Salpornitinae (Sittidae), 9
Scolopacinae (Charadriidae), 2
Sittidae, 9
Sittinae (Sittidae), 9
Stercorariidae, 3
Strigidae, 3
Sturnidae, 5
Sulidae, 1
Sylviinae (Muscicapidae), 8

Threskiornithidae, 1

Tichodromadinae (Sittidae), 9
Timaliinae (Muscicapidae), 6–7
Troglodytidae, 9
Trogonidae, 4
Turdinae (Muscicapidae), 8–9
Turnicidae, 2
Tytoninae (Strigidae), 3

Upupidae, 4

Zosteropidae, 10

HANDBOOK OF THE
BIRDS
OF INDIA
AND PAKISTAN

*TOGETHER WITH THOSE OF BANGLADESH,
NEPAL, SIKKIM, BHUTAN AND SRI LANKA*

SÁLIM ALI
AND
S. DILLON RIPLEY

With contributions by T. J. Roberts

Volume 7
LAUGHING THRUSHES TO THE MANGROVE WHISTLER
Synopsis Nos. 1272–1470
Colour Plates 77–82

SECOND EDITION

*Sponsored by the
Bombay Natural History Society*

DELHI
OXFORD UNIVERSITY PRESS
BOMBAY   CALCUTTA   MADRAS
1996

Oxford University Press, Walton Street, Oxford OX2 6DP

Oxford New York
Athens Auckland Bangkok Bombay
Calcutta Cape Town Dar es Salaam Delhi
Florence Hong Kong Istanbul Karachi
Kuala Lumpur Madras Madrid Melbourne
Mexico City Nairobi Paris Singapore
Taipei Tokyo Toronto

and associates in
Berlin Ibadan

© Oxford University Press, 1972, revised 1990

Sálim ALI 1896–1987
Sidney Dillon RIPLEY 1913
Second edition 1996

ISBN 0 19 563590 6

Printed in India at Rekha Printers Pvt. Ltd., New Delhi 110020
and published by Manzar Khan, Oxford University Press
YMCA Library Building, Jai Singh Road, New Delhi 110001

SYSTEMATIC INDEX

Order PASSERIFORMES (cont.)

Family MUSCICAPIDAE (cont.)

Subfamily TIMALIINAE : Babblers (cont.)

1272	Ashyheaded Laughing Thrush, *Garrulax cinereifrons* Blyth	3
1273	Whitethroated Laughing Thrush, *Garrulax albogularis whistleri* Baker	4
1274	ssp. *albogularis* (Gould)	5
1275	Necklaced Laughing Thrush, *Garrulax monilegerus monilegerus* (Hodgson)	6
1276	ssp. *badius* Ripley	7
1277	Blackgorgeted Laughing Thrush, *Garrulax pectoralis pectoralis* (Gould)	8
1278	ssp. *melanotis* Blyth	8
1279	Striated Laughing Thrush, *Garrulax striatus striatus* (Vigors)	10
1280	ssp. *vibex* Ripley	12
1281	*sikkimensis* (Ticehurst)	12
1282	*cranbrooki* (Kinnear)	13
1283	Whitecrested Laughing Thrush, *Garrulax leucolophus leucolophus* (Hardwicke)	14
1284	ssp. *patkaicus* Reichenow	16
1285	Chestnutbacked Laughing Thrush, *Garrulax nuchalis* Godwin-Austen	17
1286	Yellowthroated Laughing Thrush, *Garrulax galbanus galbanus* Godwin-Austen...	18
1287	Rufousvented Laughing Thrush, *Garrulax delesserti delesserti* (Jerdon)	19
1288	ssp. *gularis* (McClelland)	21
1289	Variegated Laughing Thrush, *Garrulax variegatus similis* (Hume)	22
1290	ssp. *variegatus* (Vigors)...	23
1291	Ashy Laughing Thrush, *Garrulax cineraceus cineraceus* (Godwin-Austen)..	24
1292, 1293	Rufouschinned Laughing Thrush, *Garrulax rufogularis occidentalis* Hartert)..	26
1294	ssp. *rufogularis* (Gould)	27
1295	*rufitinctus* (Koelz)	28
1295a	*rufiberbis* (Koelz)	28
1296	Assam Rufouschinned Laughing Thrush, *Garrulax rufogularis assamensis* (Hartert)	29
1297	Whitespotted Laughing Thrush, *Garrulax ocellatus maximus* (Verreaux)..	29
1298	ssp. *griseicauda* Koelz	31
1299	*ocellatus* (Vigors)...	32
1300	Greysided Laughing Thrush, *Garrulax caerulatus caerulatus* (Hodgson)	33
1301	ssp. *subcaerulatus* Hume	34
1302	*livingstoni* Ripley	35
1303	Rufousnecked Laughing Thrush, *Garrulax ruficollis* (Jardine & Selby)	36
1304	Spottedbreasted Laughing Thrush, *Garrulax merulinus merulinus* Blyth	37
1305	ssp. *toxostominus* (Koelz)	38
1306	Whitebrowed Laughing Thrush, *Garrulax sannio albosuperciliaris* Godwin-Austen..	39
1307, 1308	Nilgiri Laughing Thrush, *Garrulax cachinnans* (Jerdon)	40
1309	Whitebreasted Laughing Thrush, *Garrulax jerdoni jerdoni* Blyth	42
1310	ssp. *fairbanki* (Blanford)	42
1311	*meridionale* (Blanford)	43
1312	Streaked Laughing Thrush, *Garrulax lineatus bilkevitchi* (Zarudny)	44
1313	ssp. *gilgit* (Hartert)	46
1314	*lineatus* (Vigors)..	46
1315	*setafer* (Hodgson)	48
1316	*imbricatus* Blyth...	48
1317	Manipur Streaked Laughing Thrush, *Garrulax virgatus* (Godwin-Austen)...	49

SYSTEMATIC INDEX

1318	Browncapped Laughing Thrush, *Garrulax austeni austeni* Godwin-Austen)	50
1319	Bluewinged Laughing Thrush, *Garrulax squamatus* (Gould)	51
1320	Plaincoloured Laughing Thrush, *Garrulax subunicolor subunicolor* (Blyth)	52
1321	Prince Henry's Laughing Thrush, *Garrulax henrici* (Oustalet)	54
1322	Blackfaced Laughing Thrush, *Garrulax affinis affinis* Blyth	55
1323	ssp. *bethelae* Rand & Fleming	55
1324	Redheaded Laughing Thrush, *Garrulax erythrocephalus erythrocephalus* (Vigors)	57
1325	ssp. *kali* Vaurie	59
1326, 1327	ssp. *nigrimentum* (Oates)	59
1328	ssp. *chrysopterus* (Gould)	60
1329	*godwini* (Harington)	61
1330	*erythrolaema* (Hume)	61
1331	Crimsonwinged Laughing Thrush, *Garrulax phoeniceus phoeniceus* Gould)	62
1332	ssp. *bakeri* (Hartert)	63
1333, 1334	Silvereared Mesia, *Leiothrix argentauris argentauris* (Hodgson)	64
1333a	ssp. *aureigularis* (Koelz)	66
1335	Redbilled Leiothrix, *Leiothrix lutea kumaiensis* Whistler	66
1336, 1337	ssp. *calipyga* (Hodgson)	67
1338	Firetailed Myzornis, *Myzornis pyrrhoura* Blyth	68
1339	Nepal Cutia, *Cutia nipalensis nipalensis* Hodgson	70
1340	Rufousbellied Shrike-Babbler, *Pteruthius rufiventer* Blyth	72
1341	Redwinged Shrike-Babbler, *Pteruthius flaviscapis validirostris* Koelz	73
1342	Green Shrike-Babbler, *Pteruthius xanthochlorus occidentalis (Harington)*	75
1343	ssp. *xanthochlorus* Gray	76
1344	*hybrida* Harington	77
1345	Chestnut-throated Shrike-Babbler, *Pteruthius melanotis melanotis* Hodgson	77
1346	Chestnutfronted Shrike-Babbler, *Pteruthius aenobarbus aenobarbulus* Koelz	79
1347	Whiteheaded Shrike-Babbler, *Gampsorhynchus rufulus rufulus* Blyth	80
1348	Speckled Barwing, *Actinodura egertoni egertoni* Gould	81
1349	ssp. *lewisi* Ripley	83
1350	*khasiana* Godwin-Austen	83
1351	*ripponi* Ogilvie-Grant	84
1352, 1353	Hoary Barwing, *Actinodura nipalensis nipalensis* (Hodgson)	85
1354	ssp. *daflaensis* Godwin-Austen	86
1355	*waldeni* Godwin-Austen	87
1356	*poliotis* (Rippon)	88
1357	Redtailed Minla, *Minla ignotincta ignotincta* Hodgson	88
1358	Barthroated Siva, *Minla strigula simlaensis* (Meinertzhagen)	90
1359	ssp. *strigula* (Hodgson)	91
1360	*yunnanensis* (Rothschild)	93
1361	*cinereigenae* (Ripley)	93
1362	Bluewinged Siva, *Minla cyanouroptera cyanouroptera* (Hodgson)	94
1363	Whitebrowed Yuhina, *Yuhina castaniceps rufigenis* (Hume)	96
1364	ssp. *plumbeiceps* (Godwin-Austen)	97
1365	*castaniceps* (Moore)	98
1366	Whitenaped Yuhina, *Yuhina bakeri* Rothschild	98
1367	Yellownaped Yuhina, *Yuhina flavicollis albicollis* (Ticehurst & Whistler)	100
1368, 1369	ssp. *flavicollis* Hodgson	100
1370	*rouxi* (Oustalet)	102
1371	Stripethroated Yuhina, *Yuhina gularis vivax* Koelz	102
1372	ssp. *gularis* Hodgson	103
1373	Slatyheaded or Rufousvented Yuhina, *Yuhina occipitalis occipitalis* Hodgson	104
1374	Blackchinned Yuhina, *Yuhina nigrimenta nigrimenta* Hodgson	105

SYSTEMATIC INDEX

1375	Whitebellied Yuhina, *Yuhina zantholeuca zantholeuca* (Hodgson)..........	106
1376	Goldenbreasted Tit-Babbler, *Alcippe chrysotis chrysotis* (Blyth)...........	109
1377	ssp. *albilineatus* (Koelz)...	110
1378	Dusky Green or Yellowthroated Tit-Babbler, *Alcippe cinerea* (Blyth)	110
1379	Chestnut-headed Tit-Babbler, *Alcippe castaneceps castaneceps* (Hodgson).........	111
1380	Whitebrowed Tit-Babbler, *Alcippe vinipectus kangrae* (Ticehurst & Whistler).................	113
1381	ssp. *vinipectus* (Hodgson)..	113
1382	*chumbiensis* (Kinnear)..	114
1382a	*perstriata*..	115
1383	*austeni* (Ogilvie-Grant)...	115
1384	Brownheaded Tit-Babbler, *Alcippe ludlowi* (Kinnear).....................	116
1385	Greyheaded Tit-Babbler, *Alcippe cinereiceps manipurensis* (Ogilvie-Grant).............	116
1385a	Streakthroated Tit-Babbler, *Alcippe striaticollis* (Verreaux)...............	117
1386	Redthroated Tit-Babbler, *Alcippe rufogularis rufogularis* (Mandelli).....	118
1387	ssp. *collaris* Walden...	119
1388	Rufousheaded Tit-Babbler, *Alcippe brunnea mandelli* (Godwin Austen)	120
1389	Quaker Babbler, *Alcippe poioicephala brucei* Hume...........................	121
1390	ssp. *poioicephala* (Jerdon)...	122
1391	*fusca* Godwin-Austen...	123
1392, 1393	Nepal Quaker Babbler, *Alcippe nipalensis nipalensis* (Hodgson)....	124
1394	ssp. *stanfordi* Ticehurst...	126
1395	Chestnutbacked Sibia, *Heterophasia annectens annectens* (Blyth)...........	126
1396	Blackcapped Sibia, *Heterophasia capistrata capistrata* (Vigors).............	128
1397	ssp. *nigriceps* (Hodgson)..	130
1398	*bayleyi* (Kinnear)..	130
1399	Grey Sibia, *Heterophasia gracilis* (Harsfield)................................	131
1400	Beautiful Sibia, *Heterophasia pulchella* (Godwin-Austen)..................	132
1401	Longtailed Sibia, *Heterophasia picaoides picaoides* (Hodgson)..............	133

Subfamily MUSCICAPINAE : Flycatchers

1402	Olive Flycatcher, *Rhinomyias brunneata nicobarica* Richmond.............	135
1403, 1404	Spotted Flycatcher, *Muscicapa striata sarudnyi* Snigirewski.........	138
1405	Sooty Flycatcher, *Muscicapa sibirica gulmergi* (Baker)......................	140
1406	ssp. *cacabata* Penard..	142
1407	Brown Flycatcher, *Muscicapa latirostris* Raffles..............................	143
1408	Brownbreasted Flycatcher, *Muscicapa muttui muttui* (Layard).........	146
1409	Rufoustailed Flycatcher, *Muscicapa ruficauda* Swainson.....................	148
1410	Ferruginous Flycatcher, *Muscicapa ferruginea* (Hodgson)..................	151
1411	Redbreasted Flycatcher, *Muscicapa parva parva* Bechstein.................	153
1412	ssp. *albicilla* Pallas...	154
1413	Kashmir Redbreasted Flycatcher, *Muscicapa subrubra* Hartert & Steinbacher.................	156
1414	Orangegorgeted Flycatcher, *Muscicapa strophiata strophiata* (Hodgson)	158
1415	Whitegorgeted Flycatcher, *Muscicapa monileger monileger* (Hodgson) ...	159
1416	ssp. *leucops* (Sharpe)..	160
1417	Rufousbreasted Blue Flycatcher, *Muscicapa hyperythra hyperythra* Blyth	161
1418	Rustybreasted Blue Flycatcher, *Muscicapa hodgsonii* (Verreaux)..........	163
1419	Little Pied Flycatcher, *Muscicapa westermanni collini* Rothschild..........	164
1420	ssp. *australorientis* Ripley...	165
1421	Ultramarine Flycatcher, *Muscicapa superciliaris superciliaris* Jerdon.....	166
1422	ssp. *aestigma* Gray..	168
1423	Slaty Blue Flycatcher, *Muscicapa leucomelanura leucomelanura* (Hodgson).............	169
1424	ssp. *minuta* (Hume)...	170
1425	*cerviniventris* (Sharpe)...	171
1426	Sapphireheaded Flycatcher, *Muscicapa sapphira* (Blyth)..................	172
1427	Black-and-Orange Flycatcher, *Muscicapa nigrorufa* (Jerdon).............	174
1428	Large Niltava, *Muscicapa grandis grandis* (Blyth)............................	175

SYSTEMATIC INDEX

1429	Small Niltava, *Muscicapa macgrigoriae macgrigoriae* (Burton)	177
1430	ssp. *signata* (Horsfield)	177
1431	Rufousbellied Niltava, *Muscicapa sundara whistleri* (Ticehurst)	178
1432	ssp. *sundara* (Hodgson)	180
1433	Rufousbellied Blue Flycatcher, *Muscicapa vivida oatesi* (Salvadori)	181
1434	Whitetailed Blue Flycatcher, *Muscicapa concreta cyanea* (Hume)	182
1435	Whitebellied Blue Flycatcher, *Muscicapa pallipes* Jerdon	183
1436	Brooks's Flycatcher, *Muscicapa poliogenys poliogenys* (Brooks)	185
1437	ssp. *cachariensis* (Madarász)	186
1438	*vernayi* (Whistler)	187
1439	Pale Blue Flycatcher, *Muscicapa unicolor unicolor* (Blyth)	188
1440	Bluethroated Flycatcher, *Muscicapa rubeculoides rubeculoides* (Vigors)	189
1441	Largebilled Blue Flycatcher, *Muscicapa banyumas magnirostris* (Blyth)	191
1442	Tickell's Redbreasted Blue Flycatcher, *Muscicapa tickelliae tickelliae* (Blyth)	192
1443	ssp. *jerdoni* (Holdsworth)	195
1444	Dusky Blue Flycatcher, *Muscicapa sordida* (Walden)	195
1445	Verditer Flycatcher, *Muscicapa thalassina thalassina* Swainson	197
1446	Nilgiri Verditer Flycatcher, *Muscicapa albicaudata* Jerdon	198
1447	Pygmy Blue Flycatcher, *Muscicapella hodgsoni hodgsoni* (Moore)	200
1448	Greyheaded Flycatcher, *Culicicapa ceylonensis calochrysea* Oberholder	201
1449	ssp. *ceylonensis* (Swainson)	203
1450	Yellowbellied Fantail Flycatcher, *Rhipidura hypoxantha* Blyth	205
1451	Whitebrowed Fantail Flycatcher, *Rhipidura aureola aureola* Lesson	206
1452	ssp. *compressirostris* (Blyth)	209
1453	*burmanica* (Hume)	209
1454	Whitethroated Fantail Flycatcher, *Rhipidura albicollis canescens* (Koelz)	210
1455	ssp. *albicollis* (Vieillot)	212
1456	*stanleyi* Baker	213
1457	*orissae* Ripley	213
1458	*albogularis* (Lesson)	214
1459	*vernayi* (Whistler)	215

Subfamily MONARCHINAE : Monarch Flycatchers

1460	Paradise Flycatcher, *Terpsiphone paradisi leucogaster* (Swainson)	216
1461	ssp. *paradisi*	218
1462	*ceylonensis* (Zarudny & Härms)	220
1463	*saturatior* (Salomonsen)	221
1464	*nicobarica* Oates	222
1465	Blacknaped Monarch Flycatcher, *Hypothymis azurea styani* Hartlaub	223
1466	ssp. *ceylonensis* Sharpe	225
1467	*tytleri* (Beavan)	226
1468	*idiochroa* Oberholser	226
1469	*nicobarica* Bianchi	227

Subfamily PACHYCEPHALINAE : Thickheads or Shrikebilled Flycatchers

1470	Grey Thickhead or Mangrove Whistler, *Pachycephala grisola* (Blyth)	228

COLOUR PLATES

Synopsis numbers in brackets
(SE) = species extralimital

Plate 77

1 *Garrulax subunicolor* Plaincoloured Laughing Thrush (1320)
2 *Garrulax lineatus* Streaked Laughing Thrush (1314)
3   ssp. *nigrimentum* of 1324 (1326)
4 *Garrulax erythrocephalus* Redheaded Laughing Thrush (1324)
5 *Garrulax squamatus* Bluewinged Laughing Thrush (1319)
6 *Garrulax austeni* Browncapped Laughing Thrush (1318)
7 *Garrulax virgatus* Manipur Streaked Laughing Thrush (1317)
8 *Garrulax merulinus* Spottedbreasted Laughing Thrush (1304)
9 *Garrulax rufogularis* Rufouschinned Laughing Thrush (1294)
10 *Garrulax striatus* Striated Laughing Thrush (1279)
11 *Garrulax monilegerus* Necklaced Laughing Thrush (1275)
12 *Babax lanceolatus* Chinese Babax (1270)
13 *Garrulax pectoralis* Blackgorgeted Laughing Thrush (1277)
14 *Babax waddelli* Giant Tibetan Babax (1271)
15 *Garrulax ocellatus maximus* Giant Laughing Thrush (1297)
16 *Garrulax ocellatus* Whitespotted Laughing Thrush (1299)

Plate 78

1 *Garrulax phoeniceus* Crimsonwinged Laughing Thrush (1331)
2 *Garrulax cachinnans* Nilgiri Laughing Thrush (1307, 1308)
3 *Garrulax jerdoni* Whitebreasted Laughing Thrush (1309)
4 *Garrulax cineraceus* Ashy Laughing Thrush (1291)
5 *Garrulax galbanus* Yellowthroated Laughing Thrush (1286)
6 *Garrulax delesserti* Rufousvented Laughing Thrush (1287)
7   ssp. *gularis* of 1287 (1288)
8 *Garrulax sannio* Whitebrowed Laughing Thrush (1306)
9 *Garrulax ruficollis* Rufousnecked Laughing Thrush (1303)
10 *Garrulax nuchalis* Chestnutbacked Laughing Thrush (1285)
11 *Garrulax henrici* Prince Henri's Laughing Thrush (1321)
12 *Garrulax affinis* Blackfaced Laughing Thrush (1322)
13 *Garrulax variegatus* Variegated Laughing Thrush (1290)
14 *Garrulax caerulatus* Greysided Laughing Thrush (1300)
15 *Garrulax albogularis* Whitethroated Laughing Thrush (1274)
16 *Garrulax leucolophus* Whitecrested Laughing Thrush (1283)

Plate 79

1 *Gampsorhynchus rufulus* Whiteheaded Shrike-Babbler (1347)
2 *Actinodura nipalensis* Hoary Barwing (1352, 1353)
3 *Alcippe brunnea* Rufousheaded Tit-Babbler (1388)
4 *Actinodura egertoni* Spectacled Barwing (1348)
5   ssp. *waldeni* of 1352, 1353 (1355)
6 *Alcippe castaneceps* Chestnut-headed Tit-Babbler (1379)
7 *Alcippe cinerea* Dusky-Green or Yellowthroated Tit-Babbler (1378)
8 *Alcippe rufogularis* Redthroated Tit-Babbler (1386)
9 *Alcippe nipalensis* Nepal Quaker Babbler (1392, 1393)
10 *Alcippe cinereiceps* Greyheaded Tit-Babbler (1385)
11 *Alcippe poioicephala* Quaker Babbler (1390)
12 *Yuhina xantholeuca* Whitebellied Yuhina (1375)
13 *Yuhina castaniceps* Whitebrowed Yuhina (1365)

Terms used in the description of a bird's plumage and parts

Topography of a sparrow

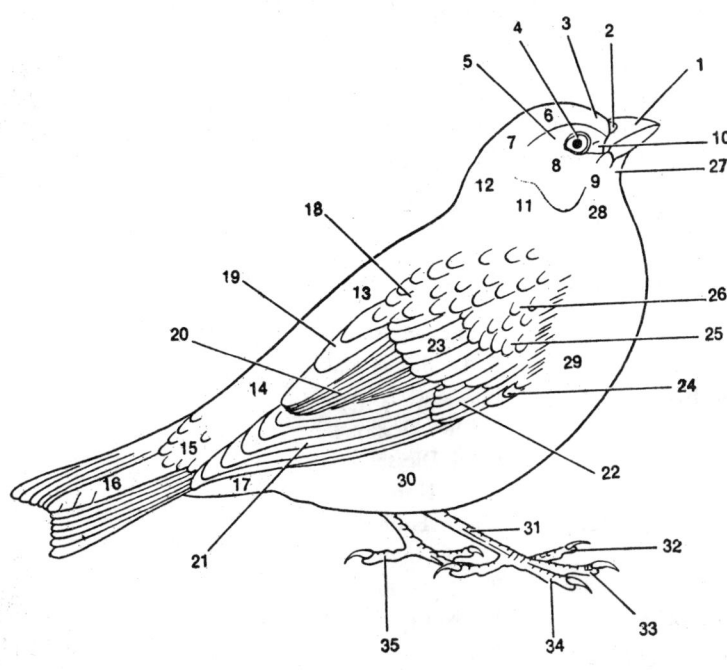

1A Maxilla (=upper mandible)
1B Mandible (=lower mandible)
2 Nostril
3 Forehead
4 Iris
5 Supercilium
6 Crown
7 Hind crown
8 Ear-coverts
9 Malar region (malar stripe, moustache)
10 Lores
11 Side of neck
12 Hindneck (=nape)
13 Back
14 Rump
15 Upper tail-coverts
16 Tail (rectrices)
17 Under tail-coverts
18 Scapulars
19 Tertials
20 Secondaries ⎫ (remiges)
21 Primaries ⎭
22 Primary coverts
23 Greater coverts
24 Bastard wing (alula)
25 Median coverts
26 Lesser coverts
27 Chin
28 Throat
29 Breast
30 Belly (abdomen)
31 Tarsus
32 Inner toe
33 Middle toe
34 Outer toe
35 Hind toe (hallux)

ABBREVIATIONS

Bull. BOC	*Bulletin of the British Ornithologists' Club*
Dementiev	*Birds of the Soviet Union*
FBI	*Fauna of British India, Birds*
Ind. Hb.	*Handbook of the Birds of India and Pakistan*
J. Orn.	*Journal für Ornithologie*, Berlin
JBNHS	*Journal of the Bombay Natural History Society*
PZS	*Proceedings of the Zoological Society*, London
SF	*Stray Feathers*
SZ	*Spolia Zeylanica*
Witherby	*The Handbook of British Birds*

Abbreviated references to persons frequently quoted

Abdulali or HA	Humayun Abdulali
Baker	E. C. Stuart Baker
Biswas or BB	Biswamoy Biswas
Desfayes or MD	Michel Desfayes
Diesselhorst or GD	Gerd Diesselhorst
Hartert	Ernst Hartert
Hume	A. O. Hume
Inglis	C. M. Inglis
Jerdon	T. C. Jerdon
Kinnear or NBK	Sir Norman B. Kinnear
Meinertzhagen or RM	Col. R. Meinertzhagen
Murphy	R. C. Murphy
Osmaston or BBO	B. B. Osmaston
Paynter or RAP	R. A. Paynter
Peters	J. L. Peters
Phillips	W. W. A. Phillips
Ripley or SDR	S. Dillon Ripley
Sálim Ali or SA	Sálim Ali
TJR or T. J. Roberts	Tom J. Roberts
Stresemann	Erwin Stresemann
Ticehurst or CBT	Claud B. Ticehurst
Whistler or HW	Hugh Whistler

Order PASSERIFORMES (cont.)
Family MUSCICAPIDAE (cont.)
Subfamily TIMALIINAE: Babblers (cont.)

Whitecrested Laughing Thrush (1283)

Genus GARRULAX Lesson

Garrulax Lesson, 1831 (June), Traité d'Orn.: 647. Type, by subsequent designation, Gray, 1846, *Garrulax Belangeri* Lesson, but *nomen nudum*, type designated by Ripley (Synopsis: 380), *Garrulax rufifrons* Lesson

Ianthocincla Gould, 1835, Proc. Zool. Soc. London: 48. Type, by original designation, *Cinclosoma ocellatum* Vigors

Trochalopteron Blyth, 1843, Jour. Asiat. Soc. Bengal 12: 952, *ex* Hodgson MS. Type, by subsequent designation, *Trochalopteron subunicolor* Hodgson

Grammatoptila Gray, 1855, Cat. Gen. Subgen. Bds.: 48. Type, by original designation, *Garrulus striatus* Vigors

Stactocichla Sharpe, 1883, Cat. Bds. Brit. Mus. 7: 328, 449. Type, by monotypy, *Garrulax merulinus* Blyth

Dryonastes Sharpe, 1883, Cat. Bds. Brit. Mus. 7: 453. Type, by original designation, *Ianthocincla ruficollis* Jardine & Selby

Bill strong and straight. Legs strong. Wing short and rounded, often edged with a brighter colour. Sexes usually alike. Most species noisy and very gregarious outside the breeding season.

TIMALIINAE

Key to the Species

	Page
I Under surface of tail tipped with white or rufous	
A Size large, tail over 150 mm; upperparts conspicuously spotted with white or buff	
1 Crown and throat black .... *G. ocellatus*	31
2 Crown brown, throat cinnamon .... *G. ocellatus maximus*	29
B Size medium, tail under 150 mm; upperparts not spotted with white	
3 Underparts yellow .... *G. galbanus*	18
4 Underparts not yellow	
i Throat white or buff	
a A black collar	
a^1 Larger, tarsus over 40 mm; a black cheek stripe *G. pectoralis*	8
a^2 Smaller, tarsus under 40 mm; no cheek stripe *G. monilegerus*	6
b No black collar	
a^3 Crown plain olive-brown .... *G. albogularis*	4
a^4 Crown finely barred with black *G. caerulatus* (*subcaerulatus*)	34
ii Chin and throat black .... *G. variegatus*	21
iii Chin and throat neither black nor white	
c Tip of tail rufous	
a^5 Edge of wing pale blue .... *G. squamatus*	51
a^6 Edge of wing whitish .... *G. rufogularis*	25
a^7 Wing crimson .... *G. phoeniceus*	62
d Tip of tail white or greyish	
a^8 Primary-coverts black	
b^1 Wing tipped with black and white .... *G. cineraceus*	24
b^2 Wing almost uniform slaty .... *G. henrici*	54
a^9 Primary-coverts not black	
b^3 Plumage striated .... *G. lineatus*	44
b^4 Plumage squamated .... *G. subunicolor*	52
b^5 Underparts barred whitish and brown .... *G. austeni*	50
II Under surface of tail not tipped with white or rufous	
C Plumage striated	
5 Tail narrowly cross-barred .... *G. virgatus*	49
6 Tail not cross-barred .... *G. striatus*	10
D Plumage not striated	
7 Chin and throat white	
iv Crown white .... *G. leucolophus*	14
v Crown squamated .... *G. caerulatus*	33
(except *subcaerulatus*)	
vi Crown slaty .... *G. delesserti*	19
8 Throat and breast grey .... *G. jerdoni*	42
9 Chin black	
vii Ear-coverts white .... *G. nuchalis*	17
viii Ear-coverts black	
e A rufous patch on sides of neck .... *G. ruficollis*	36
f A white patch on sides of neck .... *G. affinis*	55
ix Ear-coverts tawny olive-brown; a white supercilium *G. cachinnans*	40
x Chin neither black nor white	

		Page
g A whitish line behind eye		
a^{10} Throat spotted	G. merulinus	37
a^{11} Throat not spotted	G. sannio	39
h No line behind eye		
a^{12} Crown slate-grey, back unspotted	G. cinereifrons	3
a^{13} Crown chestnut, back spotted	G. erythrocephalus	57

1272. **Ashyheaded Laughing Thrush.** *Garrulax cinereifrons* Blyth[1]

Garrulax cinereifrons Blyth, 1851, Jour. Asiat. Soc. Bengal 20: 176 (Ceylon)
Baker, FBI No. 190, Vol. 1: 196
Plate 76, fig. 14 (Vol. 6)

LOCAL NAMES. *Alu-demalichchā* (Sinhala); *Vēlaikkāra-kūrūvi* (Tamil).

SIZE. Myna; length *c.* 23 cm (9 in.).

FIELD CHARACTERS. *Above*, forehead, crown and ear-coverts grey; rest of upperparts dark reddish brown. *Below*, chin and throat whitish (ochraceous buff); rest of underparts rufous-brown. Sexes alike.
Distinguished from the Rufous Babbler (1266) mainly by its black bill, dark grey legs and grey head. It is also somewhat less unkempt-looking.

STATUS, DISTRIBUTION and HABITAT. Resident in Sri Lanka; moderately plentiful in the low-country Wet zone and Hill zone to *c.* 1500 m. Affects dense, humid forest.

GENERAL HABITS. Lives in flocks of ten to twenty individuals working steadily through the damp undergrowth, fluttering from tree to tree and exploring the mossy recesses of fallen trunks. Keeps up a constant flow of squeaks and chattering which can be mistaken for those of the Rufous Babbler which inhabits the same jungles. Such flocks are often accompanied by one or two jungle squirrels.

FOOD. Chiefly insects.

VOICE and CALLS. Breaks out constantly into a harsh chattering which is taken up in turn by all the members of the troop; this chattering is usually finished up with a hurried sort of scream (Legge). Also a flow of squeaks and chatterings while working through the undergrowth.

BREEDING. *Season*, appears to be between March and July (full-fledged nestlings seen in April and August). *Nest*, an untidy mass of leaves, placed in the fork of a small sapling, 15 feet above ground, on a fairly steep hillside with quite tall grass. Nest cup 3.5 inches in diameter, made largely of pliable twigs, rootlets, horsehair, lichen with a few *Cullinia ceylonica* leaves at base. *Eggs*, 3 turquoise blue; incubation begins with last egg; period, 17–18 days. Three individuals observed working at nest building (Judy & John Banks, JBNHS 84: 682–3).

MUSEUM DIAGNOSIS. See Field Characters. Young as adult, but first primary soft and blunt and rectrices narrower and pointed.

---

[1] As already pointed out by Ripley (*Spolia Zeylanica* 24: 223, 1946), this species is most closely related to *Garrulax delesserti*.

TIMALIINAE

MEASUREMENTS

	Wing	Bill (from feathers)	Tail
4 ♂♂	114–118	22–24	100–104 mm
2 ♀♀	110, 116	22, 23	95, 105 mm (SDR)

Bill from skull 27 mm. Tarsus 36 mm (MD)
Weight 1 ♂ 70 g (SDR).

COLOURS OF BARE PARTS. Iris pale yellowish white or whitish. Bill black, base of lower mandible greyish or greyish white. Legs bluish brown or bluish grey-brown.

GARRULAX ALBOGULARIS (Gould): WHITETHROATED LAUGHING THRUSH

Key to the Subspecies

Belly more richly coloured, ferruginous .... *G. a. albogularis*
Belly paler, ochraceous .... *G. a. whistleri*

1273. *Garrulax albogularis whistleri* Baker

Garrulax albogularis whistleri Baker, 1921, Bull. Brit. Orn. Cl. 42: 29 (Simla)
Baker, FBI No. 138, Vol. 1: 154

LOCAL NAMES. None recorded.

SIZE. Myna +; length *c.* 28 cm (11 in.).

FIELD CHARACTERS. *Above*, olive-brown with tawny forehead, black lores and eye-rim. Outer edge of wing grey. Tail olive-brown, graduated, the rectrices broadly tipped with white (except the central pair) showing up as a prominent white terminal band, especially when spread in flight and while alighting. *Below*, throat conspicuously white bordered on breast by an olive-brown band. Belly and under tail-coverts ochraceous. Terminal half of tail white. Sexes alike.

STATUS, DISTRIBUTION and HABITAT. Resident, common but rather locally distributed especially in the western part of its range; subject to some vertical movements. The western Himalayan outer ranges. Formerly from the Murree Hills east to Kumaon but not recorded from Pakistan since early 1960s and believed to be locally extirpated (T. J. Roberts). From 1800 to 2900 m in summer (up to tree-line in Garhwal), and from 1200 to at least 2400 m in winter, rarely or occasionally descending to 450 m in the cold weather. Affects dense forest of oak, fir, deodar; also scrub and light jungle, sometimes coming into gardens.

GENERAL HABITS. Gregarious even in the breeding season; keeps in flocks of six to twelve birds and in winter up to thirty or more, commonly in association with other laughing thrushes, tree pies, jays and Blue Magpies. Feeds a good deal on the ground but on the whole keeps more to the trees than *G. leucolophus*, searching the crevices of bark, tearing off lumps of moss, and slipping away through the branches with rapidity. Sometimes ventures into harvested fields, hopping on the ground like a Jungle Babbler, digging with its bill for food. Less wary of man than other laughing thrushes.

FOOD. Chiefly insects; also berries and seeds.

VOICE and CALLS. While feeding keeps up a soft, low *teh, teh,* rather like a common note of the Black Bulbul and very reminiscent of a flock of tits though louder (HW). When alarmed, bursts into an extraordinary series of sibilant squeals and hisses which develop into choruses of shrill squeaky 'laughter' when birds really excited. For renditions see 1274.

BREEDING. *Season,* March to July. *Nest,* a cup of coarse grass, bamboo leaves, roots, moss, lichen or other material, usually thickly lined with rootlets. Placed in a bush, at the top of a sapling or near the extremity of a horizontal branch, between one and four metres from the ground, sometimes up to six metres. *Eggs,* normally 3, exceptionally 4, longish ovals, glossy intense blue, deeper than those of most Indian laughing thrushes. Average size of 60 eggs 29 × 21.1 mm (Baker).

MUSEUM DIAGNOSIS. Differs from *Albogularis* (1274) in having the upperparts paler (more greyish), the ochraceous of underparts duller, and in being slightly larger.

Young like adult but upperparts tinged rufescent; flanks paler; olive-brown pectoral band less distinct.

MEASUREMENTS

Wing, ♂ ♀ 132 (worn) to 144 mm (Baker).

	Wing	Bill (from skull)	Tarsus	Tail
1 ♀	137	23	43	149 mm (MD)

COLOURS OF BARE PARTS. As in 1274.

1274. *Garrulax albogularis albogularis* (Gould)

Ianthocincla albogularis Gould, 1836 (8 April), Proc. Zool.
Soc. London: 187 (Nepal)
Baker, FBI No. 137, Vol. 1: 153
Plate 78, fig. 15

LOCAL NAME. *Karriam-pho* (Lepcha).

FIELD CHARACTERS. As in 1273, q.v.

STATUS, DISTRIBUTION and HABITAT. Common resident subject to some vertical movements. The Himalayas from western Nepal east through Sikkim, Bhutan and Arunachal Pradesh at least to Shergaon *c.* 27°N., 92°15′ E. (Whistler, *Ibis* 1941: 172); eastern limit unknown. Also Assam in North Cachar ['Very rare everywhere'; recorded twice in the Baril range; nest found (Baker, JBNHS 8: 174 and FBI 1: 153)]. From 1800 to 3500 m (Inskipp & Inskipp) descending to *c.* 900 m in winter. Affects dense forest but does not shun thin and open scrubby hillsides, apparently frequenting the latter in winter.

Extralimital. The species extends to southwestern Sichuan, Yunnan and northwestern Vietnam; also Taiwan. Introduced on Kauai (Hawaii).

GENERAL HABITS and FOOD. As in 1273. In winter large loose flocks of between fifty and a hundred not uncommon. Fond of rummaging on the ground amongst bonfire and kitchen ashes on the site of pilgrim and muleteer bivouacs.

VOICE and CALLS. As in 1273; very noisy. A continual musical chattering *chip chip chip chip*; alarm *quoik, tsueeeeee*(Fleming). Voice more subdued and less harsh than that of *G. leucolophus*, but shriller.

BREEDING. As in 1273.

MUSEUM DIAGNOSIS. Differs from *whistleri* (1273) in being more richly coloured, especially on underparts.

MEASUREMENTS

	Wing	Bill (from skull)	Tarsus	Tail	
♂♂	123–136	25–30	45–48	125–141	mm
♀♀	122–135	24–30	45–48	125–137	mm

(BB, Fleming, SA)

Weight 17 ♂♂ 97–114; 7 ♀ ♀ 78–105 g (GD, SA).

COLOURS OF BARE PARTS. Iris dull white to bluish white. Bill horny black; mouth yellow. Legs and feet plumbeous; claws grey; soles yellowish white.

GARRULAX MONILEGERUS (Hodgson): NECKLACED LAUGHING THRUSH

Key to the Subspecies

Paler, less richly coloured .... *G. m. monilegerus*
Darker, more saturated with rufous .... *G.m.badius*

1275. ***Garrulax monilegerus monilegerus*** (Hodgson)

Cinclosoma monilegera Hodgson, 1836, Asiat. Res. 19: 147 .
(Nepal)
Baker, FBI No. 134 (part), Vol. 1: 151
Plate 77, fig. 11

LOCAL NAMES. As for 1278; *Chhōta pengā* (Bengal).

SIZE. Myna +; length *c.* 27 cm (11 in.).

FIELD CHARACTERS. At first sight confusingly like *G. pectoralis* (1278) from which it may be distinguished by the following: Lack of cheek-stripe between throat and ear-coverts. Necklace much narrower on the breast, nearly covered by the rufous of the lower throat; chin and throat white (*G. pectoralis* has a buff throat bordered with white along the broad black necklace). White of belly extends up to the nuchal collar along and below the necklace. Shoulders (primary-coverts) olive-brown, concolorous with upperparts and rest of wing (*contra* blackish brown in *pectoralis*). It is also smaller in size with less strong bill, legs and feet, the latter paler coloured (yellowish brown as against slate-grey); the bill is entirely dark while in *pectoralis* it is paler at the base of the lower mandible.

STATUS, DISTRIBUTION and HABITAT. Resident, fairly common except in the western part of its range. From west-central Nepal east through Sikkim, Darjeeling, Jalpaiguri district, Bhutan, Arunachal Pradesh to the range of *G. m. badius*, the hills of NE. India (except the Patkai Range), Nagaland, Manipur and Bangladesh in the northeastern hills and the Chittagong region; from the edge of the plains to *c.* 1000 m, locally up to 1400 m. Affects thick evergreen and moist-deciduous forest and secondary jungle with undergrowth of rattan brakes, etc. Geographical and vertical distribution of this species coincides with that of *G. pectoralis*.

Extralimital. Burma. The species extends east to Vietnam, Hainan and Anhui.

GENERAL HABITS. Highly gregarious. Keeps in flocks outside the breeding season, often in company with its larger 'double' *G. pectoralis* or other laughing thrushes. Feeds on the ground, turning over dead leaves with much rustling and scratching. Behaves exactly as *G. pectoralis* (1278) with which it shares the same ecological niche. Flocks break up at the end of March.

FOOD. Insects, snails, small lizards, etc.; also berries, seeds and other vegetable matter.

VOICE and CALLS. Usually silent and secretive if people are about, but once they fancy the danger has passed, utter a noisy chorus of hollow-sounding musical whistles (SA). Voice indistinguishable from that of *G. pectoralis* (1278) q.v.

BREEDING. *Season*, March to July, chiefly April and May. *Nest* and *eggs* (4 or 5) very similar to those of *G. pectoralis*; nest placed in identical situations. Average size of 100 eggs 28.4 × 21.3 mm (Baker). Commonly brood-parasitized by the cuckoos *Clamator jacobinus*, *C. coromandus* and presumably also *Cuculus sparverioides*.

MUSEUM DIAGNOSIS. *G. m. badius* (1276) is darker, more saturated with rufous. For distinguishing from *G. pectoralis*, see Field Characters.

Young very similar to adult but paler. Necklace dusky, underparts less fulvous. Distinguished from the young of *pectoralis* by the olive-brown primary-coverts (*v.* black). Postjuvenal moult complete.

MEASUREMENTS

	Wing	Bill (from skull)	Tarsus	Tail	
♂♂	119–132	29–30	41–44	121–132	mm
♀♀	116–126	—	—	—	mm

(SA, Rand & Fleming, Heinrich)

Weight 5 ♂♀ 77–91 g (SA).

COLOURS OF BARE PARTS. Iris yellow, orange-yellow or reddish yellow. Bill dark horn, tip paler. Legs and feet yellowish brown or brownish flesh.

1276. *Garrulax monilegerus badius* Ripley

Garrulax moniliger (sic) *badius* Ripley, 1948, Proc. Biol. Soc. Washington 61: 102
(Tezu, Mishmi Hills, NE. Assam)
Baker, FBI No. 134 (part), Vol. 1:151

LOCAL NAMES. None recorded.

SIZE. Myna +; length 27 cm (11 in.).

FIELD CHARACTERS. As in 1275, q.v.

STATUS, DISTRIBUTION and HABITAT. Common resident. Arunachal Pradesh from the Mishmi Hills to the Patkai Range; from the edge of the plains to *c.* 900 m. Affects tropical evergreen forest.

GENERAL HABITS, FOOD and VOICE. As in 1275.

BREEDING. As in 1275.

MUSEUM DIAGNOSIS. Differs from *G. m. monilegerus* (1275) in being darker, more saturated with rich rufous, particularly on the nuchal collar and on the underparts.
MEASUREMENTS
3 ♂♂ Wing 120–124 mm (SDR).
COLOURS OF BARE PARTS. Iris whitish orange to orange-yellow. Bill black, tip light grey. Legs light grey.

GARRULAX PECTORALIS (Gould): BLACKGORGETED LAUGHING THRUSH
Key to the Subspecies

Paler, less richly coloured .... *G. p. pectoralis*
Darker, more richly coloured .... *G. p. melanotis*

1277. *Garrulax pectoralis pectoralis* (Gould)

Ianthocincla pectoralis Gould, 1836 (8 April), Proc. Zool. Soc. London: 186 (Nepal)
Baker, FBI No 132 (part), Vol. 1: 150

Plate 77, fig. 13

LOCAL NAMES. None recorded.
SIZE. Myna +; length *c.* 29 cm (12 in.).
FIELD CHARACTERS. As in 1278, q.v.
STATUS, DISTRIBUTION and HABITAT. Common resident. Central Nepal to eastern Nepal, from the edge of the terai up to *c.*1650 m. Affects secondary growth, cut-over scrub and forests of sal, pine, etc. with dense undergrowth.
GENERAL HABITS, FOOD and VOICE. As in 1278.
BREEDING. As in 1278.
MUSEUM DIAGNOSIS. Paler and less richly coloured than *melanotis* (1278). This species has a fine fleshy yellow eye-rim which is absent in *G. monilegerus*.
MEASUREMENTS
Wing, 4 ♂♂ 144–149; 4 ♀♀ 139–143 mm (Rand & Fleming).
COLOURS OF BARE PARTS. As in 1278.

1278. *Garrulax pectoralis melanotis* Blyth

Garrulax melanotis Blyth, 1843, Jour. Asiat. Soc. Bengal 12: 949 (Arracan)
Garrulax McClellandi Blyth, 1843, Jour. Asiat. Soc. Bengal 12: 949 (Assam)
Garrulax uropygialis Bonaparte, 1850, Consp. Gen. Av. 1: 371 (Assam)
Garrulax waddelli Ogilvie-Grant, 1894, Bull. Brit. Orn. Cl. 3: xxix
[Rungeet (Rangit) River, Sikkim]
Baker, FBI No. 132 (part), Vol. 1: 150

LOCAL NAMES. *Ol-pho* (Lepcha); *Piang-kam* (Bhutia); *Poreri* or *Purirhi* (Dafla); *Bădā pengā* (Bengali).

SIZE. Myna+; length *c.* 29 cm (12 in.).

FIELD CHARACTERS. *Above*, olive-brown with a rufous nuchal collar and white supercilium extending to the collar. Shoulder (primary-coverts) blackish brown; outer edge of wing whitish. Tail graduated, the rectrices, except central pair, black, broadly tipped with white, showing in flight as a broad terminal band. Ear-coverts varying from striped black and white to entirely white or black, bordered above and behind by black, and below by a black cheek stripe. *Below*, throat mostly buff bordered by white along the broad necklace. Centre of belly white, flanks ochraceous. Sexes alike.

Confusingly similar to Necklaced Laughing Thrush (*G. monilegerus*); for distinguishing from it see 1275 Field Characters.

STATUS, DISTRIBUTION and HABITAT. Common resident. From Darjeeling, Sikkim, and Jalpaiguri district east through Bhutan and Arunachal Pradesh to the Mishmi Hills; the hills of Assam, Nagaland and Manipur, and Bangladesh and the Chittagong region (Cox's Bazar); from the edge of the plains to *c.* 1200 m, locally up to 1700 m. Affects dense forest, secondary growth and bamboo jungle, often bordering cultivation.

Extralimital. Burma to Arakan and western Yunnan. The species ranges east to Vietnam, Hainan and Anhwei.

GENERAL HABITS. Highly gregarious; goes about in troops of 10 to 25, often in association with *G. monilegerus* and other laughing thrushes. To cross an open space, the members of a flock glide one after another (never all together unless scared), in a continuous flowing motion. Flight rather clumsy and jay-like, yet stronger than of most laughing thrushes. Feeds much on the forest floor, proceeding by long hops. On alarm, flies up into bushes and mounts into trees, hopping rapidly from branch to branch, then gliding off as described. Its ability to disappear into cover when in danger is remarkable. The birds often display, in spring and at other times as well, hopping about on the ground, flirting and spreading their wings, bowing and performing like circus contortionists, all the while uttering loud calls (Baker).

Food. Mostly insects.

VOICE and CALLS. Very noisy. Keeps up an incessant querulous conversational squeaking, a nasal *week, week, week* (SA); also described as a strange, very human piping to which are usually added several short high whistles. The several individuals simultaneously and confusedly uttering these strange calls sound like an orchestra of mournful piping (Heinrich). According to the same observer, its voice is indistinguishable from that of *G. monilegerus*. Also has some harsh, grating calls.

BREEDING. *Season*, March to August, mainly April to June. *Nest*, a broad, rather shallow but bulky saucer, untidily and rather loosely put together; made mostly of bamboo leaves, with other dead leaves, roots, scraps of moss, bracken, etc., bound by weed stems and tendrils, lined with rootlets and fine grass stems. Placed in bushes or small trees, from near the ground to about six metres up. *Eggs*, 3 to 5 (rarely up to 7), most often 4, deep blue. Average size of 100 eggs 31.4 × 22.7 mm (Baker). The incubating bird sits close but is adept in slipping off without giving a clear view and hiding in the undergrowth with no sign of its presence. Commonly brood-parasitized by *Clamator coromandus* and *Cuculus sparverioides*.

MUSEUM DIAGNOSIS. Differs from *G. p. pectoralis* (1277) in being darker and more richly coloured.

Young, as adult but paler. Primary-coverts black with pale tips. Postjuvenal moult complete.

MEASUREMENTS

	Wing	Bill (from skull)	Tarsus	Tail	
♂♂	130–152	32–36	46–48	122–144	mm
♀♀	130–148	c. 35	46–49	118–135	mm
			(SA, HW, SDR, Heinrich)		

Weight 1 ♂ 156; 2 ♀ ♀ 135, 135 g (SDR, SA).

COLOURS OF BARE PARTS. Iris reddish- or orange-brown; fleshy circumorbital ring chrome yellow. Bill: upper mandible blackish brown, lower mandible basally grey, distally brown; extreme bill-tip whitish. Legs and feet slate-grey; claws horny white.

GARRULAX STRIATUS (Vigors): STRIATED LAUGHING THRUSH

Key to the Subspecies

		Page
A	A broad black stripe from eye to nape .... *G. s. cranbrooki*	13
B	No black stripe	
1	Paler, less olive; larger .... *G. s. striatus*	10
2	Darker, more olive; smaller .... *G. s. vibex*	12
3	Darker than 2, more red-brown .... *G. s. sikkimensis*	12

1279. *Garrulax striatus striatus*(Vigors)

Garrulus striatus Vigors, 1831, Proc. Zool. Soc. London: 7
(Himalaya Mountains, restricted to Naini Tal by Baker, 1920, JBNHS 27: 245)
Baker, FBI No. 177 (part), Vol. 1: 184

Plate 77, fig. 10

LOCAL NAMES. None recorded.
SIZE. Myna +; length *c.* 28 cm (11 in.).
FIELD CHARACTERS. A large umber-brown laughing thrush conspicuously white-streaked, with a short thick bill and loose, mop-like crest. *Above*, crest dark brown, white-streaked in front. Back umber-brown with fine white streaks. Tail chestnut-brown, with minute white tips to outer rectrices. *Below*, throat and sides of head densely streaked. Breast and belly brownish grey with paler streaks. Sexes alike.

× *c.* 1

Distinguished from the species *virgatus* mainly by its larger size, crest, and lack of supercilium. *G. lineatus* is much smaller with dusky streaks.

Garrulax striatus

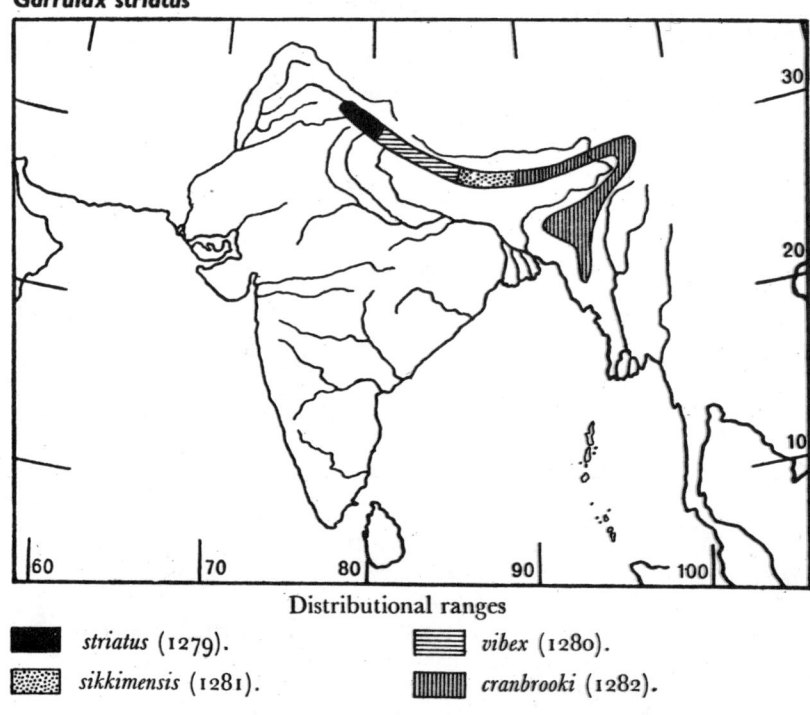

Distributional ranges

▪ *striatus* (1279). ≡ *vibex* (1280).
▦ *sikkimensis* (1281). ||||| *cranbrooki* (1282).

STATUS, DISTRIBUTION and HABITAT. Resident, locally common, subject to vertical movements. The western Himalayas from Kulu and Mandi to Garhwal; from 1200 to 2700 m, mostly between 2000 and 2700 m in summer. Affects dense forest and better-wooded nullahs and ravines; local in its choice of habitat.

GENERAL HABITS. More arboreal than most laughing thrushes, feeding in the canopy of tall trees as well as in undergrowth and lower branches. Keeps in pairs in the breeding season, otherwise in small noisy parties of five to eight birds—sometimes singly—frequently in association with other laughing thrushes and trees pies; rather parochial, frequenting the same patch of forest day after day. Often found at fruiting trees in company with bulbuls, barbets and fruit pigeons.

FOOD. Insects, berries and seeds.

VOICE and CALLS. A variety of loud, discordant cackling notes resembling those of a domestic fowl that has laid an egg, and sometimes a harsh unmusical chorus of chattering laughter. A lively, rich whistling call rendered as *oh see-saw-oh-whitey oh-white* (Frome), which usually betrays the presence of the birds a long way off. A frequent call-note is a double whistle *whe-ho*, easily imitated (Stanford). Alarm, a harsh *oick-oick-oick-oick*. Other calls, doubtless variants of above, under 1281.

BREEDING. *Season*, April to July, mostly May and June. *Nest*, a broad, shallow cup strongly made of coarse grass, twigs, rootlets, dead leaves, moss or green ferns, and lined with rootlets. Placed in a sapling or among climbers, from one to six metres above the ground. *Eggs*, almost invariably 2, pale blue.

Average size of 35 eggs 33.3 × 23.3 mm (Baker). Brood-parasitized by *Clamator coromandus*.

MUSEUM DIAGNOSIS. Differs from *vibex* (1280) in being paler, less olive. Young, like adult but streaks on underparts not so sharp.

MEASUREMENTS

	Wing	Bill (from skull)	Tail
♂♀	145–165	27–28	139–148 mm (SDR, CBT)

COLOURS OF BARE PARTS. As in 1280.

1280. *Garrulax striatus vibex* Ripley

Garrulax striatus vibex Ripley, 1950, Proc. Biol. Soc. Washington 63: 103 (Godavari, Central Valley, Nepal)
Baker, FBI No. 177 (part), Vol. 1: 184

LOCAL NAME. *Bhiakura* (Pahari, all laughing thrushes).
SIZE. Myna +; length *c.* 28 cm (11 in.).
FIELD CHARACTERS. As in 1279; see Museum Diagnosis.

STATUS, DISTRIBUTION and HABITAT. Common resident, subject to some vertical movements. The central Himalayas from eastern Kumaon to eastern Nepal; from *c.* 1500 to 2850 m (Inskipp & Inskipp, p. 308) descending to *c.* 1000 m in winter. Affects dense forest and thickets.

GENERAL HABITS, FOOD and VOICE. As in 1279.
BREEDING. As in 1279.

MUSEUM DIAGNOSIS. Differs from *striatus* by smaller size and generally darker more olive-tinted upper surface and flanks. From *sikkimensis* it differs by being distinctly lighter, less red-brown, particularly on the back and lower parts, the dark edging to the median streaks noticeably darker.

MEASUREMENTS

	Wing	Bill (from skull)	Tarsus	Tail
♂♂	138–151	26–30	*c.* 42	128–138 mm
♀♀	132–142	27–30	*c.* 42	126–139 mm

(BB, SDR, Rand & Fleming)

Weight 5 ♂ ♂ 126–148; 1 ♀ 138 g (GD).

COLOURS OF BARE PARTS. Iris brownish pink to dull brick-red with a thin yellow inner ring. Bill dark horny, paler on base of lower mandible. Legs and feet pale slate; claws horny; soles yellowish grey.

1281. *Garrulax striatus sikkimensis* (Ticehurst)

Grammatoptila striata sikkimensis Ticehurst, 1924,
Bull. Brit. Orn. Cl. 44: 104 (Sikkim)
Baker, FBI No. 177 (part), Vol. 1: 184

LOCAL NAMES. *Nampiok-pho* (Lepcha); *Kopiam* (Bhutea).
SIZE. Myna +; length *c.* 28 cm (11 in.).

FIELD CHARACTERS. As in 1279; see Museum Diagnosis.

STATUS, DISTRIBUTION and HABITAT. Common resident, subject to some vertical movements. Eastern Nepal from the Arun Kosi river to eastern Bhutan. Breeds between 1500 and 2700 m, optimum zone 1800–2400 m; in winter descends to *c.* 750 m in Sikkim and to the foothills of Bhutan. Recorded also as high as 2700 m in winter. Affects dense rain forest with heavy undergrowth; also secondary and scrub jungle.

GENERAL HABITS and FOOD. As in 1279.

VOICE and CALLS. As in 1279. Flocks keep up a sharp conversational squealing, reminiscent of the Pariah Kite's (*Milvus migrans*). Song, recorded in Sikkim and Bhutan—doubtless variants of those under 1279—a loud rich musical whistle uttered with shaggy crest erected *O-willyou-willyou-wit* sometimes ending with *wit-witoo*. Two common variants of this are *wheeyou-you-witoo* and *white to greet you* (accent on *greet*). These phrases repeated unvaryingly every 3 to 5 seconds for many minutes from the same stance. Another call is rather reminiscent of the 'brain fever' call of hawk-cuckoo, *tiwo-wo* (first *wo* loudest) repeated singly at intervals, not in runs or crescendo.

BREEDING. As in 1279. *Eggs* said to be more pointed than those of the western races. Average size of 16 eggs 33.1 × 22.7 mm (Baker).

MUSEUM DIAGNOSIS. Differs from *vibex* in being darker, more red-brown.

MEASUREMENTS

	Wing	Bill (from skull)	Tarsus	Tail
♂♀[1]	126–152 (mostly 130–142)	23–30	39–47	121–137 mm

(SA, SDR, CBT, Stresemann)

Weight 1 ♂ 92; 3 ♀ ♀ 99–106 g (SA).

COLOURS OF BARE PARTS. Iris reddish brown (the iris of one specimen examined immediately after death was pinkish biscuit colour; two hours later brownish scarlet. SA). Bill blackish brown. Legs and feet plumbeous; claws horny brown.

1282. *Garrulax striatus cranbrooki* (Kinnear)

Grammatoptila striata cranbrooki Kinnear, 1932, Bull.
Brit. Orn. Cl. 53: 79 (Adung Valley, Burma)
Garrulax striatus brahmaputra Hachisuka, 1953, Auk 70: 92.
New name for *Grammatoptila austeni* Oates, preoccupied.
Grammatoptila austeni Oates, 1889, Fauna Brit. India, Birds 1: 104 (Dafla and Naga Hills, Assam). Not *Trochalopteron Austeni* Godwin-Austen, 1870
= *Garrulax austeni austeni* (Godwin-Austen)
Baker, FBI No. 178, Vol. 1: 185

LOCAL NAME. *Daopa* (Cachari).

SIZE. Myna +; length *c.* 28 cm (11 in.).

FIELD CHARACTERS. As in 1279 but with a broad black stripe on side of crest from eye to nape, and no shaft-streaks on crown.

---

[1] A remarkable disproportion in the sexes has been noted. Of 14 specimens collected in Sikkim and Bhutan between January and April in different years, 13 birds in a row proved to be female, only the last being a male! It is difficult to accept this as mere coincidence (SA).

STATUS, DISTRIBUTION and HABITAT. Common resident. From eastern Bhutan through Arunachal Pradesh in the Mishmi and Patkai hills and Assam in the Cachar hills, Meghalaya in the Khasi hills, Nagaland and Manipur (?); from *c.* 1400 to at least 2400 m, reaching the foothills in winter (600 m in Arunachal Pradesh). Affects deep forest of great trees with plenty of undergrowth.

Extralimital. The Chin Hills of Burma.

GENERAL HABITS, FOOD and VOICE. As in 1279, q.v.

BREEDING. *Season,* April to August, chiefly May and June. *Nest* as in 1279. *Eggs,* 2 or 3, pale blue. Average size of 34 eggs 31.5 × 23.5 mm (Baker). Both sexes share, at least, incubation.

MUSEUM DIAGNOSIS. See Field Characters.

MEASUREMENTS

	Wing	Bill (from skull)	Tail	
♂♂	134–149	29–31	120–129	mm
♀♀	133–143	*c.* 31	127–130	mm
		(Kinnear, SDR, HW)		

COLOURS OF BARE PARTS. Iris reddish brown. Bill blackish brown. Legs and feet greyish brown; soles yellowish.

GARRULAX LEUCOLOPHUS (Hardwicke): WHITECRESTED LAUGHING THRUSH

Key to the Subspecies

Abdomen paler, pectoral belt rufous . . . . *G. l. leucolophus*
Abdomen darker, pectoral belt chestnut . . . . *G. l. patkaicus*

1283. *Garrulax leucolophus leucolophus* (Hardwicke)

Corvus leucolophus Hardwicke, 1815, Trans. Linn. Soc. London 11: 208, pl. 15 (Mts above Hardwar)
Baker, FBI No. 128 (part), Vol. 1: 146
Plate 78, fig. 16

LOCAL NAMES. *Rawil-kahy* (Hindi); *Karrio-pho* (Lepcha); *Karria-goka* (Bhutea); *Puhu* (Mishmi).

SIZE. Myna +; length *c.* 28 cm (11 in.).

FIELD CHARACTERS. An unmistakable large olive-brown laughing thrush with a white crested head, throat and breast and prominent black eyemask.

Above, crown and crest white becoming slightly ashy on nape. A broad eye-stripe from lores to ear-coverts jet black. A rufous collar blending into the olive-brown back. Tail blackish. *Below,* throat, sides of neck and breast white bordered by a rufous band joining the nuchal collar. Belly olive-brown. Sexes alike.

STATUS, DISTRIBUTION and HABITAT. Common resident. The Himalayas from Chamba east through Nepal, Sikkim, N. Bengal (Jalpaiguri district), Bhutan and Arunachal Pradesh in the Mishmi Hills; from the duars and foothills to 1700 m, Pokhara (W. Nepal, 1000 m TJR), locally (Sikkim) up to 2100 m, extending to adjacent plains in the eastern half of its range (east of

Nepal). Optimum breeding zone 600 to 1500 m. Observed at 1700 m in winter; there may be some downward movement in the cold season. Affects forest with dense undergrowth and secondary scrub and bamboo jungle—especially broken foothills country with wooded ravines and bordering terraced cultivation.

× *c*. 1

GENERAL HABITS. A very noisy species, gregarious at all seasons, keeping in flocks of six to twelve, sometimes up to forty individuals; often accompanied by other laughing thrushes, tree pies, Green Magpies and occasionally Red Junglefowl. Feeds mostly on the ground, turning over and flicking aside dead leaves, uttering soft single contact chuckles continuously. Shy as a rule but may also be very inquisitive. Very boisterous when disturbed, exploding into choruses of loud cackling 'laughter'. The flocks move about in scattered follow-my-leader style from tree to tree, three or four birds sometimes perching huddled up affectionately on a twig for a brief moment. On the ground they progress in long bouncing or volplaning hops. Like its white-throated congener (1274) often seen rummaging for scraps among wood ashes on the site of pilgrims' or muleteers' bivouacs.

FOOD. Chiefly insects, berries and seeds; also small reptiles and flower-nectar. Larger items held under foot and torn with bill or hacked by vigorous hammer blows of the bill.

VOICE and CALLS. One of the noisiest birds, always calling one another with a variety of notes, bursting out every now and again into a cacophony of cackling choruses 'in which each member tries to outshout the rest' (Baker). One bird leads with a loud, pleasant *pick* or *pick-wo* and the whole flock chimes in with a tumult of discordant cackling; throughout the performance the leader beats time with his *pick-wo* refrain at regular intervals, bill raised skyward and his half-drooped wings fluttering—as if conducting an orchestra! (SA). On the ground these outbursts are often accompanied by dancing, posturing and flapping of wings by the entire company. In the distance the clamour sounds not unlike the yelping of a pack of hounds in full cry. It has also been aptly rendered by one observer (Zafar Futehally) as *Rē-rē-rē, mărigio, mărigio, mărigio*, etc. (= 'I am dead' or 'Help! murder!' in Gujarati).

BREEDING. *Season*, end of March to September. Mostly April to June. *Nest*, a large but shallow cup roughly made of grass, bamboo leaves, roots, moss or other material, loosely bound with creepers and tendrils, and lined with rootlets. Generally placed in shrubs within reach of the hand among low, dense jungle, sometimes as high as six metres. *Eggs*, 3 to 6, normally 4, white with innumerable tiny pits over the whole surface. Average size of 30 eggs

29.2 × 23.5 mm (Baker). Both sexes incubate; incubation period about 14 days. Frequently brood-parasitized by the cuckoos *Clamator jacobinus* and *C. coromandus*.

MUSEUM DIAGNOSIS. Differs from *patkaicus* in being paler; nuchal collar and pectoral belt rufous.

Young, like adult but crest shorter; nuchal feathers ashy brown. More ferruginous on upperparts; coverts and outer edges of wing more rusty; underparts suffused with dull vinaceous brown. Postjuvenal moult complete.

MEASUREMENTS

	Wing	Bill (from skull)	Tarsus	Tail	
♂♂	124–136	28–30	46–48	125–131	mm
♀♀	129–138	27–30	45–49	125–135	mm

(BB, SA, Rand & Fleming)

♂♀ 130–140 (SDR)

Weight 3 ♂♂ 123–129; 5 ♀♀ 119–123 g (GD, SA).

COLOURS OF BARE PARTS. Iris reddish brown; orbital skin pale bluish slate. Bill black. Legs and feet dull black or plumbeous; claws dark horny; sole yellowish grey.

1284. *Garrulax leucolophus patkaicus* Reichenow

Garrulax patkaicus Reichenow, 1913, Jour. f. Orn. 61: 557
(Patkai Mountains, Upper Burma)
Garrulax leucolophus hardwickii Ticehurst, 1926, Bull.
Brit. Orn. Cl. 46: 113 (Naga Hills)
Baker, FBI No. 128 (part), Vol. 1: 146

LOCAL NAMES. *Naga-dhaopuleka* (Assam); *Dao-flantu* (Cachari); *Ngo* (Naga).

SIZE. Myna +; length *c*. 28 cm (11 in.).

FIELD CHARACTERS. As in 1283; see Museum Diagnosis.

STATUS, DISTRIBUTION and HABITAT. Common resident. The hills of eastern Arunachal Pradesh (Noa Dihing Watershed), Nagaland and Manipur, and Bangladesh in the northeastern hills and the Chittagong region; from the base of the hills to *c*. 1800 m, most common between 400 and 800 m. Affects deep forest, dense secondary growth on abandoned cultivation and, less often, bamboo-jungle.

Extralimital. Northern and western Burma. Other races in Thailand, the Indochinese countries and western Sumatra.

GENERAL HABITS, FOOD and VOICE. As in 1283.

BREEDING. As in 1283.

MUSEUM DIAGNOSIS. Differs from the nominate race (1283) in being darker, especially on belly; nuchal collar and pectoral belt chestnut.

MEASUREMENTS and COLOURS OF BARE PARTS. As in 1283.

Weight ♂ ♀ 104–130 g (SDR).

1285. Chestnutbacked Laughing Thrush. *Garrulax nuchalis*
Godwin-Austen

Garrulax nuchalis Godwin-Austen, 1876, Ann. Mag. Nat. Hist. 18: 411
(Khasi-Naga Hills, North Bengal)
Baker, FBI No. 121, Vol. 1: 140
Plate 78, fig. 10

LOCAL NAME. *Pak-chi-loka* (Trans-Dikku Naga).
SIZE. Myna; length *c.* 23 cm (9 in.).
FIELD CHARACTERS. *Above*, forehead, lores, eye-rim and a short stripe behind eye black. Ear-coverts and sides of neck white. Crown slaty with a few small white feathers in front. A broad rufous-chestnut nuchal collar. Rest of upperparts olive-brown, the outer edge of wing light grey and the tip of tail black. *Below*, chin and throat black. Breast pale grey. Belly olive-brown. Sexes alike.

The large white cheek-patch in combination with the black throat and rufous collar identifies this species.

STATUS, DISTRIBUTION and HABITAT. Resident, locally common. From the Mishmi Hills south through eastern Assam (East Lakhimpur), the Patkai Range, Nagaland, east Manipur; from the base of the hills to *c.* 900 m. Affects thick scrub jungle on broken ground, and rocky scrub-clad ravines; also high grass.

Extralimital. Extends to northern Burma. The species ranges east to Vietnam, Hainan and Kwangtung.

GENERAL HABITS. Keeps in small parties sometimes in company with other laughing thrushes, feeding on the ground in thick scrub, each bird every now and then clambering up to the top of a bush and uttering loud calls and soon joined in chorus by the others.

FOOD. Insects (ants etc.) recorded. Presumably also berries and seeds.

VOICE and CALLS. An unmistakable *churr* when alarmed; a rich, loud whistling song of four or five notes which at once attracts attention (Smythies). Call-note, a soft *chip* (Stanford).

BREEDING. *Season,* March to July, mostly May and June. *Nest,* a neat and compact cup of bracken with an inner layer of dead leaves and broad grass-blades, lined with moss, rootlets and fibres. Placed in dense-foliaged bushes within a metre or so from the ground. *Eggs,* 3 sometimes 2, very pale blue (an abnormal clutch of white eggs has been taken). Average size of 40 eggs 28.5 × 20.7 mm (Baker). Both birds incubate. Sits very close, but slips away quietly into the low jungle when approached, uttering a low chuckle as it disappears.

MUSEUM DIAGNOSIS. *G. propinquus* of southern Burma (Vol. 6, plate 72) lacks the rufous collar.

MEASUREMENTS

	Wing	Bill (from skull)	Tarsus	Tail
♂♀	106–117	26–27	41(♀)	106–115 mm (HW, Baker)

Weight ♀ 71 g.

COLOURS OF BARE PARTS. Iris reddish brown to brick red; orbital skin grey. Bill black. Legs and feet pale fleshy or fleshy grey.

1286. Yellowthroated Laughing Thrush. *Garrulax galbanus galbanus* Godwin-Austen

Garrulax galbanus Godwin-Austen, 1874, Proc. Zool. Soc. London: 44, pl. 10 (Manipur Valley, NE. Bengal)
Garrulax galbanus galbanatus Koelz, 1954, Contrib. Inst. Regional Exploration, No. 1: 2 (Blue Mountain, Lushai Hills)
Baker, FBI No. 127, Vol. 1: 107
Plate 78, fig. 5

LOCAL NAMES. None recorded.

SIZE. Myna; length *c.* 23 cm (9 in.).

FIELD CHARACTERS. A striking-looking laughing thrush with a black face and chin and yellow underparts. *Above,* crown and nape ashy-brown with a thin whitish supercilium; rest of head black. Remainder of upperparts ochraceous brown. Outer rectrices with broad white tips preceded by black. *Below,* chin black. Rest of underparts pale yellow, washed with olive on flanks. Under tail-coverts white. Sexes alike.

In flight, the white of the rectrices and the under tail-coverts make it very conspicuous.

STATUS, DISTRIBUTION and HABITAT. Resident, locally common. Nagaland, Manipur, Cachar Hills of Assam, Lushai Hills of Mizoram, and Bangladesh in the Chittagong region; from *c.* 800 to 1800 m. Affects open jungle, tall grass intermixed with trees and shrubs and outskirts of dense evergreen forest.

Extralimital. The Chin Hills of Burma. Another race in northeastern Kiangsi.

GENERAL HABITS. Keeps in pairs or small parties of up to six individuals, sometimes in larger flocks ('50–80'—Hume), often in association with *G. ruficollis.* Feeds on the ground; always on the move in follow-my-leader style through the long grass; when flushed flies up into trees and threads its way through the branches.

FOOD. Chiefly insects; also small seeds.

VOICE and CALLS. A feeble chirping call frequently uttered.

BREEDING. *Season,* April to June. *Nest,* a cup roughly made of grass stems, bamboo leaves, creepers and fine twigs, lined with yellow grass-seed stems or rootlets; the yellow lining seems to be a distinguishing feature. Placed in the fork of a bush between 30 cm and 3 m above the ground. *Eggs,* normally 3, sometimes 2, exceptionally 4, white or 'occasionally very pale blue'. Average size of 80 eggs 25.8 × 18.6 mm (Baker).

MUSEUM DIAGNOSIS. See Field Characters.

MEASUREMENTS.

	Wing	Bill (from feathers)	Tarsus	Tail
♂♀	92–96	20–22	*c.* 35	*c.* 105–110 mm (Baker)

Weight 2 ♂ ♂ 56, 57; 1 ♀ 55 g (SDR).

COLOURS OF BARE PARTS. Iris pinkish brown to reddish brown; orbital skin blue. Bill black. Legs and feet blackish grey.

BABBLERS

GARRULAX DELESSERTI (Jerdon): RUFOUS-VENTED LAUGHING THRUSH

Key to the Subspecies

All rectrices blackish brown .... *G. d. delesserti*
Four outer rectrices rufous .... *G. d. gularis*

1287. *Garrulax delesserti delesserti* (Jerdon)

Crateropus delesserti Jerdon, 1839 (Oct.), Madras Jour.
Lit. Sci. 10: 256 (Wynaad, S. India)
Baker, FBI No. 131, Vol. 1: 149
Plate 78, fig. 6

LOCAL NAMES. *Patungan kili* (Malayalam); *Pūn kūrūvi* (Tamil).
SIZE. Myna; length *c.* 23 cm (9 in.).
FIELD CHARACTERS. An uncrested laughing thrush, chiefly chestnut-brown above, ashy and chestnut-below. Crown and nape brownish slate. Lores, eye-rim and ear-coverts black. Bill mostly yellow. Back chestnut-brown. Tail blackish brown. *Below*, throat white; breast ashy grey. Lower belly and under tail-coverts rufous. Legs dusky yellow. Sexes alike.
STATUS, DISTRIBUTION and HABITAT. Resident, locally common. Range disjunct from that of the subspecies *gularis* of northeastern India: the hills of southwest India from Goa and Belgaum south through western Karnataka, Kerala and western Tamil Nadu; from the base of the hills to the highest elevations. Affects humid rain forest with dense undergrowth of *Strobilanthes*, thorny cane-brakes and cardamom sholas.

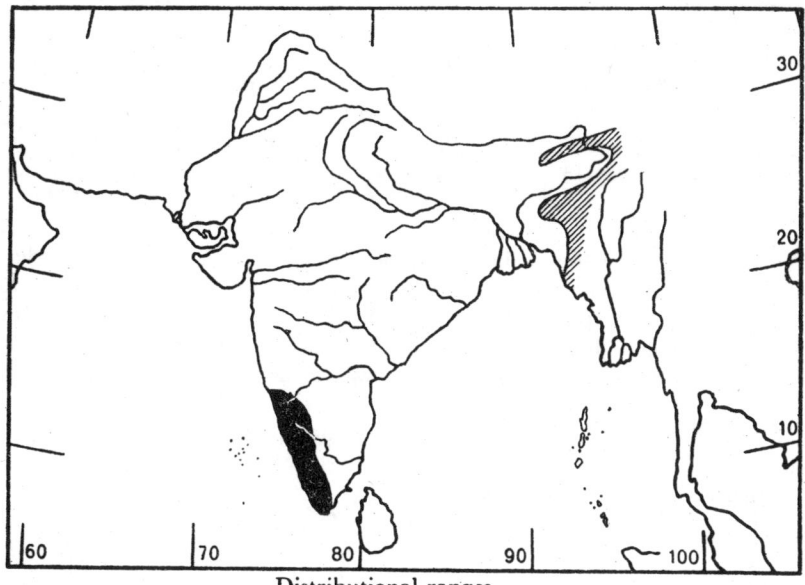

Distributional ranges
■ *delesserti* (1287)   ▨ *gularis* (1288).

GENERAL HABITS. Very gregarious. Keeps in flocks of six to fifteen and sometimes up to forty or fifty individuals. Feeds mostly on the ground, rummaging among the mulch, turning over or flicking aside dead leaves, and occasionally ascending small trees. A half-dozen birds may often be seen huddled affectionately on the same branch, preening each other. A great skulker; scuttles into cover on the least disturbance, the birds hopping swiftly from bush to bush, uttering a chorus of squeaky shrieks as they disappear.

FOOD. Mostly insects; also berries and seeds.

VOICE and CALLS. Characteristic shrill chattering and cackling of the genus, starting with one individual, followed one after another by the rest of the flock till it finally ends up in a chorus of loud, discordant 'laughter' (SA). Also an occasional low harsh churring and a call-note similar to the *chirp* of a fledgling thrush. Voice very similar to that of *Turdoides subrufus*, 1260 (La Personne).

BREEDING. *Season*, ill-defined; mostly during the monsoon: April to August in Kerala, July to September in Kanara. There is also evidence of breeding in December, February and March. *Nest*, a bulky, untidy semi-domed cup of twigs, creepers and roots, lined with rootlets. Placed in bushes, saplings or *Strobilanthes* plants, within a couple of metres from the ground. *Eggs*, 3 or sometimes 4, rarely more, white. Average size of 50 eggs 27.5 × 21.3 mm (Baker).

MUSEUM DIAGNOSIS. Differs from *gularis* in having the crown and nape darker, the feathers of the forehead tipped with black. Back a darker chestnut-brown; tail darker with the under surface blackish brown, not rufous. Underparts white, not yellow; less grey on breast.

J u v e n i l e, and first-winter birds similar to adult but slightly duller above; coverts and edges of wings not so deep chestnut; showing the normal differences of the subfamily Timaliinae, namely the possession of a soft blunt first primary and narrow, pointed rectrices. Postjuvenal moult complete.

MEASUREMENTS

	Wing	Bill (from skull)	Tarsus	Tail	
♂♂	100–113	30–31	*c.* 39	97–107	mm
♀♀	108–113	*c.* 31	—	101–106	mm
				(SA)	

COLOURS OF BARE PARTS. A d u l t : Iris scarlet or 'maroon-brown'. Gape and upper mandible dark horny brown, lower mandible pale yellowish flesh; palate yellow, gullet pink. Legs, feet and claws pinkish flesh with grey tinge. J u v e n i l e and i m m a t u r e : Iris pale pinkish buff. Eye-rim bright yellow, orbital skin paler yellow. Bill: upper mandible horny brown except tip, nostrils, lores and a spot on culmen near forehead, which are yellow; gape and lower mandible bright yellow. Legs and feet dusky yellow, claws paler.

One specimen (age?): Iris brownish orange. A bare post-orbital patch bluish slate. Bill: upper mandible horny brown, lower pale yellow or cream. Legs and feet dirty brownish grey; claws creamy white.

1288. *Garrulax delesserti gularis* (McClelland)

Ianthocincla gularis McClelland, 1839 (1840),
(Oct. 22, 1839 = March 1840), Proc. Zool. Soc. London: 159 (Assam)
Garrulax gularis gratior Koelz, 1954, Contrib. Inst.
Regional Exploration, No. 1: 2 (Sangau, Lushai Hills)
Baker, FBI No. 136, Vol. 1: 152

LOCAL NAMES. None recorded.

SIZE. Myna; length *c.* 23 cm (9 in.).

FIELD CHARACTERS. As in 1287 but white of underparts replaced by yellow, the four outer rectrices rufous and the bill entirely black. Seen from below, the tail is rufous, not blackish. Sexes alike.

STATUS, DISTRIBUTION and HABITAT. Resident, locally common. The Himalayan foothills of eastern Bhutan and Arunachal Pradesh (mostly confined to the base of the hills), the hills of Assam, Nagaland and Manipur, and Bangladesh in the northeastern hills and the Chittagong region; from *c.* 1000 to 1800 m, descending locally to base of hills in winter. Affects thick evergreen undergrowth, dense secondary growth, less often bamboo and scrub jungle.

Extralimital. Northern Burma and northern Laos. Range widely disjunct from that of the nominate race (southwest India).

GENERAL HABITS and FOOD. As in 1287, q.v.

VOICE and CALLS. A loud, rather sweet whistle in addition to the usual cackling notes of its kind (Baker).

BREEDING. *Season*, April to July, mostly May. *Nest*, as in 1287. *Eggs*, generally 3, sometimes 2, white or pale blue. Average size of 100 eggs 29.1 × 20.5 mm (Baker).

MUSEUM DIAGNOSIS. Differs from *delesserti* in having the back olive-brown tinged with rufous instead of chestnut-brown. Crown and nape slate grey. Tail less blackish above, rufous below, the four outer rectrices being rufous. Sides of breast a darker grey. White of underparts replaced by primrose yellow in fresh specimens. This colour however fades rapidly in museum skins and becomes white. The resemblance between the two forms then becomes so striking that they may casually be confused. However, wing in *gularis* usually *under* 100 mm, in *delesserti* usually 100 mm or over.

Young, like adult but wings richer in colour; ashy grey of breast mixed with rusty; crown black fringed at base with rufous.

MEASUREMENTS

	Wing	Bill (from skull)	Tarsus	Tail	
♂♀	95–104	29–32	38–42	90–94	mm

(Baker, MD, SA)

Weight 1 ♂ 92 g (SDR); 1 ♀ 72 g (SDR).

COLOURS OF BARE PARTS. Iris reddish brown. Bill black. Legs and feet yellowish orange.

GARRULAX VARIEGATUS (Vigors) VARIEGATED LAUGHING THRUSH

Key to the Subspecies

Edges of primaries grey.......................... *G. v. similis*
Edges of primaries yellow .... *G. s. variegatus*

1289. *Garrulax variegatus similis* (Hume)

Trochalopteron simile Hume, 1871, Ibis: 408 (Far Northwest = Gilgit)
Baker, FBI No. 163, Vol. 1: 174

LOCAL NAMES. None recorded.
SIZE. Myna ±; length *c.* 24 cm (10 in.).
FIELD CHARACTERS. *Above*, forehead tawny; lores black. Crown, nape and ear-coverts dark grey with a short white streak behind eye. Back, rump and upper tail-coverts olive-brown. Wings silvery grey with a black and a rufous shoulder-patch and a larger black patch on centre of wing; secondaries blackish tipped with white. Tail black, distally grey with white tip, the latter colour more in evidence when tail is spread. *Below*, chin and centre of throat black, broadly bordered with buff. Breast and flanks pale olive-brown. Lower belly and under tail-coverts rufous. Sexes alike.

The black and grey pattern of wing and tail, whitish cheeks and black band down centre of throat identify this species.

STATUS, DISTRIBUTION and HABITAT. Common resident, subject to vertical movements. The western Himalayas from Kohat, Chitral and Gilgit east through the mountains of northern Pakistan, the outer ranges of Kashmir to Lahul and Chamba, meeting the nominate race in Saraj. In Kangra both subspecies may be seen on the same ground in winter, sometimes in the same flock. Breeds between 1800 and 3800 m, mostly above 2400 m. Found in winter from 1200 to at least 2100 m. In Fulton's time (1904-5, JBNHS 16: 44-64; 744), it was common in Chitral 'throughout the year up to 1800 m',

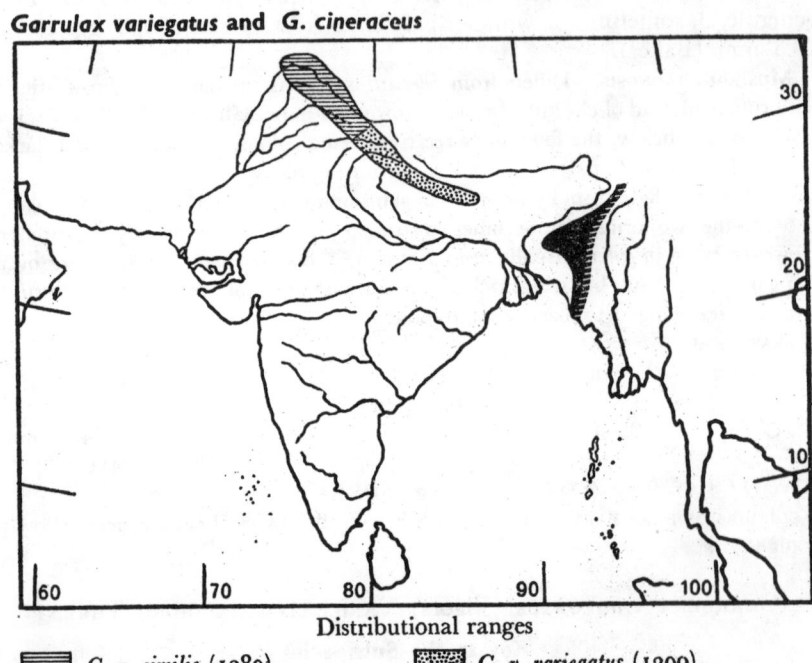

Garrulax variegatus and *G. cineraceus*

Distributional ranges

▤ *G. v. similis* (1289).   ▨ *G. v. variegatus* (1290).
■ *G. c. cineraceus* (1291).

but is now confined to deodar (*cedrus*) forest in southern Chitral (T. J. Roberts). Affects open forest of fir and birch with dense rhododendron and ringal bamboo growth, patches of *Viburnum* at the forest edges as well as various types of dense jungle, especially in winter. Occasionally enters gardens. Has adapted itself to open willow groves in the Chandra and Bhaga valleys in Lahul, Himachal Pradesh (Alexander, H. G., 1951, JBNHS 49: 609).

GENERAL HABITS. A great skulker. Keeps in pairs during the breeding season, otherwise in flocks up to twenty or more. Feeds among bushes but often ascends trees, hopping energetically from branch to branch close to the trunk, diving into cover on alarm. Flight weak, interspersed with short sailings, tail spread wide.

FOOD. Insects, berries and fruits.

VOICE and CALLS. Loud musical whistles rendered as *weet-a-weer* or *weet-a-woo-weer*, far-reaching and unmistakable (Bates); also transcribed as *zdrip-diu-i-wiuh* and *dio-pi-wiah* (*i* pronounced as *ee*) often well represented by the words *choky william* 'which is immediately taken up by hidden accomplices in the bushes around' (Magrath). In the non-breeding season, when in flocks, often call with a *pte-weer* similar to that of the Streaked Laughing Thrush but louder and clearer; when alarmed, utters subdued muttering and squealing notes which sound like a nestful of young chicks clamouring for food (Bates).

BREEDING.' *Season*, April to August, chiefly May and June. *Nest*, a large and somewhat untidy cup of grass with some strips of birch bark, leaves or moss, lined with rootlets, finer grass or pine needles. Usually placed in bushes, about one metre off the ground, sometimes in the fork of a small tree up to six metres or so. *Eggs*, normally 3, sometimes 2, rarely 4, blue or blue-green profusely blotched and spotted, especially near the large end, with liver-brown, red-brown, dark brown or brownish black. Average size of 60 eggs 27·8 × 21 mm (Baker). Both sexes incubate. Sometimes brood-parasitized by Indian Cuckoo (*Cuculus micropterus*, 576).

MUSEUM DIAGNOSIS. Differs from nominate *variegatus* (1290) in having the outer webs of wing-feathers grey and the yellow on tail replaced by grey.

MEASUREMENTS. As in 1290.
Weight 10 ♂ ♀ (Apr.-May) 59–72 (av. 64·5) g—SA.

COLOURS OF BARE PARTS. Iris peridot-green (Meinertzhagen). Bill black. Legs and feet flesh-colour.

1290. *Garrulax variegatus variegatus* (Vigors)

Cinclosoma variegatum Vigors, 1831, Proc. Zool. Soc. London: 56
(Himalayas = Simla-Almora area, according to Ticehurst & Whistler, 1924, Ibis: 471)
Baker, FBI No. 162, Vol. 1: 173
Plate 78, fig. 13

LOCAL NAME. *Ganza* (Nepal).

SIZE. Myna ±; length *c.* 24 cm (10 in.).

FIELD CHARACTERS. As in 1289 but grey of wings and tail replaced by yellow. Grey head, white cheeks, black band down centre of throat, yellow in wings and tail and rusty underparts are leading pointers. Sexes alike.

STATUS, DISTRIBUTION and HABITAT. Common resident, subject to vertical movements. From the Kareri Lake between Dharmsala and Dalhousie (*c.* 76°E.) east to east-central Nepal (Inskipp & Inskipp). One of the highest altitude laughing thrushes; breeds from 2100 m to 3300 m on the Duala Dhar and around Simla, generally above 2400 in Garhwal, and from 2700 m to tree-line (4100 m) in Nepal. From October to March, usually below 2100 m, down to 1000 m but also recorded as high as 2700 m at this season in Nepal. Affects forests of oak, fir or birch with dense undergrowth of rhododendron and other bushes, or dwarf rhododendron patches at or above timber-line. In winter frequents steep hillsides with dense ringal bamboo, briar and other undergrowth.

GENERAL HABITS, FOOD and VOICE. As in 1289. Shrill musical whistles *p'ti-pieeyou* or *pitēē-whēē* as contact calls (SA).

BREEDING. As in 1289.

MUSEUM DIAGNOSIS. For details of plumage see Baker, loc. cit. Colouring of edgings of primaries and of rectrices very variable. Typically these are yellow, varying from golden to orange and even olive-yellow, but specimens with reddish orange, orange-brown and pink have been recorded.

Young, a dull version of the adult, a little darker above and less bright below.

MEASUREMENTS

	Wing	Bill (from skull)	Tarsus	Tail	
♂♂	101–109	*c.* 23	*c.* 38	*c.* 120	mm
♀♀	98–103	—	—	—	mm
		(Rand & Fleming, MD)			
		(from feathers)			
♂♀	102–112	*c.* 20	*c.* 38	*c.* 130	mm
				(Baker)	

Weight 2 ♂♂ 67, 69 g (GD).

COLOURS OF BARE PARTS. Iris pale yellow-green, brown, raw sienna-brown, pale yellowish brown. Bill black. Legs and feet pale reddish orange-brown (Hume) *or* Iris pale yellow. Bill dark brown, yellowish at base of lower mandible. Legs and feet pale brown (HW).

1291. Ashy Laughing Thrush. *Garrulax cineraceus cineraceus* (Godwin-Austen)

Trochalopteron cineraceum Godwin-Austen, 1874, Proc. Zool. Soc. London: 45
(Manipur Valley, NE. Bengal)
Baker, FBI No. 141, Vol. 1: 156
Plate 78, fig. 4

LOCAL NAME. *Lehu* (Angami Naga).

SIZE. Myna; length *c.* 22 cm (9 in.).

FIELD CHARACTERS. *Above*, forehead and centre of crown to nape black. Lores, orbital area and ear-coverts whitish. A short black eye-stripe and black moustachial stripe breaking into short streaks on sides of head. Sides of crown and back umber-brown; tail graduated, of same colour with a black subterminal band and tipped with white, especially conspicuous when tail is spread. Wing edged with ashy, the secondaries with broad subterminal black band tipped with a thin white crescent. *Below*, throat buffish with short dark streaks. Breast pinkish grey. Belly and under tail-coverts tawny-olive. Sexes alike.

The white crescents on black wing-tip and streaked throat in conjunction with the white tips of tail identify this species.

STATUS, DISTRIBUTION and HABITAT. Common resident, subject to seasonal vertical movements. Assam in the Cachar hills, Nagaland and Manipur (see map, p. 22). Breeds from 1500 m upwards, generally above 1800 m. Tytler (*apud* Baker) found it breeding between 2100 and 2400 m near Kohima (Nagaland) while Ripley (JBNHS 50: 496) met with it between 1200 and 1500 m in winter in the same region. Affects thick bushes in damp forest and thick scrub and secondary growth near villages or cultivation. Avoids high or shady forest.

Extralimital. The Chin Hills of Burma. The species extends to northern Burma, western Yunnan and southeastern Sichuan.

GENERAL HABITS. Keeps in pairs in the breeding season, otherwise in small parties. Feeds mostly on the ground scratching and turning over dead leaves, or even cattle dung in search of insects.

FOOD. Insects and berries.

VOICE and CALLS. A variety of low, rather musical calls. Alarm, a thrush-like call. Song, usually uttered from a bush-top, a loud *dü-düuid* reminiscent of call of *Pomatorhinus erythrogenys* (Schäfer).

BREEDING. *Season*, April to June. *Nest*, a cup of moss, leaves, rootlets, grass and twigs, lined with rootlets or fine stems. Placed in thick bushes within two metres of the ground. *Eggs*, 2 or 3, unspotted blue. Average size of 150 eggs 25·3 × 18·6 mm (Baker). Brood-parasitized by Hawk-Cuckoo (*Cuculus varius*).

MUSEUM DIAGNOSIS. See Field Characters.

MEASUREMENTS

	Wing	Bill (from feathers)	Tarsus	Tail	
♂♀	86–89	*c.* 20	*c.* 32	*c.* 100	mm
				(Baker)	

Wing, 2 ♂♂ 90, 92; 2 ♀♀ 88, 88 mm (Heinrich).
Weight 4 ♂♀ 47–51 g (SDR).

COLOURS OF BARE PARTS. Iris creamy yellow to pinkish cream. Bill: upper mandible brownish horn, lower yellowish or whitish horn. Legs and feet pale brownish flesh.

GARRULAX RUFOGULARIS (Gould): RUFOUSCHINNED LAUGHING THRUSH

Key to the Subspecies

		Page
A	Crown dark brown or blackish brown	
	1 Ear-coverts rufous .... *G. r. occidentalis*	26
	2 Ear-coverts black .... *G. r. rufogularis*	27
B	Crown black, tail chestnut-brown	
	3 Rusty of throat extending to breast .... *G. r. rufitinctus*	28
	4 Rusty of throat more restricted	
	a Greyer below, paler above .... *G. r. rufiberbis*	28
	b Less grey below, darker above .... *G. r. assamensis*	29

1292, 1293. ***Garrulax rufogularis occidentalis*** (Hartert)

Ianthocincla rufogularis occidentalis Hartert, 1909, Vög. pal. Fauna 1: 635
(Dehra Dun)
Garrulax rufogularis grosvenori Ripley, 1950, Proc. Biol. Soc. Washington 63: 104
(Rekcha, Dailekh Dist., western Nepal)
Baker, FBI No. 145, Vol. 1: 159

LOCAL NAMES. None recorded.

SIZE. Myna; length *c.* 22 cm (9 in.).

FIELD CHARACTERS. As in nominate *rufogularis* (1294) but ear-coverts rufous, white apical crescents on secondaries thinner and less apparent, upperparts, especially tail, more olive and spotting on underparts lighter.

STATUS, DISTRIBUTION and HABITAT. Resident, locally common. The Himalayan foothills from Simla westwards. Specimens were collected from the Murree foothills (Jhelum Valley, Pakistan) by Biddulph in 1873 and from Lolab Valley, Kashmir, by Ward prior to 1906, but it has not been recorded from these regions since, and is believed to be extinct (TJR). East to central Nepal; from *c.* 600 to 2135 m (Inskipp & Inskipp), mostly 900 to 1200 m. Affects dense thickets and scrub jungle especially on the edges of cultivation.

GENERAL HABITS, FOOD and VOICE. As in 1294.

BREEDING. *Season*, April to August. *Nest* and *eggs* (c/2 or 3) as in 1294. Average size of 20 eggs 26·1 × 19·4 mm (Baker).

Distributional ranges

| occidentalis (1292). | rufogularis (1294). | rufiberbis (1295a). |
| assamensis (1296). | rufitinctus (1295). |

MUSEUM DIAGNOSIS. As *rufogularis* (1294) but paler, more olive, less rufous above; ear-coverts rusty red. See also Field Characters.

MEASUREMENTS and COLOURS OF BARE PARTS. As in 1294.

1294. *Garrulax rufogularis rufogularis* (Gould)

Ianthocincla rufogularis Gould, 1835, Proc. Zool. Soc. London: 48
(Himalayas = Sikkim)
Cinclosoma rufimenta Hodgson, 1836, Asiat. Res. 19: 148 (Nepal = Kathmandu)
Baker, FBI No. 143, Vol. 1: 158
Plate 77, fig. 9

LOCAL NAME. *Narbigivan-pho* (Lepcha).

SIZE. Myna; length *c.* 22 cm (9 in.).

FIELD CHARACTERS. *Above*, forehead and centre of crown to nape black. Lores buff. Ear-coverts black. Sides of head and back olive-brown, latter scaled with crescentic black marks. Tail graduated, chestnut, with a broad black subterminal band and rufous tip showing as a broad rufous band when it is spread. Wings black and grey with whitish outer edges, the secondaries with subterminal black band tipped with a white crescent. *Below*, chin and under tail-coverts rufous. Throat whitish with black sides mingled with white. Breast pale greyish, flanks olive-brown, both with black spots. Belly whitish. For sex differences see Museum Diagnosis.

The black-spotted back and rufous chin identify this species.

STATUS, DISTRIBUTION and HABITAT. Resident, locally common, subject to some seasonal vertical movements. The Himalayan foothills from central Nepal east through Sikkim, Darjeeling, Bhutan and Arunachal Pradesh at least to the Miri Hills; from *c.* 600 to 2135 m (Inskipp & Inskipp), optimum zone 1000–1800 m. Reported from as low as the Buxa duars (Inglis) and exceptionally as high as 3500 m in December in Sikkim (Meinertzhagen). Elevations below 900 m are winter records. Affects dense undergrowth in oak and rhododendron forest and forest edges, scrub jungle and secondary growth near cultivation.

GENERAL HABITS. Less gregarious than most laughing thrushes. Keeps in pairs or small family parties haunting low bushes and feeding mostly on the ground. Takes to wing with reluctance; flight weak and ill-sustained. A great skulker more often heard than seen and doubtless often overlooked.

FOOD. Insects, berries and seeds.

VOICE and CALLS. Not very noisy but has the usual range of chuckles and low conversational chatter of the genus, and some loud squealing alarm-notes.

BREEDING. *Season*, April to September. *Nest*, a rather deep cup made mostly of tendrils, with some twigs, roots, leaves, scraps of bracken and grass-bents, some nests being made entirely of one or two of these materials to the exclusion of the rest; the lining is almost always of rootlets. Usually built in a fork of a bush or tree at heights varying from *c.* 60 cm to 6 metres. *Eggs*, normally 3, sometimes 2 or 4, white. Average size of 15 eggs 26·2 × 19·4 mm (Baker). Both sexes take part in incubation. Period undetermined. Brood-parasitized by the cuckoos *Cuculus sparverioides* and *Clamator coromandus*.

MUSEUM DIAGNOSIS. Female greyer, less olive-brown on upperparts than male; black spots smaller and paler; white tips of flight-feathers narrower. Sexes differ particularly in the colour of the underparts; greyer in male, browner in female; olive-brown of breast and flanks more extensive in male (Diesselhorst).

Young, upperparts unspotted, except crown which is slightly spotted with blackish; underparts spotted as in adult; chin pale rusty.

Differs from *occidentalis* (1292) in being more rufous above and in having the tail chestnut-brown with broader subterminal black band and deeper rufous tips; ear-coverts black.

MEASUREMENTS

	Wing	Bill (from skull)	Tarsus	Tail	
♂♂	88–98	23–24	c. 35	98–110	mm
♀♀	87–99	23–25	c. 37	97–103	mm

(BB, SA, Rand & Fleming)

Weight 5 ♂♂ 58–68; 2 ♀ ♀ 58, 58 g (SA).

COLOURS OF BARE PARTS. Iris crimson-brown. Bill: upper mandible horny brown, lower pale whitish horn, darker at gape. Legs and feet greyish horny brown; claws pale horny brown.

1295. *Garrulax rufogularis rufitinctus* (Koelz)

Ianthocincla rufogularis rufitincta Koelz, 1952, Jour. Zool. Soc. India 4: 37
(Pynursla, Khasia Hills)
Baker, FBI No. 144 (part), Vol. 1: 159

LOCAL NAMES. None recorded.
SIZE. Myna; length *c.* 22 cm (9 in.).
FIELD CHARACTERS. As in 1294, q.v.
STATUS, DISTRIBUTION and HABITAT. Resident. Meghalaya in the Khasi and Garo hills, between 900 and 1800 m. Habitat as in 1296.
GENERAL HABITS, FOOD and VOICE. As in 1294.
BREEDING. As in 1294.
MUSEUM DIAGNOSIS. Differs from *rufiberbis* (1295a) in being paler; feathers of crown have much less pronounced black edging; black markings of back reduced in size; rusty of throat extends to breast; a rusty wash over the belly.
MEASUREMENTS and COLOURS OF BARE PARTS. As in 1294.

1295a. *Garrulax rufogularis rufiberbis* (Koelz)

Ianthocincla rufogularis rufiberbis Koelz, 1954, Contrib. Inst. Regional Exploration
No. 1: 3 (between Langyang and Htawgaw, Kachin State, Upper Burma)
Not in Baker, FBI

LOCAL NAMES. None recorded.
SIZE. Myna; length *c.* 22 cm (9 in.).
FIELD CHARACTERS. As in 1294, q.v.
STATUS, DISTRIBUTION and HABITAT. Resident. The Patkai Range and eastern Nagaland (Ripley, JBNHS, 58: 281), at *c.* 1200 m. Habitat as in 1296.

Extralimital. Northern Burma. Another subspecies in northern Vietnam.
GENERAL HABITS, FOOD and VOICE. As in 1294.
BREEDING. Unrecorded.
MUSEUM DIAGNOSIS. Differs from nominate *rufogularis* (1294) in being more uniformly grey, the rufous coloration of the chin paler and more restricted. Colour of back brighter, black edges of feathers somewhat larger. Crown black rather than dark brown of Himalayan races. Greyer below than *assamensis* with reduced and paler rufous on chin; paler above and on lores.
MEASUREMENTS and COLOURS OF BARE PARTS. As in 1294.
Weight ♀ 55 g.

1296. **Assam Rufouschinned Laughing Thrush.** *Garrulax rufogularis assamensis* (Hartert)

Ianthocincla rufogularis assamensis Hartert, 1909, Vög. pal. Fauna 1: 635
(Margherita, Assam)
Baker, FBI No. 144 (part), Vol. 1: 159

LOCAL NAME. *Mi-pa-pita* (trans-Dikku Naga).
SIZE. Myna; length *c.* 22 cm (9 in.).
FIELD CHARACTERS. As in 1294, q.v.
STATUS, DISTRIBUTION and HABITAT. Resident. Assam from Margherita and Cachar south to the Chin Hills of Burma, and Bangladesh in the Chittagong Hill Tracts. Affects heavy undergrowth in forest, and cut-over scrub.
GENERAL HABITS, FOOD and VOICE. As in 1294.
BREEDING. As in 1294.
MUSEUM DIAGNOSIS. Like *rufogularis* (1294) but chin and part of ear-coverts pale rust; crown black; black on sides of throat more extensive. Not so fulvous on upper-parts. On the whole a somewhat more saturated form than 1294.
MEASUREMENTS and COLOURS OF BARE PARTS. As in 1294.
Weight 1 ♂ 67; 1 ♀ 64 g (SDR).

GARRULAX OCELLATUS (Vigors):
WHITESPOTTED LAUGHING THRUSH

Key to the Subspecies

Tail chestnut-brown .... *G. o. ocellatus*
Tail with much grey .... *G. o. griseicauda*
Tail very long, graduated, white-tipped .... *G. o. maximus*

1297. *Garrulax ocellatus maximus* (Verreaux)

Pterorhinus maximus Verreaux, 1871, Nouv. Arch. Mus. Paris 6,
Bull. 36, pl. 3, fig. 1 (Montagnes du Thibet chinois = Mouping)
Not in Baker, FBI
Plate 77, fig. 15

LOCAL NAME. *Gya tra* (Tibetan).
SIZE. Pigeon ±; length *c.* 35 cm (14 in.).

FIELD CHARACTERS. A large laughing thrush conspicuously spotted with white on the back, and with a very long tail.

Above, crown brown, lores whitish, ear-coverts and supercilium cinnamon-rufous. Upper back and sides of neck grey. Back brown and black with conspicuous, round white spots. Rump and upper tail-coverts rufous brown with same spots. Outer edge of wing ashy; all flight-feathers and their coverts tipped with white. Tail graduated; central rectrices greyish brown, outer rectrices dark brown tipped with white. *Below*, throat and upper breast cinnamon-rufous finely barred with pale buff on the latter. Rest of underparts buff, darker on belly and under tail-coverts, sides of breast barred with blackish. Under surface of tail dark brown with white tips. Sexes alike.

Easily distinguished from *G. ocellatus* (1299) by the rufous throat and brown cap, and longer tail.

STATUS, DISTRIBUTION and HABITAT. Common resident. Southeast Tibet in the lower Tsangpo Valley, between 2200 and 2900 m (up to 4900 m in Sichuan). Affects dense, dry subalpine forest with plenty of glades and undergrowth.

Appears to intergrade with *ocellatus* between the upper Subansiri and the Tsangpo Valley (*vide* Ludlow, 1944, *Ibis* 86: 74). The ranges of these two species are complementary.

Extralimital. Ranges from southeast Tibet and northwestern Yunnan to southern Kansu.

GENERAL HABITS. A secretive species but conspicuous by its loud calls. Very gregarious, often associating with other laughing thrushes. Hops on the

Garrulax maximus and G. ocellatus

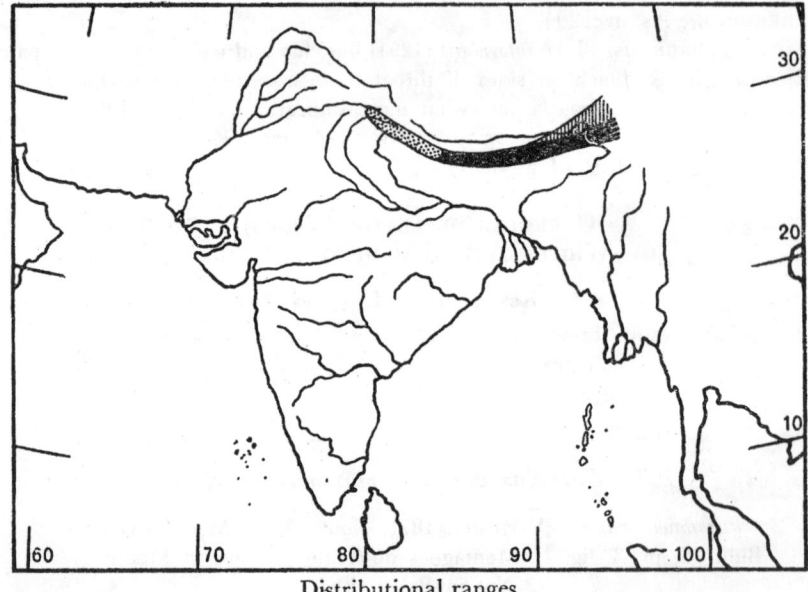

Distributional ranges

G. o. griseicauda (1298).   *G. o. ocellatus* (1299).

G. maximus (1297).

ground where it finds most of its food, tossing over leaves with the long bill.

VOICE and CALLS. Shrill, far-carrying notes reminding one of *Cuculus sparverioides*. Often calls in chorus: an individual starts with shrill screams *gno gnoit gno gnoit*, the others accompanying it with short, jerky, rapid and oft-repeated rattles *tscherr, tscherr, tscherr*. Voice very similar to that of *G. ocellatus* (Schäfer).

FOOD. Unrecorded. Probably as in 1299.

BREEDING. *Season*, presumably May to July (eggs found in June). *Nest*, (one described) a shallow, cup-shaped structure, composed of a foundation of small twigs, lined with dry grass, and placed in a clump of bamboo *c.* 120 cm above the ground. The two eggs measured 36 × 22·5 and 35·2 × 22 mm (Ludlow).

MUSEUM DIAGNOSIS. See Field Characters.

Young. *Above*, as in adult but spots on back less numerous, and triangular, not round. Rump and upper tail-coverts unspotted. *Below*, rufous of throat and breast duller, unbarred; rest of underparts uniform buff, no bars on breast or flanks. Tail shorter.

MEASUREMENTS

	Wing	Bill (from skull)	Tarsus	Tail	
11 ♂♂	133–134	31–36	[*c.* 43]	170–186	mm
5 ♀♀	134–141	32–35	—	177–183	mm
				(Ludlow)	

COLOURS OF BARE PARTS. Unrecorded.

1298. *Garrulax ocellatus griseicauda* Koelz

Garrulax ocellatus griseicauda Koelz, 1950, Amer. Mus. Novit., No. 1452: 7 (Wan, Garhwal, United Provinces, India)

Not in Baker, FBI

LOCAL NAMES. None recorded.

SIZE. Pigeon ±; length *c.* 32 cm (13 in.).

FIELD CHARACTERS. As in 1299, q.v.

STATUS, DISTRIBUTION and HABITAT. Resident. Garhwal, Kumaon and west Nepal, intergrading with *ocellatus* in Nepal (Fleming & Traylor, 1964, *Fieldiana*, Zool. 35: 534). Habitat as in 1299.

GENERAL HABITS, FOOD and VOICE. As in 1299.

BREEDING. As in 1299.

MUSEUM DIAGNOSIS. Differs from *ocellatus* (1299) in having much grey on the tail (however, Ludlow, 1944, *Ibis* 86: 74, states that the colour of the tail seems to be a variable character); also appears to have a longer tail.

MEASUREMENTS

	Wing	Bill (from skull)	Tail	
♂♂	135 (type)	33	163, 167	mm
♀♀	132–136	31–32	157–164	mm
		(Koelz, Fleming & Traylor)		

COLOURS OF BARE PARTS. As in 1299.

1299. *Garrulax ocellatus ocellatus* (Vigors)

Cinclosoma ocellatum Vigors, 1831, Proc. Zool. Soc. London: 55
(Himalaya Mts. restricted to Darjeeling by Baker, FBI 1: 156)
Baker, FBI No. 140, Vol. 1: 155
Plate 77, fig. 16

LOCAL NAMES. *Moonali bhiakora* (Pahari); *Lho-karreum-pho* (Lepcha).

SIZE. Pigeon; length *c.* 32 cm (13 in.).

FIELD CHARACTERS. Resembles *G. o. maximus* (1297) but has black crown, ear-coverts and throat. Upper back fulvous, lower chestnut-brown, both spotted with black and buff. Tail chestnut with black subterminal band and white tip, the outer rectrices dark brown tipped white. Sides of neck ashy, sides of lower throat cinnamon-rufous. Underparts buff, mottled with black on breast and flanks. Sexes alike.

STATUS, DISTRIBUTION and HABITAT. A high-elevation species. Resident, fairly common. Nepal (intergrades with *griseicauda*), Darjeeling, Sikkim, Bhutan, Arunachal Pradesh (Tawang, upper Subansiri, and presumably eastwards from 2100 to 3660 m (Inskipp & Inskipp, p. 309) at all seasons, breeding mostly above 2800 m. Affects light forest with undergrowth, thick rhododendron scrub and bushes at the edges of fields, in sunny situations.

Extralimital. Chumbi Valley and southeast Tibet, possibly intergrading with *G. o. maximus* between the upper Subansiri and the Tsangpo (Ludlow, 1944, *Ibis* 86: 74). The species extends to northeastern Burma, northwestern Yunnan and southwestern Szechuan.

GENERAL HABITS. Much as those of *G. o. maximus*. Not a conspicuous bird in spite of its size, but inquisitive and easily observed. Keeps in pairs and small parties of five to eight, working through the bushes, hopping on the ground and pecking noisily among dead leaves. Often associated with *G. affinis* (1323).

FOOD. Insects; seeds and fruits. Appears to be almost entirely vegetable even in the breeding season. Nestlings are fed with insects.

VOICE and CALLS. Very similar to those of *G. o. maximus*. In breeding season a piercing 8-syllabled call (Ludlow). Song, a beautiful, very human whistle *tu wee, tu wee, tu witty-o* (Proud). The birds not only answer one another but respond readily to an imitation of this call. Also: *Q-twe-twe-tweee-koi-koi* emitted from the top of a bush (in winter) and a subdued *pie, pie, pie, pie* (Fleming). Smythies describes its 'call note' as a loud *cacreee-creee-creee-creee-rrrr-cacreee-creee* at the rate of two *creee* per second; if suspicious, keeps uttering a low interrogative note which rises to a squawk of alarm if the bird is suddenly startled.

BREEDING. Little known. *Season*, May and June. *Nest*, a large, loose cup about 17 cm across, made of twigs, dry grass or bamboo leaves, lined with rootlets, placed in bushes or small trees within a couple of metres from the ground. *Eggs*, normal clutch probably 2, deep blue-green, spotless or with a few chocolate-brown specks near the broad end. Size of three eggs 30 × 21·8; 32·7 × 21·5 mm (Hume) and 30·9 × 21·2 mm (Osmaston, both *in* Baker). As in *G. leucolophus* and *G. albogularis*, the young leave the nest long before they are full-grown (Diesselhorst).

MUSEUM DIAGNOSIS. See Field Characters and Key to the Subspecies; for details of plumage Baker, loc. cit.

Young, like adult but head browner; white spots on upperparts smaller, absent on rump and upper tail-coverts; throat browner and edgings less sharp. Probably not distinguishable from young of *G. o. maximus*. Timaline juvenile characters of wing and tail (i.e. soft blunt first primary and narrow, pointed rectrices) present. Postjuvenal moult apparently not complete.

MEASUREMENTS

	Wing	Bill (from skull)	Tarsus	Tail	
♂♂	122–135	27–30	—	146–162	mm
♀♀	117–137	28–29	45	148–161	mm

(Rand, Fleming, Traylor, Stresemann, SA)

Weight 1 ♂ 110; 4 ♀ ♀ 110–113; 1 o? 114g (Diesselhorst, SA). 1 ♂ (September) 121 g — SDR.

COLOURS OF BARE PARTS. Iris stone-yellow. Bill horny, darker on culmen and tip. Legs and feet fleshy, dusky in front (Stevens). Or: Iris greyish khaki. Legs, feet and claws pale flesh-colour (SA).

GARRULAX CAERULATUS (Hodgson): GREYSIDED LAUGHING THRUSH

Key to the Subspecies

		Page
A	Outer rectrices tipped with white .... *G. c. subcaerulatus*	34
B	Outer rectrices not tipped with white	
1	More rufous above; crown-feathers more broadly edged with black .... *G. c. livingstoni*	35
2	More olive, less rufous above; black edgings of crown-feathers narrower... *G. c. caerulatus*	33

1300. *Garrulax caerulatus caerulatus* (Hodgson)

Cinclosoma caerulatus Hodgson, 1836, Asiat. Res. 19: 147 (Nepal)
Baker, FBI No. 123 (part), Vol. 1: 141
Plate 78, fig. 14

LOCAL NAMES. *Tarma-pho* (Lepcha); *Piang-kam* (Bhutea).
SIZE. Myna ±; length *c*. 25 cm (10 in.).
FIELD CHARACTERS. The only laughing thrush in its range with white underparts and grey flanks. *Above*, forehead and orbital area black. Earcoverts black tipped with whitish. Crown rufous-brown, nape olive-brown fulvous on the sides, both finely barred with black giving the head a scaly appearance. Sides of throat and back olive-brown. Wings rufous brown. Tail chestnut. *Below*, throat, breast, belly and under tail-coverts white, the latter tinged with pink. Sides slaty. Sexes alike.

× *c*. 1

STATUS, DISTRIBUTION and HABITAT. Resident, locally distributed; some downward movement in winter. From central Nepal to Darjeeling, Sikkim, Bhutan and (?) Arunachal Pradesh. Breeding zone not satisfactorily determined: between 1500 and 2400 m in Bhutan (Ludlow); up to 2700 m in Sikkim (Stevens) and possibly as low as 1000 m (Gammie); in Nepal, 2745 m (Inskipp & Inskipp); Proud found it very common on Sheopuri and Phul Chowk (= Phulchauki Danda) in summer, while Biswas noted it at c. 2000 m in April and May; breeds between 1500 and 2000 m (Diesselhorst). Recorded in winter at c. 1500 m in the Nepal Valley and as low as 600 m in the Bhutan duars. Affects undergrowth in forest, and ringal bamboo and scrub-covered hillsides.

GENERAL HABITS. Keeps in parties of three to twelve or more, feeding in low bushes and on the ground. Makes a great disturbance when feeding on the latter, 'hurling earth and dead leaves in all directions with the vigour of a puppy digging for rats' (Proud). Timid in winter but bolder in spring and then in taller trees (Fleming). Flight weak and ill-sustained. Usually escapes by hopping from branch to branch in the undergrowth.

FOOD. Berries, seeds and other vegetable matter (Diesselhorst). Probably also insects.

VOICE and CALLS. Loud musical calls and liquid whistles, some of them rendered by Fleming as *ovik-chorr, brain fever* and *new jericho* (do, si, fa, la). Also keeps up a constant flow of conversational, soft and pleasant notes, now and then breaking out in loud, discordant calls. One of its call-notes is a loud *oh dear dear*, and the alarm-note a very sharp and distinctive chitter (Harington).

BREEDING. *Season*, May to July. *Nest*, cup-shaped, made of bamboo leaves, twigs and stems of creepers, lined with rootlets. Placed in bushes, bamboo clumps or trees between one and three metres from the ground, rarely higher. *Eggs*, 3, sometimes 2, pale blue, unmarked. Average size of 15 eggs 30·5 × 22·1 mm (Baker).

MUSEUM DIAGNOSIS. See Key to the Subspecies.

Young, like adult but lacks the black edges to crown-feathers; flanks tinged with brown and under tail-coverts with buff (brown according to Ludlow). Postjuvenal moult complete. Postnuptial moult completed in late October.

MEASUREMENTS

	Wing	Bill (from skull)	Tarsus	Tail	
♂♂	106–115	24–28	41–43	124–133	mm
♀♀	103–110	25–29	40–42	(114) 122–131	mm

(Rand & Fleming, SDR, BB, SA)

Weight 4 ♂♂ 82–99; 6 ♀♀ 79–84 g (GD, SA).

COLOURS OF BARE PARTS. Iris red or red-brown; orbital skin livid. Bill horny black, paler and greyer at the base. Legs and feet pale fleshy.

1301. *Garrulax caerulatus subcaerulatus* Hume

Garrulax subcaerulatus Hume, 1878, Stray Feathers 7: 140 (Shillong)
Baker, FBI No. 124, Vol. 1: 142

LOCAL NAMES. None recorded.

SIZE. Myna ±; length *c*. 25 cm (10 in.).
FIELD CHARACTERS. As in 1300 but three outer rectrices broadly tipped with white.
STATUS, DISTRIBUTION and HABITAT. Resident. Meghalaya in the Khasi Hills, between 1200 and 1800 m. Keeps almost entirely to undergrowth in pine forest, occasionally mixed oak and rhododendron.
GENERAL HABITS, FOOD and VOICE. As in 1300.
BREEDING. *Season*, May and June. *Nest* and *eggs* as in 1300. Average size of 40 eggs 29·3 × 20·8 mm (Baker).
MUSEUM DIAGNOSIS. Similar to *caerulatus* but upperparts paler; ear-coverts and cheeks above and below them white, with slight black tips. Three outer rectrices broadly tipped with white.
MEASUREMENTS and COLOURS OF BARE PARTS. As in 1300.

1302. *Garrulax caerulatus livingstoni* Ripley

Garrulax caerulatus livingstoni Ripley, 1952, Jour. Bombay nat. Hist. Soc. 50: 497
(Mt Japvo, Naga Hills, Assam)
Dryonastes caerulatus biswasi Koelz, 1953, Jour. Zool. Soc. India 4: 153
(Kohima, Naga Hills)
Baker, FBI No. 123 (part), Vol. 1: 141

LOCAL NAMES. None recorded.
SIZE. Myna ±; length *c*. 25 cm (10 in.).
FIELD CHARACTERS. As in 1300, q.v.
STATUS, DISTRIBUTION and HABITAT. Resident, fairly common. Assam in north Cachar, easternmost Arunachal Pradesh (Noa Dihing Watershed), Nagaland, Manipur and adjacent Burma; from 1200 to 2300 m. Habitat as in 1300.
GENERAL HABITS, FOOD and VOICE. As in 1300.
BREEDING. As in 1300.
Extralimital. The species extends to northern Burma and Yunnan; also the mountains of Fujian and Taiwan.
MUSEUM DIAGNOSIS. Differs from *caerulatus* in having the crown more rufous-brown, the feathers more broadly edged with black; upperparts richer, more saturated with rufous. Differs from *subcaerulatus* in lacking the broad white tipping of the three outer rectrices.
MEASUREMENTS

	Wing	Bill (from skull)	Tail
♂ (type)	118	24	125 mm (SDR)

Weight 1 ♂ 98 g (SDR).
COLOURS OF BARE PARTS. Iris brown; orbital skin dark blue. Bill black. Legs and feet pale bluish white.

1303. Rufousnecked Laughing Thrush. *Garrulax ruficollis*
(Jardine & Selby)

Ianthocincla ruficollis Jardine & Selby, 1838, Ill. Orn. 2, pl. 21
(Himalayas = Sikkim, restricted by Meinertzhagen, 1928, Ibis: 515)
Baker, FBI No. 120, Vol. 1: 139
Plate 78, fig. 9

LOCAL NAMES. *Rapchen-pho* (Lepcha); *Pobduya, Hath gurri-gurri* (Bengali); *Doopooleeka* (Assam); *Dao-popalika* (Cachari).

SIZE. Myna; length *c.* 23 cm (9 in.).

FIELD CHARACTERS. A dark babbler with black forehead, ear-coverts, throat and upper breast. A large rufous patch on sides of neck and rufous vent. Crown and nape slaty. Back, rump, belly and wings dark olive-brown, the last with pale outer edges. Tail black. Sexes alike.

The only laughing thrush with a large rufous patch or collar on sides of neck. While calling with upraised bill this rufous collar becomes very prominent.

STATUS, DISTRIBUTION and HABITAT. Common resident. Central Nepal (sparingly—Inskipp & Inskipp, p. 310), Sikkim, N. Bengal (the Darjeeling and Jalpaiguri districts), Bhutan, Arunachal Pradesh to the Mishmi Hills, Bangladesh in the northeastern hills and the Chittagong region, Assam, Tripura, Mizoram, Nagaland and Manipur. A bird of the foothills and adjacent plains, ascending to 1500 m in the Himalayas and to 600 m (rarely up to 1200 m) south of the Brahmaputra.

Frequents a variety of habitats: scrub and grass, bamboo jungle, high grass and reeds, edges of cultivation, hedgerows, scrub pastures, secondary growth, tea gardens, outskirts of forest and less commonly humid evergreen forest.

Extralimital. Western and northern Burma.

GENERAL HABITS. In pairs of parties of three to twenty birds according to the season. Hops about on the ground or in low bushes, rummaging among the mulch, flicking aside or turning over dead leaves in search of food. Will often flutter up into a larger bush, gradually hopping further up towards the top, flirting its tail, and then fly down among lower bushes again. At times indulges in a noisy game of 'follow-my-leader' across open spaces, 'each venture into an opening being a good and sufficient reason for an outburst of raucous cackling' (Baker). Flight rather heavy and ungainly. Parties break up in March.

FOOD. Insects, molluscs, seeds and berries.

VOICE and CALLS. A noisy species. The birds constantly utter a running chorus of sharp rather musical notes reminiscent of the Pied Myna (*Sturnus contra*) as the flock threads its way through the thickets, often bursting out in a tumult of loud squeals or calling and answering each other. Has a three-noted mellow whistle, the first of which sounds as if produced by a broken reed (SDR). Two of the more usual calls may be rendered as: *wiweeit-witoo*, repeated for several minutes, and *weeeoo-wihoo-wick* (pause) *weeeoo-withoo-wick* (pause) and so on unvaryingly for maybe a quarter-hour at a stretch. Possibly two separate individuals are involved in this. The song is varied and pleasing, one or two phrases forming the basis of the theme, upon which many variations are grafted and often combined into phrases of fair length. While singing with upraised bill flicks wings and the depressed expanded tail and pivots from side to side on the perch.

BREEDING. *Season*, overall March to August, chiefly April and May. *Nest*, cup-shaped, rather untidy exteriorly, made of leaves, grass, roots, weed stems, occasionally dry moss and lichen, all bound together with long weed stems and tendrils. Placed in bushes, generally between one and two metres up, less often up to six metres. *Eggs*, 3 or 4, pale skim-milk blue, very rarely white. Average size of 200 eggs 25.7 × 20 mm (Baker).

MUSEUM DIAGNOSIS. See Field Characters.

MEASUREMENTS

	Wing	Bill (from skull)	Tarsus	Tail	
♂♀	96–100	23–24	36–37	101–111	mm
				(SA)	

Weight 6 ♂ ♀ 60–73 g (SDR, SA); 1 ♀ 59 G (SDR).

COLOURS OF BARE PARTS. Iris crimson or brown; eye-rim yellow. Bill, legs, feet and claws dark horny brown.

GARRULAX MERULINUS Blyth: SPOTTEDBREASTED LAUGHING THRUSH

Key to the Subspecies

More richly coloured; spots more numerous .... *G. m. toxostominus*
Less richly coloured; spots less numerous .... *G. m. merulinus*

1304. *Garrulax merulinus merulinus* Blyth

Garrulax merulinus Blyth, 1851, Jour. Asiat. Soc. Bengal 20: 521
(Cherra Punji, Khasia Hills)
Baker, FBI No. 179 (part), Vol. 1: 186
Plate 77, fig. 8

LOCAL NAMES. None recorded.
SIZE. Myna; length *c*. 22 cm (9 in.).
FIELD CHARACTERS. A very thrush-like babbler with a buff throat and breast spotted with dark brown, and a whitish stripe behind the eye. Centre of belly buff, rest of plumage brown.
STATUS, DISTRIBUTION and HABITAT. Resident, locally common; Meghalaya in the Khasi and Mizo (Lushai) Hills, between 900 and 1800 m. Affects deep and damp forest with heavy undergrowth of *Rubus* etc. and bamboo jungle. Densely overgrown abandoned shifting-cultivation clearings are favourite haunts.
GENERAL HABITS. A true laughing thrush in its habits, noisy, gregarious and an inveterate skulker, usually extremely difficult to approach on account of its wellnigh impenetrable habitat. 'Never will you find them in any place in which it is possible even to creep about, without cutting your way' (Hume). Keeps in parties of ten to twenty individuals, feeding on the ground among fallen leaves. Flies better than most laughing thrushes.
FOOD. Unrecorded; presumably insects for the most part.
VOICE and CALLS. Very noisy. Has a great variety of clear, beautiful notes and a coughing, oft-repeated chuckle (Hume).

BREEDING. *Season*, from the end of April to July, mostly in June and July. *Nest*, a rather bulky cup of roots, grass, bamboo and other leaves, more or less mixed with bracken fronds and moss, lined with rootlets, or occasionally with fine creeper stems or tendrils. Materials vary according to location of nest, whether in evergreen forest or bamboo jungle. Nest placed low down in thick shrubs or well inside bamboo clumps. *Eggs*, generally 2, often 3, blue with a tinge of green, unmarked. Average size of 50 eggs 28.7 × 21.2 mm (Baker). Both sexes share in incubation; period not known. The bird is a close sitter, slinking off the nest at the last moment, then skulking quietly in the cover.

MUSEUM DIAGNOSIS. For distinction from *toxostominus* see 1305.

Young, like adult but more rufous above and on flanks, wings and tail. Postjuvenal moult complete.

MEASUREMENTS

	Wing	Bill (from feathers)	Tarsus	Tail
♂♀	93–99	24	40	*c*. 96–100 mm (Baker)

Bill from skull 29 mm

COLOURS OF BARE PARTS. Iris pale yellowish or pinkish; brown in young birds. Bill dark horny brown, black at tip and on culmen, greyish on lower mandible. Legs and feet pale to dark brown, claws darker, soles paler.

1305. *Garrulax merulinus toxostominus* (Koelz)

Stactocichla merulina toxostomina Koelz, 1952, Jour. Zool. Soc. India 4: 38 (Karong, Manipur)

Baker, FBI No. 179 (part), Vol. 1: 186

LOCAL NAME. *Moh mepeh* (Angami Naga).

SIZE. Myna; length *c*. 22 cm (9 in.).

FIELD CHARACTERS. As in 1304, q.v.

STATUS, DISTRIBUTION and HABITAT. Resident, locally common. Assam in north Cachar, Arunachal Pradesh, Nagaland and Manipur, from 900 to 2400 m. Affects dense secondary scrub and undergrowth in damp evergreen forest.

Extralimital. The adjacent hills of western Burma, and northern Burma. The species extends to Vietnam.

GENERAL HABITS, FOOD and VOICE. As in 1304.

MUSEUM DIAGNOSIS. Differs from *Merulinus* (1304) in being more richly coloured. Below, the feather bases are deeper in colour, especially on the throat. Black spots average rounder, smaller and more numerous. Under tail-coverts more ferruginous. Crown usually differentiated from back. For details of plumage see Hume, SF 11: 162.

MEASUREMENTS. As in 1304.

COLOURS OF BARE PARTS. Iris pale pinkish buff; orbital skin pale leaden. Bill: upper mandible blackish, lower mandible and gape pale greyish. Legs and feet pale brown.

1306. Whitebrowed Laughing Thrush. *Garrulax sannio albosuperciliaris* Godwin-Austen[1]

Garrulax albosuperciliaris Godwin-Austen, 1874, Proc. Zool. Soc. London: 45
(Manipur Valley, NE. Bengal)
Baker, FBI No. 126, Vol. 1: 144
Plate 78, fig. 8

LOCAL NAMES. None recorded.

SIZE. Myna; length *c.* 23 cm (9 in.).

FIELD CHARACTERS. A plain-coloured laughing thrush with buff lores continued above the eye by a white supercilium and below by a buffish white cheek-patch. Rest of plumage olive-brown, darker on head, throat and tail. Centre of belly buff, under tail-coverts ferruginous. Sexes alike.

STATUS, DISTRIBUTION and HABITAT. Rare resident. Nagaland, Manipur, and Assam in north Cachar, from *c.* 1000 to 1800 m. Affects dense forest, scrub pastures, secondary growth or bamboo and open hillsides covered with bracken and wild raspberry scrub.

Extralimital. The species extends north to Sichuan and Hupeh, and east through northern Burma to Fujian, Guangdong and Vietnam.

GENERAL HABITS. Less shy than most laughing thrushes. Usually seen in small parties, sometimes singly or in pairs, working through scrub jungle or undergrowth. Parties break up in March.

FOOD. Seeds, berries, rice, small molluscs.

VOICE and CALLS. A noisy species. Notes more complaining and less hilarious than those of *G. leucolophus*.

BREEDING. *Season*, March to June. *Nest*, cup-shaped, similar to that of *G. ruficollis* but larger and more massive. Made of dark-coloured grass, fern, roots and bamboo leaves, bound together with weed stems and tendrils and lined with fern roots and fern stems; some nests are made almost entirely of bamboo leaves and grass. Usually placed low down in thick bushes— *Rubus* brakes etc.—or sometimes in small saplings. *Eggs*, 3 or 4, white or pale blue. Average size of 100 eggs 26 × 19.4 mm (Baker).

MUSEUM DIAGNOSIS. See Field Characters. Said to be distinguishable from other (extralimital) races by the continuous white supercilium and brown ear-coverts. See also footnote below.

Young, a little paler than adult. Upper tail-coverts and throat yellowish brown. Jevenal timaline characters of wings and tail present, namely soft, blunt first primary and narrow, pointed rectrices. Postjuvenal moult complete.

MEASUREMENTS

	Wing	Bill (from skull)	Tarsus	Tail	
♂♀	93–104	22–25	*c.* 37	105–117 (HW)	mm

Weight 2 ♂ ♂ 68, 68; 1 ♀ 56 g (SDR).

---

[1] A doubtfully tenable race, but see Deignan, H. G. 1952, *Postilla*, No. 11 (26 March). He points out that this subspecies of *G. sannio*, wholly isolated from the Chinese populations, is distinct in having a deep brown pileum, a cold dark olivaceous brown mantle and a strong vinaceous wash over the entire underparts. The post-ocular stripe is blackish brown in the adult *v.* pale brown to dark in birds from China.

1307, 1308. **Nilgiri Laughing Thrush.** *Garrulax cachinnans* (Jerdon)[1]

Crateropus cachinnans Jerdon, 1839, Madras Jour. Lit. Sci. 10: 255 (Nilgiris)

Baker, FBI No. 165, Vol. 1: 176

Plate 78, fig. 2

LOCAL NAMES. None recorded.

SIZE. Myna ±; length *c.* 20 cm (8 in.).

FIELD CHARACTERS. An olive-brown and rufous laughing thrush with a very pronounced white eye-stripe. *Above*, crown slaty brown, nape slaty. A conspicuous white supercilium. Lores and a short streak behind eye black. An incomplete white eye-ring. Cheeks tawny olive. Rest of upperparts olive-brown. *Below*, chin black. Breast and belly ochraceous. Lower belly and flanks olive-brown. Sexes alike.

Distinguished from *G. jerdoni* by its rufous, not grey, breast.

STATUS, DISTRIBUTION and HABITAT. Resident, very common but range curiously restricted. Confined to the Nilgiri Hills (western Tamil Nadu), from *c.* 1200 m to the summit. A geographical representative of the closely related *G. jerdoni* of Kerala. Affects dense undergrowth in forest, sholas, even hill-station gardens or wherever there are a few bushes or a patch of scrub.

GENERAL HABITS. Much the same as those of *G. jerdoni* (1310), q.v. Generally seen in flocks of a dozen or more commonly in association with the itinerant mixed hunting parties in sholas. Feeds on the ground as well as in low bushes, climbing into trees when disturbed and hopping along branches lightly and with great agility.

FOOD. Insects and berries, especially wild raspberry (*Rubus* spp.), and 'hill guava' (*Rhodomyrtus tomentosa*).

VOICE and CALLS. Very noisy. Has loud squeals and laughing calls—some of the most characteristic bird sounds of the Nilgiris—similar to those of *G. jerdoni*. The spirited 'laughing' calls, *pee-ko-ko* etc., are uttered with bill raised vertically and tail depressed. Another member of the flock invariably beats time with a harsh *kē-kē-kē*, almost identical with one of the calls of the tree pie (*Dendrocitta vagabunda*).

BREEDING. *Season*, overall February to July, chiefly May and June. *Nest*, cup-shaped, made of roots, dead leaves, small twigs, moss, grass and some lichen, lined with fine grass, sometimes with a few feathers and scraps of wool. Placed in bushes or small trees between one and three metres above the ground. *Eggs*, pale blue, marked with blotches, spots and specks of pale reddish brown, and a few short lines of the same colour. Average size of 40 eggs 25.6 × 18.8 mm (Baker). Commonly brood-parasitized by Pied Crested Cuckoo (*Clamator jacobinus*).

---

[1] No. 1308, *Garrulax cachinnans cinnamomeum* (Davison), 1886, *Ibis:* 204 (locality unknown) is believed to be based on two stained specimens in Trivandrum Museum, Kerala. They differ from *cachinnans* in having the black of lores and chin replaced by dark brown, and the crown hair-brown instead of slaty brown.

Garrulax cachinnans and **G. jerdoni**

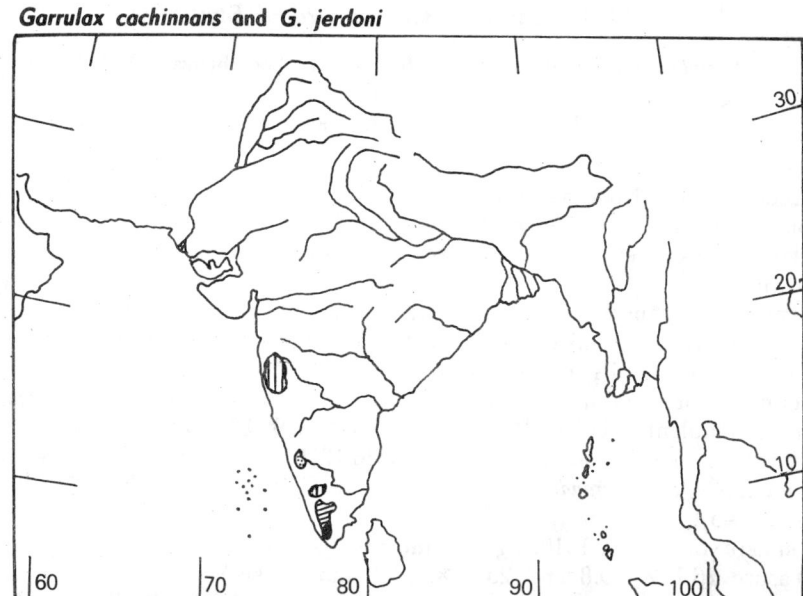

Distributional ranges

▨ *G. cachinnans* (1307).   ▥ *G. j. jerdoni* (1309).
▩ *G. j. fairbanki* (1310).   ■ *G. j. meridionale* (1311).

MUSEUM DIAGNOSIS. See Field Characters.

MEASUREMENTS

	Wing	Bill (from skull)	Tarsus	Tail	
♂♀	92–96	16–18	30–32	c. 100 (Baker)	mm
2 ♀♀	85, 89	(from skull) c. 22	c. 32	86, 94 (MD)	mm

COLOURS OF BARE PARTS. Iris red-brown to crimson. Bill black. Legs and feet greenish plumbeous.

GARRULAX JERDONI Blyth: WHITEBREASTED LAUGHING THRUSH

Key to the Subspecies

A Chin black .... *G. j. jerdoni*
B Chin grey ....
   1 Supercilium mostly grey, not extending behind eye ....
      .... *G. j. meridionale*
   2 Supercilium white, extending behind eye .... *G. j. fairbanki*

1309. *Garrulax jerdoni jerdoni* Blyth

Garrulax(?) jerdoni Blyth, 1851, Jour. Asiat. Soc. Bengal 20: 522
(Banasore Peak)
Baker, FBI No. 167, Vol. 1: 177
Plate 78, fig. 3

LOCAL NAMES. None recorded.

SIZE. Myna ±; length *c*. 20 cm (8 in.).

FIELD CHARACTERS. As in 1310 but chin and cheeks black and crown slaty brown.

STATUS, DISTRIBUTION and HABITAT. Resident. From Goa (Dudhsagar) and northern Karnataka (Castle Rock) along the Sahyadris to the south in the Brahmagiri hills and Banasore Peak at *c*. 1500 to 1800 m. Habitat, evergreen forest with *Trema orientalis, Maesa indica* and bracken *Pteridium aquilinum* aplenty (Ulhas Rane, JBNHS 80: 639; 81: 474–5).

GENERAL HABITS, FOOD and VOICE. As in 1310, q.v. Feeds on fruit of *Maesa indica, Lavinga eleutherandra* and *Trema orientalis* (Ulhas Rane, JBNHS 80: 639; 81: 474–5).

BREEDING. As in 1310. *Eggs*, similar to those of *G. cachinnans;* two eggs measure 26.1 × 19.8 and 25.1 × 19.2 mm (Baker).

MUSEUM DIAGNOSIS. See Key to the Subspecies; for details of plumage, Baker, loc. cit.

MEASUREMENTS

	Wing	Bill (from feathers)	Tarsus	Tail
♂♀	80–83	*c*. 18	*c*. 32	*c*. 90 mm (Baker)

COLOURS OF BARE PARTS. Iris crimson (adult), red-brown (immature). Bill dull black. Legs, feet and claws dark plumbeous brown (Davison).

1310. *Garrulax jerdoni fairbanki* (Blanford)

Trochalopteron fairbanki Blanford, 1869, Jour. Asiat. Soc. Bengal 38: 175
(Palni Hills)
Baker, FBI No. 168, Vol. 1: 178

LOCAL NAME. *Chiluchilăppăn* (Malayalam).

SIZE. Myna ±; length *c*. 20 cm (8 in.).

FIELD CHARACTERS. *Above*, crown and nape dark sooty brown; a white supercilium extending behind eye and a blackish stripe below it from lores through eyes; rest of upperparts olive-brown. *Below*, throat, cheeks and breast grey; belly rufous. Sexes alike.

STATUS, DISTRIBUTION and HABITAT. Common resident. The hills of Kerala and western Tamil Nadu north of the Achankovil Gap (*c*. 9°N. lat.): Cardamom, Kannan Devan and Palni hills (not recorded from the Nelliampathies), and High Wavy Mountains; from *c*. 1200 m to the summits. Confined to evergreen biotope in wild raspberry and bracken thickets lining hill streams through tea and cardamon plantations, scrub and secondary jungle near hillmen's settlements, occasionally populous hill-station gardens

(Kodaikanal, Munnar), and edges of sholas; appears to shun the deeper woods. Its occurrence coincides with that of the wild raspberry (*Rubus* spp.).

GENERAL HABITS. Keeps in parties of six to twelve birds hunting among undergrowth and in low bushes, occasionally descending to the ground. A great skulker, creeping away through cover on the least suspicion.

FOOD. Insects, berries and fruits (especially *Rubus, Maesa* and *Trema*).

VOICE and CALLS. Loud shrieks, whistles and 'laughter' uttered in chorus, and a variety of pleasant call-notes, some mellow and rather like an oriole's. Members of a party keep in touch by means of a call rendered as a rousing *pee-koko, pee-koko*: call, answer, quiet; call, answer, quiet—and so on. Other notes: *ku-hi-yu* repeated several times, a deliberate *har-har-har* and a low-pitched scolding resembling that of the Redwhiskered Bulbul (Nichols). Alarm, some squeaky shrieks rather like those of the Jungle Babbler but louder and shriller, and a low *wit-wit-wit* as the bird disappears through the undergrowth.

BREEDING. *Season* ill defined. Overall December to June, chiefly April to June. *Nest*, cup-shaped, made mostly of coarse grass occasionally mixed with moss or bracken leaves, lined with fine grass. Usually well concealed in an upright fork of a rather isolated bush, or anchored in tall brackens, within a couple of metres from the ground. *Eggs*, usually 2, pale blue with blotches, spots and speckles of pale reddish brown and dark brown; indistinguishable from those of *G. cachinnans*. Average size of 15 eggs 25.6 × 19.1 mm (Baker). Both sexes share in building, incubation and care of the young. Incubation period undetermined.

MUSEUM DIAGNOSIS. See Key to the Subspecies.

Young, like adult but crown and nape not so dark, less contrasting with mantle; lower breast tinged with rufous; ear-coverts browner.

MEASUREMENTS

	Wing	Bill (from skull)	Tarsus	Tail	
♂♂	83–91	21–23	33–35	86–97	mm
♀♀	81–86	20–23	33–35	86–92	mm
				(SA, HW)	

COLOURS OF BARE PARTS. Adult: Iris reddish brown. Bill horny brown: mouth pink. Legs and feet slaty brown; claws brown; soles greyish yellow. Young: Iris brown. Mouth pale pinkish yellow. Rest as in adult.

1311. *Garrulax jerdoni meridionale* (Blanford)

Trochalopteron meridionale Blanford, 1880, Proc. Asiat. Soc. Bengal: 184
(Mynall, S. Travancore Hills)
Baker, FBI No. 169, Vol 1: 178

LOCAL NAME. *Chiluchilăppăn* (Malayalam).

SIZE. Myna ±; length *c.* 20 cm (8 in.).

FIELD CHARACTERS. As in 1310 but supercilium shorter, not extending behind the eye; grey of underparts whiter and extending down to centre of belly. Crown, nape and rest of upperparts unicolorous dull sooty grey-brown, no contrast between them.

STATUS, DISTRIBUTION and HABITAT. Common resident. Southern Kerala ghats from the Achankovil Gap south to the Ashambu Hills, from *c.* 1100 m upwards. Habitat as in 1310. Has a parallel altitudinal distribution with the wild raspberry, *Rubus* spp., whose lower limit here is also the same.

GENERAL HABITS, FOOD and VOICE. As in 1310, q.v.

BREEDING. As in 1310.

MUSEUM DIAGNOSIS. See Field Characters.

MEASUREMENTS

	Wing	Bill (from skull)	Tarsus	Tail	
♂♂	85–88	21–22	35–36	95–96	mm
♀♀	84–85	*c.* 20	*c.* 35	*c.* 96 (SA)	mm

COLOURS OF BARE PARTS. Adult, Iris crimson. Bill horny brown; mouth pink. Legs and feet brownish slate; claws horny brown; soles yellow. Immature: Iris olive-brown. Mouth pinkish yellow. Rest as in adult. Juvenile: Iris olive-brown. Bill horny brown except commissure and extreme tip which are yellow; gape and mouth bright yellow. Legs, feet and claws as in adult.

GARRULAX LINEATUS (Vigors): STREAKED LAUGHING THRUSH

Key to the Subspecies

Page

A Subterminal band on under surface of outer rectrices over 1 cm wide, white tips to tail reduced (< 1 cm)
   1 Throat and breast dark chocolate, ear-coverts dull greyish brown
     .... *G. l. imbricatus* 48
   2 Throat, breast and ear-coverts rufous .... *G. l. setafer* 48

B Subterminal band on under surface of outer rectrices under 1 cm wide, terminal tipping extensive
   3 Lower belly and rump olive-brown
    a Throat more rufous; subterminal band of outer rectrices well marked .... *G. l. lineatus* 46
    b Throat greyer, the shafts on each side of it tipped with white; subterminal band reduced to a line .... *G. l. gilgit* 46
   4 Lower belly and rump grey .... *G.l. bilkevitchi* 44

1312. *Garrulax lineatus bilkevitchi* (Zarudny)

Trochalopteron (Ianthocincla) lineatum bilkevitchi Zarudny, 1910, Orn. Monatsb. 18: 188 (Kulyab, Tadzhikistan)

Ianthocincla lineatum ziaratensis Ticehurst, 1920, Bull. Brit. Orn. Cl. 41: 55 (Ziarat)

Baker, FBI No. 174, Vol. 1: 182

LOCAL NAMES. None recorded.

SIZE. Bulbul ±; length *c.* 20 cm (8 in.).

FIELD CHARACTERS. As in 1314, q.v.

STATUS, DISTRIBUTION and HABITAT. Resident, subject to vertical movements, fairly common. The Quetta-Pishin district of Pakistan in the Central Brahui Range (Murdar, Takatu, Zarghun and Khalifat mountains), and in

BABBLERS

Garrulax lineatus and G. virgatus

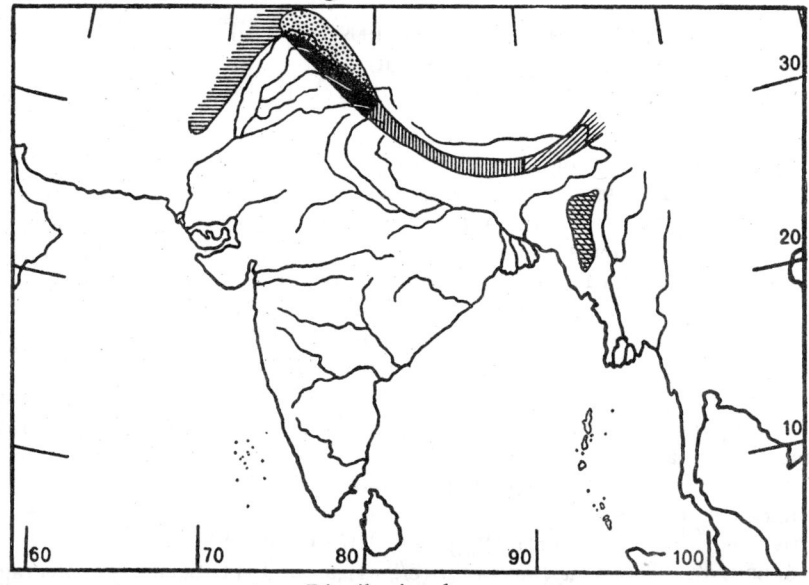

Distributional ranges

▬ G. l. bilkevitchi (1312).   ▦ G. l. gilgit (1313).
■ G. l. lineatus (1314).   ▥ G. l. setafer (1315).
▧ G. l. imbricatus (1316).   ▨ G. virgatus (1317).

the Zhol district [TJR] (Torkhan, Shingar and the Takht-i-Sulaïman). Breeds between 2400 and 3000 m. Moves down in winter and reaches Quetta during cold waves. Affects juniper scrub, bush-covered slopes and bushy nullahs, entering gardens in winter.

Extralimital. Eastern Afghanistan, Tadzhikistan and Uzbekistan in the U.S.S.R. (Flint *et al.*, 1984)-TJR. The species in other races extends east along the Himalayas to Arunachal Pradesh and southeastern Tibet.

GENERAL HABITS, FOOD and VOICE. As in 1314.

BREEDING. *Season,* May to July. *Nest,* described as a massive structure of fibrous material and bulbous plant stems, lined with fine grass stems and hair, about 120 cm up in a leafy bush. *Eggs,* unmarked turquoise-blue as in 1314; fade considerably after being blown. Two eggs measure 26.3 × 19 and 26 × 18.8 mm (Baker).

MUSEUM DIAGNOSIS. Paler than *gilgit* (1313); differs from it in having paler rusty, not chestnut, ear-coverts; grey, not olive-brown, rump and upper tail-coverts, and grey, not grey-brown belly; the markings on the breast are paler and yellower, not red-brown.

MEASUREMENTS

	Wing	Bill (from skull)	Tarsus	Tail	
♂♀	80–85	c. 18	26–27	92–97 (HW, CBT)	mm

COLOURS OF BARE PARTS. Not recorded. Probably not different from those of 1314.

1313. *Garrulax lineatus gilgit* (Hartert)

Ianthocincla lineatum gilgit Hartert, 1909, Vög. pal. Fauna 1: 636
(Gilgit)
Baker, FBI No. 173, Vol. 1: 182

LOCAL NAMES. None recorded.
SIZE. Bulbul ±; length *c.* 20 cm (8 in.).
FIELD CHARACTERS. As in 1314, q.v.
STATUS, DISTRIBUTION and HABITAT. Common resident, subject to seasonal vertical movements. Pakistan in the N.W.F.P. from Safed Koh and Samana to Chitral, east to the Indus, north to Gilgit, the Hunza Valley as far as Hunza (Baltit), the Indus and Shyok valleys as far as Khapalu (TJR), and Astor to the north side of the Burzil Pass; from 1400 to 3600 m, mostly between 1800 and 3000 m, descending to the lower valleys in winter (but recorded as high as 2700 m in this season). Affects bushes in open pine or fir forest, wild rose scrub and other thickets, shrubbery near cultivation, entering orchards at lower levels in winter. Fond of creeping about the stone walls of terraced cultivation in mountain regions.
GENERAL HABITS, FOOD and VOICE. As in 1314. Gregarious and most active in flocks on ground (P. Jones, JBNHS 76: 48).
BREEDING. *Season, nest* and *eggs* (c/2–4) as in 1314. Seven eggs average 24.9 × 18.5 mm (Baker).
MUSEUM DIAGNOSIS. Similar to *lineatus* (1314) but paler and greyer, less rufous. See also Key to the Subspecies.
MEASUREMENTS and COLOURS OF BARE PARTS. As in 1314.

1314. *Garrulax lineatus lineatus* (Vigors)

Cinclosoma lineatum Vigors, 1831, Proc. Zool. Soc. London: 56
(Himalaya Mts =NW. Himalayas *vide* Hume, 1875, Stray Feathers
3: 396 = Simla-Almora area)
Ianthocincla lineatum grisescentior Hartert, 1909, Vög. pal. Fauna 1: 636
(Western Himalayas, Simla)
Baker, FBI No. 172, Vol. 1: 142
Plate 77, fig. 2

LOCAL NAME. *Sheen-a-pi-pin* (Kashmir).
SIZE. Bulbul ±; length *c.* 20 cm (8 in.).
FIELD CHARACTERS. A small uncrested streaked laughing thrush with a greyish white terminal band on rounded tail.
Above, crown and upper back grey streaked with dark brown; middle back streaked with white; rump olive-brown. Ear-coverts and wings rufous; tail olive-brown faintly barred, rufous on sides, tipped greyish. *Below*, throat and breast streaked with rufous, each feather with a white shaft. Lower belly olive-brown. Tail graduated with black subterminal spot on outer rectrices and pale grey tip. Sexes alike.
The rather similar Striated Laughing Thrush (1279) is much larger and has a conspicuous mop-like crest.
STATUS, DISTRIBUTION and HABITAT. Common resident, subject to

seasonal vertical movements. The western Himalayas from Murree and Hazara east through Kashmir (south of the main range), Himachal Pradesh, Lahul, Kulu and Uttar Pradesh to Kumaon. Intergrades with *setafer* in Kumaon–western Nepal, and with *gilgit* in Hazara. Breeds between 1200 and 3600 m; optimum zone 1800–3000 m. Altitudinal zone appears to vary according to local conditions. From October to March, it is mostly found between 1000 and 2700 m though a few birds may be seen as high as 2700 m. In Lahul it occurs between 3000 and 3600 m. Regularly descends to 500 m (Islamabad, Pakistan) in winter, remaining until March (T. J. Roberts). Affects bush-covered slopes, *Berberis* and *Rosa* scrub (Lahul), wooded nullahs, cultivation, undergrowth in open forest, even venturing into gardens, courtyards of houses or verandas.

GENERAL HABITS. Seen in pairs or small parties of three to six according to the season. Keeps more to the ground than most other laughing thrushes, shuffling along in tangles of grass, bracken and low bushes, hopping or creeping on the ground like an accentor, flirting its wings and jerking its tail to one side, then to the other. Reluctant to fly; when disturbed, merely flies a short distance downhill and pitches into the nearest cover. Hops up a tree rapidly from branch to branch but seldom if ever ascends very high and does not sail from tree to tree as other members of the genus often do, but rather drops to the ground and scuttles hurriedly to the next bush, half hopping, half flying. A skulker when away from habitations.

FOOD. Insects, berries, seeds and even take bread-crumbs.

VOICE and CALLS. Members of a party keep up an incessant conversational squeaking in low querulous notes rendered as *chit-chit-chitrr*, *chicker-chicker*, *witty-kitty-cree* or a soft, churring *crrer-r* (HW). The oftenest heard call-note is a loud, clear whistle *p'ty-weer*, or *titty-titty-we* are uttered from a prominent perch. Alarm, a plaintive, unmistakable *twee-twee-twee* (Bates) or *sweet-pea-pea-pea* (Fleming); the Kashmiris hear it as *sheen-a-pay-pay-pay* and regard the bird as the herald of winter on first hearing it in autumn when it comes into town limits, for this call means 'oh! snow, fall, fall, fall'. Song phrases uttered during the breeding season from tree-top or branch: *trit-tew*, *tewit* and *ju-wi-ye* (Magrath); a jingling squeaky whistle of three descending rather plaintive notes *pee-pi-pi* of the timbre of those of *Zosterops* or *Muscicapa thalassina* (SA). Song period: March to September.

BREEDING. *Season*, overall March to September, chiefly May and June. At least two broods are normally reared. *Nest*, a deep cup of coarse grass and dead leaves, loose and untidy on the exterior but more compactly woven inside, lined with fine grass and rootlets. Well concealed in low bushes, sometimes at the end of low fir branches, in a pollarded tree, in thick grass on sloping banks, once in a honeysuckle climbing up a veranda; usually within a metre from the ground, sometimes up to a couple of metres. *Eggs*, 2 to 4, normally 3, turquoise-blue, unmarked. Average size of 16 eggs 25.8 × 18.7 mm (Osmaston) and of 100 eggs 25.6 × 18.4 mm (Baker). Both sexes share in, at least, incubation. Frequently brood-parasitized by the cuckoos *Clamator jacobinus*, *Cuculus micropterus* and *C. sparverioides*.

MUSEUM DIAGNOSIS. Differs from *setafer* (1315) in being paler both above and below, with broader grey edges to the feathers of the underparts. See also Key to the Subspecies.

Young, like adult but rusty markings and shaft-streaks on breast less well defined. Head and rump browner. Primary character not present (i.e 1st primary not blunt and soft), but rectrices narrower. Postjuvenal moult, only of body-feathers.

MEASUREMENTS

Wing 3 ♂ ♂ 71–81; 2 ♀ ♀ 78, 82 mm (Rand & Fleming).
Weight 25 ♂♀ (Apr.-May) 36–46 (av. 40.7) g—SA.

COLOURS OF BARE PARTS. Iris reddish brown. Bill dark brown, paler at base of lower mandible. Legs and feet fleshy brown.

1315. *Garrulax lineatus setafer* (Hodgson)

Cinc.(losoma) setafer Hodgson, 1836, Asiat. Res. 19: 148 (Nepal)
Baker, FBI No. 171, Vol. 1; 180

LOCAL NAME. *Bhekura* (Nepal).
SIZE. Bulbul ±; length *c.* 20 cm (8 in.).
FIELD CHARACTERS. As in 1314, q.v.
STATUS, DISTRIBUTION and HABITAT. Resident, locally common, subject to seasonal vertical movements. Nepal, Darjeeling and Sikkim; from *c.* 1800 to 3900 m, up to 3300 in the Langtang Valley, and 3900 m in Khumbu. Recorded in winter from 1000 to 2700 m. Affects hillside scrub undergrowth, open forest and edges of cultivation. Not confined to thick growth as most laughing thrushes are.
GENERAL HABITS, FOOD and VOICE. As in 1314.
BREEDING. As in 1314.
MUSEUM DIAGNOSIS. Differs from *lineatus* in having the head browner (less ashy), rump darker, ear-coverts and streaks of underparts darker. Black subterminal band on outer rectrices much broader.

MEASUREMENTS

	Wing	Bill (from skull)	Tarsus	Tail	
♂♂	77–81	18–19	*c.* 29	90–97	mm
♀♀	72–80	*c.* 19	*c.* 29	*c.* 90	mm

(BB, Rand & Fleming)

Weight 3 ♂ ♂ 39–45; 3 ♀ ♀ 37–43 g (GD).

COLOURS OF BARE PARTS. Iris brown. Bill horny brown, paler on commissure and lower mandible; mouth pale yellowish flesh. Legs and feet pale horny brown.

1316. *Garrulax lineatus imbricatus* Blyth

G.(arrulax) imbricatus Blyth, 1843, Jour. Asiat. Soc. Bengal 12: 951
(Bhutan)
Baker, FBI No. 175, Vol. 1: 183

LOCAL NAMES. None recorded.
SIZE. Bulbul ±; length *c.* 20 cm (8 in.).
FIELD CHARACTERS. As in 1314, q.v.
STATUS, DISTRIBUTION and HABITAT. Resident, locally common. Bhutan and western Arunachal Pradesh (Tawang region). Eastern limit unknown.

Breeds between 1500 and 2400 m, locally up to 3000 m. A specimen from the Buxa duars in May. Affects thick scrub on the edges of cultivation; also bushes and long grass in uncultivated areas.

Extralimital. Tibetan areas adjacent to its range in Bhutan and Arunachal Pradesh.

GENERAL HABITS, FOOD and VOICE. As in 1314. Usually tame and confiding. On occasion behaves very like Jungle Babbler, hopping about picking kitchen scraps around, and even entering, occupied labourers' hutments; scuttling off with squeaky whistles into adjoining shrubbery in alarm at inmates' movements, but soon returning to feed when all is quiet again.

BREEDING. Unrecorded; probably as in 1314.

MUSEUM DIAGNOSIS. Streaks darker than in other subspecies and occupying the whole feather. Rump darker than in *setafer*. Upperparts uniform dark chocolate colour with dark shafts on head and white shafts on mantle and scapulars. Underparts dark chocolate with white shaft-streaks on breast. Ear-coverts dull greyish brown with pronounced white shaft-tips. Subterminal band of tail blacker and wider.

MEASUREMENTS

	Wing	Bill (from skull)	Tarsus	Tail	
♂♂	77–82	17–21	29–34	84–102	mm
♀♀	74–80	18–20	30–33	94–102	mm
				(SA, Kinnear)	

Weight 3 ♂ ♂ 46–50; 4 ♀ ♀ 42–45; 8 ♂ ♀ 44–52 g (SA).

COLOURS OF BARE PARTS. As in 1315.

1317. **Manipur Streaked Laughing Thrush.** *Garrulax virgatus* (Godwin-Austen)

Trochalopteron virgatum Godwin-Austen, 1874, Proc. Zool. Soc. London: 46 (Razami under Kopamedza Ridge, Naga Hills)
Trochalopteron virgatum querulum Koelz, 1952, Jour. Zool. Soc. India 4: 38 (Hmuntha, Lushai Hills)
Baker, FBI No. 170, Vol. 1: 179
Plate 77, fig. 7

LOCAL NAME. *Dao-phéré* (Cachari).

SIZE. Myna +; length *c.* 25 cm (10 in.).

FIELD CHARACTERS. *Above,* forehead and supercilium to nape white. Lores rufous-chestnut; crown and back dark brown with conspicuous white shaft-streaks. Rump olive-brown with pale streaks. Tail olive-brown finely cross-rayed. Wings rufous-chestnut with whitish shoulders. *Below,* chin and throat chestnut shading into ochraceous-buff on rest of underparts, all except chin with white shaft-streaks. Flanks olive-brown. Sexes alike.

The larger Striated Laughing Thrush (1279) has a loose mop-like crest and no supercilium.

STATUS, DISTRIBUTION and HABITAT. Resident, locally common. Assam hills in Cachar (the whole of the hill-ranges, according to Baker), Meghalaya, Nagaland, Manipur and Mizoram (see map, p. 45); from *c.* 900

to 2400 m. Affects damp evergreen forest with heavy undergrowth, stunted oak forest with undergrowth of ferns and other plants, and thick secondary growth.

Extralimital. The Chin Hills of Burma.

GENERAL HABITS. Not gregarious and not as noisy as other laughing thrushes, shy and retiring. Generally keeps in pairs to thick undergrowth and on the ground, seldom showing itself.

FOOD. Chiefly insects; doubtless also berries and seeds.

VOICE and CALLS. A peculiar, soft, single-noted call and some sweet conversational notes.

BREEDING. *Season,* April to July. *Nest,* a deep, stoutly built cup of tendrils, dead leaves, grasses, roots, fine bents, bamboo-leaves, bracken and moss, lined with rootlets. Well concealed in thick bushes, from near the ground to about two metres up. *Eggs,* 2 or 3, more often the former, clear blue with sometimes a greenish tinge, indistinguishable from those of *G. lineatus*. Average size of 100 eggs 26 × 19.2 mm (Baker).

MUSEUM DIAGNOSIS. See Field Characters; for details of plumage Baker, loc. cit.

MEASUREMENTS

	Wing	Bill (from feathers)	Tarsus	Tail
♂♀	85–89	c. 17	c. 31	110–115 mm (Baker)
		(from skull)		
6 ♂♂	82–87	— mm		
1 ♀	81	21 mm (Heinrich)		

COLOURS OF BARE PARTS. Iris brown, sometimes yellowish; orbital skin dusky plumbeous. Bill very dark to blackish brown. Legs and feet fleshy; claws brownish; sole yellowish.

1318. Browncapped Laughing Thrush. *Garrulax austeni austeni* (Godwin-Austen)

Trochalopteron austeni Godwin-Austen, 1870, Jour. Asiat. Soc. Bengal 39: 105 (Hengdan Peak, N. Cachar)
Baker, FBI No. 146, Vol. 1: 160
Plate 77, fig. 6

LOCAL NAME. *Deo-gajao-i-ba* (Cachari).

SIZE. Myna; length *c.* 22 cm (9 in.).

FIELD CHARACTERS. A rufous-brown laughing thrush with underparts barred and mottled with whitish and brown. A pale wing-bar and pale grey outer edge of wing. Sides of neck finely streaked with white. Tail rufous-brown above, blackish below with white tips. Sexes alike.

STATUS, DISTRIBUTION and HABITAT. Scarce resident. The hills of Meghalaya, Assam in the Cachar Hills and Mizoram, Nagaland and Manipur; from 1800 to 2700 m, less often down to 1500 m, rarely 1200 m. Affects oak and rhododendron forest, bushes in ravines and clearings, and bamboo thickets along forest margins.

Extralimital. Another subspecies on Mt Victoria, Chin Hills.

GENERAL HABITS. Keeps in pairs or small family parties, moving along the ground under cover of thickets, occasionally climbing in low bushes. Reluctant to fly and when forced to, does so with the usual fluttering and sailing.

FOOD. Insects and seeds.

VOICE and CALLS. A liquid three-noted whistle *to meet you* very reminiscent of that of *Pellorneum ruficeps* but much louder; also a soft *tick* when working in the underbush and some loud harsh calls like a wounded tree pie (SDR). Song, a pleasant, flute-like phrase, easy to imitate, *ti-ti-ti-tia-tui-ti* the fourth note longer and lower, with several variations; call-note, a double, loud, resounding whistle *krrü-krrü*; alarm-note, a subdued *krüpp* . . . *krüpp* . . . *krüpp* frequently and quickly repeated (Heinrich).

BREEDING. *Season*, April to August, mostly the latter half of April and May. *Nest*, cup-shaped, made of tendrils, roots, leaves, moss, grass, etc., lined with rootlets and a few bents. Usually placed in bushes within a couple of metres from the ground, sometimes up to three metres in small trees. *Eggs*, usually 3, sometimes 4, occasionally 2, white. Average size of 46 eggs 26.3 × 19 mm (Baker).

MUSEUM DIAGNOSIS. See Field Characters; for details of plumage Baker, loc. cit.

MEASUREMENTS

	Wing	Bill (from feathers)	Tarsus	Tail	
4 ♂♂	90–100	c. 20	c. 35	c. 120	mm
1 ♀	90				

(Baker, ♀ wing by SDR)

Weight 4 ♂ ♂ 63–74; 1 ♀ 59 g (SDR).

COLOURS OF BARE PARTS. Iris brown to pale whitish brown. Bill black. Legs and feet brown.

1319. Bluewinged Laughing Thrush. *Garrulax squamatus* (Gould)

Ianthocincla squamata Gould, 1835, Proc. Zool. Soc. London: 48 (Sikkim)
Trochalopteron squamatum subsquamatum Koelz, 1952, Jour. Zool. Soc. India 4: 38
(Pynursla, Khasia Hills)
Baker, FBI No. 164, Vol. 1: 174
Plate 77, fig. 5

LOCAL NAMES. *Tarmal-pho* (Lepcha); *Nabom* (Bhutea).

SIZE. Myna +; length *c.* 25 cm (10 in.).

FIELD CHARACTERS. A dark olive-brown laughing thrush with black scaly markings over the whole body, especially on back. A black supercilium and white eye, very conspicuous. Wing black with pale blue outer edge and a large rufous shoulder-patch. Both upper and under tail-coverts chestnut; tail blackish with a rufous terminal band. Sexes alike.

The similarly scaly species *subunicolor* (1320) lacks the black supercilium, has buff wings and belly, and the outer rectrices tipped white.

STATUS, DISTRIBUTION and HABITAT. Resident, uncommon in the Himalayas, more common south of Brahmaputra R. From Central Nepal (Biswas, JBNHS 59: 215) east through Darjeeling, Sikkim, Bhutan and Arunachal

Pradesh to the Mishmi and the Patkai hills; the hills of Assam, Nagaland, Manipur and Meghalaya. In the Himalayas breeds between 1000 and 2400 m, mostly 1300–2200 m (recorded at 3400 m in winter in the Darjeeling district by Meinertzhagen). In the Khasi and Cachar hills, breeds mostly between 900 and 1200 m, more sparingly to 1500 m, occasionally to 1800 m. Affects humid dense bushes, ringal bamboo and rhododendron, especially along the banks of streams and rivers.

Extralimital. Extends to western and northern Burma, and western Yunnan; also northern Vietnam.

GENERAL HABITS. Found singly or more usually in pairs or small family parties keeping close to the ground in dense cover. An inveterate skulker, seldom taking flight when alarmed, diving into thicker cover and escaping silently and unseen.

FOOD. Insects, berries and seeds.

VOICE and CALLS. Call-note, a thrush-like *chuck* (SDR); some rich and full conversational notes (Baker); clamouring loudly when excited (Meinertzhagen). Song rendered as *cur-white-to-go* and *free-for-you* (Fleming).

BREEDING. *Season,* April to July. *Nest,* a compact cup of leaves, fine twigs, grass, roots and some moss, bound with a few tendrils and long roots, lined with rootlets. Usually placed in low bushes within a couple of metres from the ground. *Eggs,* normally 3, often 2, seldom 4, deep blue, unmarked. Average size of 50 eggs from Assam 29·4 × 20.7 mm (Baker). Eggs taken in Sikkim by Osmaston differ in being much paler and bigger, nine eggs averaging 29·9 × 22·2 mm.

MUSEUM DIAGNOSIS. See Field Characters. Appears to be dimorphic, some birds having a bronze-coloured tail with rufous tip and the crown concolorous with the back, others a black tail and ashy crown. Sides of head and lores may be rufous or olive-brown. For details of plumage, see Baker loc. cit.

MEASUREMENTS

	Wing	Bill (from skull)	Tarsus	Tail	
♂♀	99–105	24–25	39	97–103	mm (HW)
♂♂	100–106	—	—	102–107	mm
♀♀	98–103	—	—	98 (one)	mm

(Heinrich, Mayr, BB, SDR)

Weight 1 ♂ 79 g (SDR); 1 ♀ 84 g (SA).

COLOURS OF BARE PARTS. Iris ivory white (brown in juvenile). Bill dark brown. Legs and feet brownish flesh.

1320. Plaincoloured Laughing Thrush. *Garrulax subunicolor subunicolor* (Blyth)

Trochalopteron subunicolor Blyth, 1843, Jour. Asiat. Soc. Bengal 12: 952 (Nepal)
Ianthocincla subunicolor griseata Rothschild, 1921, Novit. Zool. 28: 33
(Shweli-Salween Divide, Yunnan)
Baker, FBI No. 160, Vol. 1: 171
Plate 77, fig. 1

LOCAL NAMES. *Tarmal-pho* (Lepcha); *Nabom* (Bhutea).

SIZE. Myna; length *c.* 23 cm (9 in.).

FIELD CHARACTERS. An olive-brown laughing thrush with dark scaly markings over the whole body plumage. Crown and nape washed with grey. Bill very short. Central rectrices olive-brown; the rest largely black, the outer three tipped white. A large straw-coloured patch on wing, the outer edge very pale grey, some flight-feathers with pale tips. Abdomen buff, also scaly. Sexes alike.

Distinguished from *G. squamatus* by lack of the black supercilium; also by the straw-coloured wing-patch and the white-tipped outer rectrices.

STATUS, DISTRIBUTION and HABITAT. Resident, subject to some vertical movements; locally common. Eastern Nepal [Ilam district; Ting Sang, *c.* 86°E. long. (Diesselhorst); also a probable sight record from the Chautara district, central Nepal (Biswas, JBNHS 59: 215)], Darjeeling, Sikkim, Bhutan and Arunachal Pradesh to the Mishmi Hills. Breeds from 1800 to at least 3600 m (from 2100 to 3900 m in northern Burma). Recorded in winter from 800 to 3400 m; a withdrawal from the highest zone in the cold season is probable. Affects thickets of *Rubus* and dwarf rhododendron, bamboo, undergrowth in mixed deciduous forest, and secondary growth. Occurs both in thick forest and open areas ('over alpine meadow', Cranbrook, JBNHS 37: 353).

Extralimital. Extends to southeast Tibet, northeast Burma and northwest Yunnan; another race in northern Vietnam.

GENERAL HABITS. Keeps in flocks of ten to twenty birds in the non-breeding season, moving through tangles of bushes and vines or on the ground.

FOOD. Insects (beetles, grasshoppers, etc.), snails, centipedes, berries and green vegetable matter.

VOICE and CALLS. A clear whistle of four notes, a sharp alarm-note and some squeaky conversational notes.

BREEDING. *Season,* April to June. *Nest,* cup-shaped, made of grass, moss, lined with soft bamboo leaves. Placed in bushes or on a low branch within a metre or so from the ground. *Eggs,* 3 or 4, pale greenish blue, unmarked. Three eggs measure 29·5 × 23 to 30·3 × 23 mm (Baker).

MUSEUM DIAGNOSIS. See Field Characters; for details of plumage Baker, loc. cit.

Young, like adult but upperparts, especially rump, more rusty with the black scalloping barely indicated. Throat dark brown, breast more rusty brown, scalloping absent or faintly indicated. Rectrices narrower.

MEASUREMENTS

	Wing	Bill (from feathers)	Tarsus	Tail
♂♀	90–95	18	*c.* 35	*c.* 100–105 mm (Baker)
♂♀	90–100	(from skull) 19–20	36–40	95–120 mm (Mayr, SDR, SA)
7 ♂♀	89–96 mm (Stresemann).			

Weight 3 ♂ ♂ 63–69 g (GD, SA).

COLOURS OF BARE PARTS. Iris straw. Bill blackish brown. Legs and feet horny brown.

1321. **Prince Henry's Laughing Thrush.** *Garrulax henrici* (Oustalet)

Trochalopteron henrici Oustalet, 1892, Ann. Sci. Nat. 12: 274
(South Tibet = Aio and Soutu)
Baker, FBI No. 176, Vol. 1: 183
Plate 78, fig. 11

LOCAL NAME. *Jomo* (= the lady. Tibetan).

SIZE. Myna +; length *c.* 25 cm (10 in.).

FIELD CHARACTERS. *Above,* crown and back mouse grey. Lores and ear-coverts dark chestnut-brown, with fine white supercilium above and white crescentic cheek-patch below. Folded wings ashy with a blackish shoulder-patch and a larger rufous patch. Tail blackish grey tipped with white. *Below,* buffish grey diffusely dark streaked, with chestnut vent and under tail-coverts. Sexes alike.

STATUS, DISTRIBUTION and HABITAT. Common resident, subject to vertical movements. Southeastern Tibet and probably Arunachal Pradesh in Tibetan facies (upper Subansiri and Tsangpo basins).[1] Also locally in the upper Arun Valley, just north of the Nepal border (Wollaston, *Ibis* 1922: 506). The dominant laughing thrush between 2700 and 4500 m, descending to 2000 m in winter. Affects scrub in the dry as well as the wet zone.

GENERAL HABITS. Keeps in pairs during the breeding season, otherwise in small parties, always on the move, seldom showing itself, flying low from cover to cover.

FOOD. Unrecorded. Doubtless berries and insects.

VOICE and CALLS. Has the typical noisy chatterings of the genus, and a fluty call *whoh-hee* (Waddell *in* FBI). Its call-notes resemble those of *G. affinis* with which at times it associates (Ludlow).

BREEDING. *Season,* May and June. *Nest,* a rather untidy structure of dried grass, dead leaves, strips of bark, moss, etc. lined with dry twisted grass. Placed in low bushes and hedges. *Eggs,* 2 or 3, indistinguishable from those of *G. erythrocephalus.* Average size of 13 eggs, 30·5 × 21 mm (Ludlow).

MUSEUM DIAGNOSIS. See Field Characters.

Young, like adult but browner; ear-coverts blackish instead of brown.

MEASUREMENTS

Wing 3 ♂♂ 103, 106, 111 mm; 10 ♀♀ 101–107 mm (Ludlow. The figure 130 given in *Ibis* 1944: 77 is a *lapsus* for 103 mm.)

Bill (from skull) 23; tarsus 37; tail 137 mm (MD).

Hartert (p. 632) gives

	Wing	Bill (from feathers)	Tarsus	Tail
♂♂	115	22–23	*c.* 37	140–150 mm
♀♀	110			

COLOURS OF BARE PARTS. Iris crimson. Bill nut-brown. Legs and feet dark plumbeous.

---

[1] The locality Shoaka (=Showa) given in FBI 1: 183 is not in the Mishmi Hills but in Tibet as pointed out by Ludlow, *Ibis* 1944: 77.

GARRULAX AFFINIS Blyth: BLACKFACED LAUGHING THRUSH

Key to the Subspecies

Edgings of breast-feathers greyish white and conspicuous ....
.... *G. a. affinis*
Edgings of breast-feathers dull grey and relatively inconspicuous
.... *G. a. bethelae*

1322. *Garrulax affinis affinis* Blyth

Garrulax affinis Blyth, 1843, Jour. Asiat. Soc. Bengal 12: 950
(Nepal = central Nepal)
Garrulax affinis flemingi Rand, 1953, Nat. Hist. Miscellanea, No. 116: 2
(Lete, Baglung Dist., West Nepal)
Baker, FBI No. 161 (part), Vol. 1: 172
Plate 78, fig. 12

LOCAL NAMES. None recorded.
SIZE. Myna ±; length *c.* 25 cm (10 in.).
FIELD CHARACTERS. As in 1323, q.v.
STATUS, DISTRIBUTION and HABITAT. Resident, subject to vertical movements; fairly common. West-central and central Nepal. Breeds between 2800 and 4100 m, winters between 1500 and 3600 m. Affects pine and birch thickets, bamboo, juniper and stunted rhododendron.
GENERAL HABITS, FOOD and VOICE. As in 1323.
BREEDING. As in 1323.
MUSEUM DIAGNOSIS. See 1323 under Museum Diagnosis.
MEASUREMENTS and COLOURS OF BARE PARTS. As in 1323.

1323. *Garrulax affinis bethelae* Rand & Fleming

Garrulax affinis betheiae Rand & Fleming, 1956, Fieldiana, Zool. 39, No. 1: 2
(Thangu, Sikkim)
Baker, FBI No. 161 (part), Vol. 1: 172

LOCAL NAMES. None recorded.
SIZE. Myna ±; length *c.* 25 cm (10 in.).
FIELD CHARACTERS. *Above*, head mostly black with a white malar patch, white sides to nape and white semi eye-ring. Back brown, finely scalloped. Wings olive-yellow with slate tip and outer edge, and a small black shoulder-spot. Upper tail-coverts rufous-brown. Tail olive-yellow with slate tip. *Below*, chin black; rest of underparts rufous-brown with grey scale-like markings. Sexes alike.
The black face with two white patches combined with yellow-and-slate wings identify this species.
STATUS, DISTRIBUTION and HABITAT. Common resident, subject to vertical movements. Eastern Nepal (Inskipp & Inskipp, p. 312), Darjeeling, Sikkim, Bhutan, Arunachal Pradesh [Tawang (Whistler, *Ibis* 1941: 173), Tsu Valley, Mishmi Hills (Bailey, JBNHS 24: 75) at suitable elevations]. One of the highest-altitude laughing thrushes, breeding between 2400 and 4200 m,

mostly above 2800 m, locally up to 4500 m. Found in winter between 1600 and at least 3600 m. Affects rhododendron bushes, scrub oak and bamboo in mixed oak and conifer, or birch and fir forest, and dwarf rhododendron above timber-line.

Extralimital. Southeastern Tibet. The species extends to southwestern Sichuan and northern Vietnam; also Taiwan.

GENERAL HABITS. Keeps in pairs during the breeding season, otherwise in small parties usually not in mixed company during the summer months, but in winter often with *Alcippe vinipectus.* Feeds mostly in bushes, often near or on the ground and also in the crowns of small trees. Runs about rat-like in the scrub over moss-covered boulders and fallen tree-trunks at great speed, seldom taking wing. Not so timid as other laughing thrushes but very noisy when excited.

FOOD. Berries (*Rubus, Viburnum, Gaultheria,* wild strawberries, etc.), crab-apples, seeds and insects.

VOICE and CALLS. Has the usual low conversational chuckles of the genus. Alarm, a long rolling *whirr-whirrer* (Stanford). Song, uttered from an elevated position, the top of a bush or rock, the bird usually remaining hidden by overhanging foliage; described as a mournful, monotonous and continually repeated four-note strophe (Cranbrook), probably the same rendered by Diesselhorst as *tsi tsitü wiü* or by SA as a melodious whistling *to-wee* or *to-wee-you.* Fleming describes a variety of clear calls *you weary, wheeooo, eee-rrr, kay-luck.* The song may be heard every month, but more regularly from March to August (Proud).

BREEDING. *Season,* May and June. *Nest,* cup-shaped, exteriorly made mostly of moss and fine twigs with an inner layer of dry rhododendron leaves, root fibres and a few grass blades, lined only with root fibres. Diameter of nest *c.* 18 cm, of cup 7·5 cm, depth of cup 5 cm. Placed in bushes, usually rhododendron, between 1 and 2·5 m from the ground. *Eggs,* 2 or 3, blue with a greenish tinge, marked with a few blackish brown spots and scrolls, mostly around the larger end. Average size of 12 eggs 28·5 × 21·2 mm (Baker). Like many species of the genus, both parents get very excited, raising intense outcries against intruders.

MUSEUM DIAGNOSIS. Differs from the nominate race (1322) in having the underparts generally darker, with the edgings of the feathers dull grey *v.* greyish white. The crown averages blacker but this not constant. For details of plumage see Baker loc. cit.

Young, a dull version of the adult; lacks grey edges to mantle and ashy edges on breast. Rectrices narrower, but juvenal timaliine character of wing (soft blunt first primary) not present (Whistler).

MEASUREMENTS

	Wing	Bill (from skull)	Tarsus	Tail
♂♂	98–115	24 (2)	36–40 (3)	134–135 (2) mm
♀♀	102–110	23 (2)	39–42 (2)	118–127 (2) mm

(Kinnear, Rand & Fleming, SA)

Wing 28 ♂ ♀ 99–109 mm (Stresemann).

Weight 8 ♂ ♂ 67–80; 6 ♀ ♀ 66–72 g (GD, SA, SDR).

COLOURS OF BARE PARTS. Iris brown or 'olive-khaki'. Bill horny black. Legs pinkish- or horny brown; claws dark brown.

GARRULAX ERYTHROCEPHALUS (Vigors): REDHEADED LAUGHING THRUSH

Key to the Subspecies
(All races intergrade along their given boundaries)

				Page
A	Chin and throat black			
	a	Crown chestnut		
		1 Paler .... *G. e. erythrocephalus*		57
		2 Darker .... *G. e. kali*		59
	b	Anterior crown streaked with black .... *G. e. nigrimentum*		59
B	Chin and throat chestnut			
	c	Back and breast with black round spots		
		3 Supercilium absent or less pronounced .... *G. e. erythrolaema*		61
		4 A well-marked grey supercilium.................. *G. e. godwini*		61
	d	Back and breast with brown oval spots .... *G. e. chrysopterus*		60

1324. *Garrulax erythrocephalus erythrocephalus* (Vigors).

Cinclosoma erythrocephalum Vigors, 1832, Proc. Zool. Soc. London: 171 (Himalayas, restricted to Chamba by Baker, 1920, 'Handlist': 15, an action which takes precedence over the restriction to Simla-Almora area by Ticehurst & Whistler, 1924, Ibis: 468–73, or to Simla by Vaurie, 1953, Bull. Brit. Orn. Cl. 73: 78)
Baker, FBI No. 148 (part), Vol. 1: 163
Plate 77, fig. 4

LOCAL NAMES. None recorded.
SIZE. Myna ±; length *c.* 28 cm (11 in.).
FIELD CHARACTERS. An olive-brown laughing thrush with rufous-chestnut crown and nape, black chin and scale-like black markings on breast, neck and upper back. Wing and sides of tail olive-yellow; a chestnut shoulder-patch. Underparts deep ferruginous. Sexes alike.
STATUS, DISTRIBUTION and HABITAT. Resident, locally common, subject to vertical movements. The Himalayas from Kangra (Himachal Pradesh) east to Kumaon—no specimens have been taken from Murree Hills range, since Stoliczka's in 1873 in Changla Gali, the bird being believedly extinct in Pakistan (T. J. Roberts, *pers. com.*). Breeds between 1800 and 3300 m, optimum zone 2100–2700 m. From mid-October to early April found mostly between 1200 and 2000 m and probably also higher, descending to *c.* 700 m during cold spells. Affects dense undergrowth of dwarf rhododendron, *Rubus, Berberis*, etc. on hillsides, along nullahs, and at the edge of terraced cultivation and upland pastures.
GENERAL HABITS. Keeps in pairs in the breeding season, otherwise in parties of four to six, sometimes up to thirty individuals, often in company with *G. albogularis* or *G. affinis*. Frequents low cover, flying restlessly from bush to bush or bouncing quickly over the ground, the long tail slightly raised, and digs under dead leaves; sometimes seen at medium heights in moss-covered trees, less often high up in trees. Shy, a great skulker and usually silent; easily overlooked and difficult to observe even when its characteristic chuckling murmurs proclaim its proximity in the undergrowth.

TIMALIINAE

Garrulax erythrocephalus

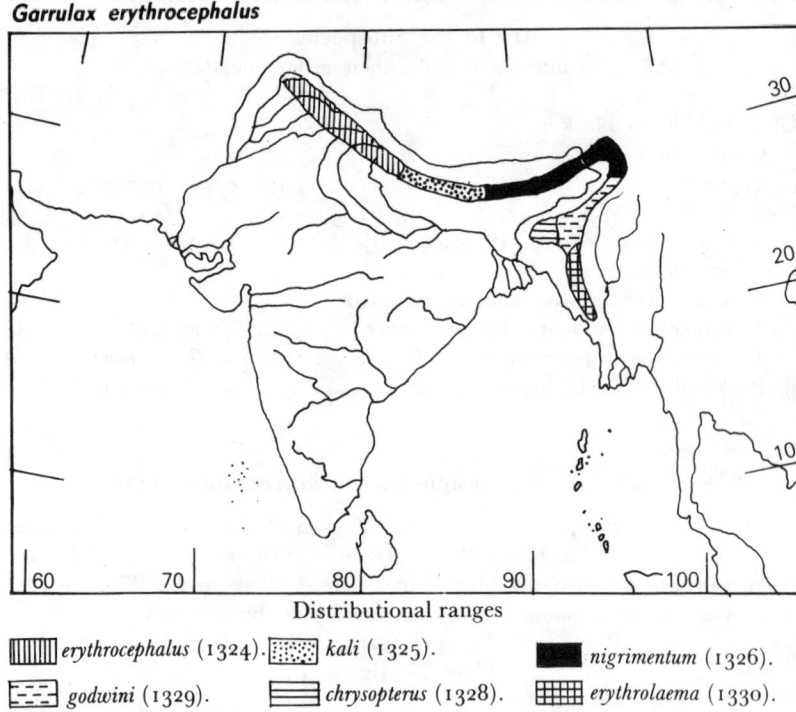

Distributional ranges

|||||| *erythrocephalus* (1324). ░░ *kali* (1325). ■ *nigrimentum* (1326).
▭ *godwini* (1329). ☰ *chrysopterus* (1328). ⊞ *erythrolaema* (1330).

FOOD. Insects, snails, leeches (once), seeds, berries and other vegetable matter. Food appears to be predominantly animal in spring and summer, and vegetal in autumn (Diesselhorst) and very likely also in winter, this being probably true for other high-altitude laughing thrushes as well.

VOICE and CALLS. A constant series of low twitters and chuckles when moving through thickets. A deep churring alarm-note *m-u-r-r-r-r*. The song consists of several short phrases given three or four times at intervals of a few seconds: *pearl-lee, to-reaper, to-real-year, you reap* (Fleming). Other calls: a clear double whistle *pheeou* or *piweep* easily imitated and to which the bird responds (HW). A loud *wee-ou-wee-whip*, the last note higher and louder [probably uttered by a different bird] (Smythies) sometimes given separately, sometimes in runs. Questing note, when looking for mate that has been shot, a soft musical single *twi-ee-you* repeated every half-second or so (once for over 10 minutes) while hopping in the shrubbery (SA).

BREEDING. *Season,* May to August. *Nest,* a substantial, cup-shaped structure of broad grass bents, fern fronds, moss and dead leaves, lined with rootlets. Usually placed in a bush within a couple of metres from the ground. *Eggs,* 2 or 3, turquoise-blue sparsely marked with liver-coloured blotches and streaks, mostly at the broad end. Average size of 100 eggs 28·4 × 21·4 mm (Baker). Commonly brood-parasitized by Pied Crested Cuckoo and hawk-cuckoos.

MUSEUM DIAGNOSIS. The palest member of an east-west cline. Crown entirely rufous. For details of plumage see Baker, loc. cit.

Young, as adult but mantle tinged brown, black spots absent; chin black; throat rusty-olive, no spots; ear-coverts less markedly chestnut. Underparts grey with a buffy tinge; edges to wings duller. Rectrices narrower.

MEASUREMENTS. As in 1325.

COLOURS OF BARE PARTS. Iris greyish hazel. Bill blackish brown. Legs and feet horn-brown.

1325. *Garrulax erythrocephalus kali* Vaurie

Garrulax erythrocephalus kali Vaurie, 1953, Bull. Brit. Orn. Cl. 73: 78
(Lete, Kali River Valley, Baglung dist., west-central Nepal)
Baker, FBI No. 148 (part), Vol. 1: 163

LOCAL NAMES. None recorded.

SIZE. Myna +; length *c*. 28 cm (11 in.).

FIELD CHARACTERS. As in 1324, q.v.

STATUS, DISTRIBUTION and HABITAT. Common resident, subject to vertical movements. Western and central Nepal. Altitudinal distribution and habitat as in 1324.

GENERAL HABITS, FOOD and VOICE. As in 1324.

BREEDING. As in 1324.

MUSEUM DIAGNOSIS. Intermediate between 1324 and 1326. Crown entirely rufous.

MEASUREMENTS

	Wing	Bill (from skull)	Tail	
♂♂	100–105	25–26	114–117	mm
♀♀	92–105	24–25	106–108	mm
		(BB, Rand & Fleming)		

COLOURS OF BARE PARTS. As in 1324.

1326, 1327. *Garrulax erythrocephalus nigrimentum* (Oates)

Trochalopteron nigrimentum Oates, 1889, Fauna Brit. India, Bds., ed. 1, 1: 91
(The Himalayas from Nepal to the Dafla Hills in Assam, inferentially restricted to Sikkim, by Kinnear, 1937, Ibis: 32)
Garrulax erythrocephalus inprudens Ripley, 1948, Proc. Biol. Soc. Wash. 61: 102 (Tidding Saddle above Dreyi, Mishmi Hills, northeast Assam).
Baker, FBI No. 150, Vol. 1: 164
Plate 77, fig. 3

LOCAL NAMES. *Tarphom-pho* (Lepcha); *Paniong* (Bhutea).

SIZE. Myna +; length *c*. 28 cm (11 in.).

FIELD CHARACTERS. As in 1324. See Museum Diagnosis.

STATUS, DISTRIBUTION and HABITAT. Common resident, subject to vertical movements. East Nepal, Darjeeling, Sikkim, Buxa duars, Bhutan, Arunachal Pradesh, Assam (Aka and Dafla Hills) and southeast Tibet; from 1500 to 3000 m (Nepal), 3300 m (Bhutan) in summer, and from 1500 to over 3000 m in winter [recorded at 3500 m in December (Meinertzhagen) and from Buxa

× c. 1

duars 600 m in the same month]. Affects forest with thick undergrowth and especially tangled bushes on steep sides of ravines.

GENERAL HABITS, FOOD and VOICE. As in 1324.

BREEDING. As in 1324. *Eggs*, usually 2, rarely 3, turquoise-blue marked with liver-coloured blotches and streaks. Average size of 32 eggs 28·5 × 20·9 mm (Baker). Size of 2 eggs 31·2 × 21·7 and 28·2 × 21·5 mm (Ludlow).

MUSEUM DIAGNOSIS. Differs from *kali* (1325) in having the sides of crown grey, the rufous restricted to nape, the anterior part of crown being broadly streaked with black; ear-coverts black with pale edges.

MEASUREMENTS

	Wing	Bill (from skull)	Tarsus	Tail	
♂♂	97–107	23–25	37–43	110–124	mm
♀♀	93–103	22–25	39–42	105–125	mm
			(Rand & Fleming, SA)		

Weight 14 ♂ ♂ 75–89; 7 ♀ ♀ 70–88 g (GD, SA).

COLOURS OF BARE PARTS. Iris olive-brown or hazel-brown. Bill horny black. Legs and feet brownish flesh; claws horny brown.

1328. *Garrulax erythrocephalus chrysopterus* (Gould)

Ianthocincla chrysoptera Gould, 1835, Proc. Zool. Soc. London: 48
(Himalayas *errore* = Khasi Hills)
Baker, FBI No. 153, Vol. 1: 166

LOCAL NAMES. None recorded.

SIZE. Myna +; length *c*. 28 cm (11 in.).

FIELD CHARACTERS. As in 1324 but with a grey supercilium and chestnut throat. See Museum Diagnosis.

STATUS, DISTRIBUTION and HABITAT. Resident, very locally distributed. Meghalaya (Khasi Hills) between 1200 and 1900 m. Affects bushes in open pine forest.

GENERAL HABITS, FOOD and VOICE. As in 1324.

BREEDING. *Season*, April to July. *Nest*, as in 1324. *Eggs*, generally 3, sometimes 2, more boldly marked than those of the Himalayan races. Average size of 50 eggs 30·6 × 21·6 mm (Baker).

MUSEUM DIAGNOSIS. Differs from other races in having brown oval lunar (gibbous) spots instead of black round spots on back and breast. A broad grey supercilium. Ear-coverts rufous, more or less tinged with grey. Chin and throat dark chestnut. Yellow on wing more golden, less greenish yellow.

Young (immature). A dull edition of the adult. Spots on breast and back obsolete. Juvenal timaline characters — soft blunt first primary, narrower rectrices — present.

MEASUREMENTS

	Wing	Bill (from feathers)	Tarsus	Tail	
♂♀	101–106	20–21	c. 38	110–115	mm (Baker)

COLOURS OF BARE PARTS. Iris yellowish or greyish brown, sometimes grey. Bill dark horny brown. Legs and feet fleshy or yellowish brown.

1329. *Garrulax erythrocephalus godwini* (Harington)

Trochalopteron erythrocephalum godwini Harington, 1914, Bull. Brit. Orn:. Cl. 33: 92
(Hengdan Peak, N. Cachar Hills)
Baker, FBI No. 151, Vol. 1: 165

LOCAL NAME. *Dau-qua-lok* (Cachari).
SIZE. Myna +; length *c.* 28 cm (11 in.).
FIELD CHARACTERS. As in 1324. See Museum Diagnosis.
STATUS, DISTRIBUTION and HABITAT. Common resident. Assam in Cachar, Nagaland and western Manipur; from *c.* 1500 to 2800 m. Affects undergrowth in evergreen forest, stunted oak and dwarf rhododendron scrub.
GENERAL HABITS, FOOD and VOICE. As in 1324.
BREEDING. Not recorded. Probably as in 1328.
MUSEUM DIAGNOSIS. Similar to *erythrolaema* (1330) and doubtfully distinct (Peters's *Check-List* 10: 376). A grey supercilium, and forehead greyer.
MEASUREMENTS
Wing ♂ ♂ 97–103; ♀ ♀ 93–106 mm (SDR).
Weight ♂ ♂ 74–92 g (SDR).
COLOURS OF BARE PARTS. Iris dark greyish brown. Bill blackish brown. Legs and feet fleshy brown.

1330. *Garrulax erythrocephalus erythrolaema* (Hume)

Trochalopteron erythrolaema Hume, 1881, Stray Feathers 10: 153
(Matchi, E. Manipur Hills)
Baker, FBI No. 149, Vol. 1: 164

LOCAL NAMES. None recorded.
SIZE. Myna +; length *c.* 28 cm (11 in.).
FIELD CHARACTERS. As in 1324. See Museum Diagnosis.
STATUS, DISTRIBUTION and HABITAT. Common resident. East Manipur hills south through the hills of Mizoram; from 1500 to 2700 m (mostly 2500–3000 m on Mt Victoria). Affects thick bushes and bamboo in forest.
Extralimital. Chin Hills and Arakan Yomas. The species extends to southwestern Yunnan and northern Vietnam.
GENERAL HABITS and FOOD. As in 1324.
VOICE and CALLS. The song consists of three loud, far-carrying, flute-like notes *tiuwiu* repeated every few seconds, the first and last notes short and on

the same tone, the middle one loud and rising; another call is rendered as *kreh-krü-krüo* the last 'ü' more accentuated; call-note a sharp and loud trill (Heinrich). See also 1324.

BREEDING. *Season*, April-May. *Nest*, as in 1324. *Eggs*, usually 2. Average size of 33 eggs 29·9 × 20·5 mm (Baker).

MUSEUM DIAGNOSIS. Differs from the nominate race (1324) in having a grey forehead and supercilium, the ear-coverts pinkish rufous, the point of chin grey and the throat and breast chestnut, nearly concolorous with crown. Spots on back browner; spots on breast much smaller and triangular, not hemispherical.

MEASUREMENTS

Wing 10 ♂ ♂ 98–106; 8 ♀ ♀ 92–102 mm (Stresemann).

COLOURS OF BARE PARTS. As in 1329.

GARRULAX PHOENICEUS (Gould): CRIMSONWINGED LAUGHING THRUSH

Key to the Subspecies

Paler; grey of belly more pronounced .... *G. p. bakeri*
Darker; grey of belly less pronounced .... *G. p. phoeniceus*

1331. *Garrulax phoeniceus phoeniceus* (Gould)

Ianthocincla phoenicea Gould, 1837, Icones Av., pl. 3 (Nepal)
Baker, FBI No. 156, Vol. 1: 168
Plate 78, fig. 1

LOCAL NAMES. *Tilji-pho* (Lepcha); *Repcha* (Bhutea).

SIZE. Myna ±; length *c*. 23 cm (9 in.).

FIELD CHARACTERS. An olive-brown laughing thrush with bright crimson sides of head, wings and tips of under tail-coverts. A black supercilium. Crown streaked with black. Tail black with reddish tip and outer rectrices. Sexes alike.

STATUS, DISTRIBUTION and HABITAT. Resident, locally common, subject to slight vertical movements or nomadism in winter. From Nepal (not recorded since Hodgson's days and probably restricted to the easternmost part) east through Darjeeling, Sikkim, Bhutan and Arunachal Pradesh to the Mishmi and Patkai hills; from *c*. 900 to 1800 m at all seasons, reaching the foothills (duars, *c*. 600 m) in winter. Affects undergrowth in evergreen forest and dense thickets of secondary growth on the edge of cultivation and along streams.

GENERAL HABITS, FOOD and VOICE. As in 1332, q.v.

BREEDING. As in 1332. Average size of 50 eggs 25·9 × 18·5 mm (Baker).

MUSEUM DIAGNOSIS. Differs from *bakeri* in being darker. For details of plumage see Baker, loc. cit.

Young, like adult but crimson of face less brilliant; upperparts less rufous; supercilium less distinct; rectrices narrower; reddish tip of tail less sharply defined.

MEASUREMENTS

	Wing	Bill (from feathers)	Tarsus	Tail	
♂♀	81–93	*c*. 18	*c*. 32	*c*. 100	mm
					(Baker)

BABBLERS

		(from skull)			
3 ♂♂	85–93	20–22	33–35	92–110	mm
1 ♀	85	21	33	102	mm
2 o?	83, 87	20, 21	33; 35	90, 95	mm
				(SDR, SA)	

Weight 1 ♂ 44 g; 1 ♀ 46 g (SDR); 2 o? 51, 52 g (SA).

COLOURS OF BARE PARTS. Iris brown (juv.) to deep crimson; orbital skin dull leaden-dusky. Bill dark brown to black. Legs and feet brown.

1332. *Garrulax phoeniceus bakeri* (Hartert)

Trochalopteron phoeniceum bakeri Hartert, 1908, Bull. Brit. Orn. Cl. 33: 10
(Laisung, North Cachar)
Trochalopteron phoeniceum khasium Koelz, 1952, Jour. Zool. Soc. India 4: 38
(Laitlyngkot, Khasi Hills)
Baker, FBI No. 157, Vol. 1: 169

LOCAL NAME. *Dao-yao-gajao* (Cachari).
SIZE. Myna ±; length *c.* 23 cm (9 in.).
FIELD CHARACTERS. As in 1331. q.v.
STATUS, DISTRIBUTION and HABITAT. Resident, fairly common. The hills of Bangladesh, Meghalaya, eastern Arunachal Pradesh (upper Noa Dihing watershed), Nagaland, Manipur and Mizoram, from 900 to 1800 m, up to 2400 m in the eastern ranges. Inhabits dense undergrowth in evergreen forest or in shady deciduous forest.
Extralimital. Western Burma. The species extends to southern Yunnan and northern Vietnam.
GENERAL HABITS. A skulker like most others of the genus. Keeps in pairs or small parties of four or five individuals according to the season, sometimes in the mixed hunting parties. Forages in undergrowth and on the ground, only occasionally going up in trees.
FOOD. Insects, seeds and berries.
VOICE and CALLS. Song, a five- to six-note phrase, the end of which is a characteristic series of three or four notes on the same tone. The second note of the song is about two tones higher than the first, the last between the second and first. Also a four-noted call, the first and last on the same tone, the second and third a half tone higher (Heinrich). It also has a variety of squeaky conversational calls both harsh and sweet, typical of the genus.
BREEDING. *Season*, April to July. *Nest*, a deep cup of dry leaves, grass, fine twigs and roots, occasionally some moss, lichen or other materials, lined with rootlets. Usually placed low down in bushes, occasionally in saplings up to two or three metres above the ground. *Eggs*, 2 or 3, rarely 4, blue with numerous dark brown scrolls and a few spots scattered irregularly over the whole surface. Average size of 100 eggs 26·1 × 18·5 mm (Baker). Both sexes take part in incubation and feeding the young. Incubation period undetermined.
MUSEUM DIAGNOSIS. Differs from the nominate race (1331) in being paler and in having the grey of the abdomen much more pronounced. Also said to be slightly smaller.

MEASUREMENTS

Wing 7 ♂ ♂ 85–90; 8 ♀ ♀ 79–87 mm (Stresemann).
Weight 2 ♂ ·♂ 45, 48 g (SDR).
COLOURS OF BARE PARTS. As in 1331.

Genus LEIOTHRIX Swainson

Leiothrix Swainson, 1832, Fauna Boreali-Americana, Bds.: 490. Type, by original designation, *Parus furcatus* Temminck = *Sylvia lutea* Scopoli
Mesia Hodgson, 1837, Ind. Rev. 2 (1): 34; (2): 88. Type, by original designation, *Mesia argentauris* Hodgson

Bill stout, about half the length of the head, slightly notched near tip, culmen curved.

Key to the Species

		Page
Crown black	*L. a. argentauris*	64
Crown greenish	*L. lutea*	66

LEIOTHRIX ARGENTAURIS (Hodgson): SILVEREARED MESIA

Key to the Subspecies

		Page
Orange on nape pale and less extensive	*L. a. argentauris*	64
Orange on nape deep and more extensive	*L. a. aureigularis*	66

1333, 1334. *Leiothrix argentauris argentauris* (Hodgson)

Mesia argentauris Hodgson, 1837, Ind. Rev. 2 (2): 88 (Nepal)
Mesia argentauris vernayi Mayr & Greenway, 1938, Proc. New England Zool. Cl. 17: 3 (Hai Bum, Upper Burma).
Leiothrix argentauris gertrudis Ripley, 1948, Proc. Biol. Soc. Washington 61: 103 (Dening, Mishmi Hills).
Baker, FBI No. 376 (part), Vol. 1: 354
Plate 80, fig. 1

LOCAL NAMES. *Dang-rapchil-pho* (Lepcha); *Jhărjhări* (Pahari).
SIZE. Sparrow; length *c.* 15 cm (6 in.).
FIELD CHARACTERS. A bright-coloured arboreal babbler with black crown and moustachial stripe and silvery ear-coverts.

× *c.* 1

Male. Forehead yellow, throat and breast bright orange-yellow. Wing edged with yellow, with a crimson patch. Both upper and under tail-coverts crimson.

Female differs in having the under tail-coverts ochraceous and the upper olive-yellow.

Young (immature) like female but crown yellowish.

Easily distinguished from the Redbilled Leiothrix by the black crown, silvery ear-coverts, and square (not forked) tail.

STATUS, DISTRIBUTION and HABITAT. Resident, fairly common but less so west of Sikkim, subject to slight vertical or erratic movements in winter. The Himalayas from Garhwal east through Nepal, Darjeeling, Sikkim, Bhutan, Arunachal, Assam in the Cachar Hills, Nagaland and Manipur; from the foothills to c. 2100 m. Affects scrub jungle, bush-clad open spaces, secondary growth, bushes in evergreen forest, especially the outskirts and more open portions, abandoned cultivation clearings and tea plantations.

GENERAL HABITS. Usually seen in parties of six to thirty birds or more outside the breeding season. Keeps to bushes in forest, sometimes going fairly high up in the canopy. Searches actively for insects among the foliage, behaving much like tits, clinging to sprigs in acrobatic positions and peering under leaves, the flocks rapidly 'flowing' from tree to tree in disorderly follow-my-leader style, all the while uttering subdued chirrups. Seemingly always in great haste. Occasionally makes fly-catching sallies after escaping insects. Pairs form in April; the birds then become very shy and hard to see.

FOOD. Insects, seeds and berries.

VOICE and CALLS. Members of a feeding party keep up a continual chirrup with occasional long-drawn, clear whistling notes *seesee-siweewee* reminiscent of the jingling call of Blackheaded Sibia (SA). A cheerful song of seven or eight notes rendered as *u-cherit, cheroi-cherit* (Stanford). Wings frequently flirted while singing.

BREEDING. *Season*, April to August. *Nest*, cup-shaped, c. 8·5 cm in diameter, 8 cm in overall depth, the depth of cup being c. 6.5 cm. Made of bamboo or other leaves, grass and moss, lined with rootlets. Placed in bushes within a couple of metres from the ground, sometimes as low as a few centimetres. Indistinguishable in size, structure and situation from that of *L. lutea*. *Eggs*, 4, sometimes 3, rarely 5, also identical with those of *L. lutea* (1336) although eggs with a white ground are perhaps more often found among those of *argentauris*. Average size of 200 eggs 20·9 × 16·1 mm (Baker). Building, incubation and care of young shared by both sexes. Nest completion takes about four days. Incubation begins with first egg; period c. 14 days. The birds sit closely, but even when not seen leaving the nest, soon give away the site by their fussy demonstrations.

MUSEUM DIAGNOSIS. For details of plumage Baker, loc. cit. In all races the back is greenish, wearing to grey. Abdulali (1983, JBNHS 80: 349) has shown that the colours of the specimens fade dramatically with age—especially the yellow pigments.

Young (immature), male and female, dull versions of their respective adults. Birds can be sexed upon fledging: male has red tail-coverts, female rusty-green. Nestlings are covered with pale buff down. Postjuvenal moult of body, wing-coverts and tertials.

MEASUREMENTS

	Wing	Bill (from skull)	Tarsus	Tail	
♂♂	70–81	17–19	24–26	63–72	mm
♀♀	70–78	17–18	23–24	63–70	mm
				(BB, CBT, SA)	

Weight ♂♂ 23–25; ♀♀ 22–25 g (SA).

COLOURS OF BARE PARTS. Iris brown to red-brown. Bill orange tinged brown at base; gape light orange in nestlings. Legs and feet yellowish flesh.

MISCELLANEOUS. This species and the Redbilled Leiothrix (1336) are commonly kept as cage-birds and under the name of Silver-ear and Peking Robin respectively largely exported to Europe and the U.S.A.

1333a. *Leiothrix argentauris aureigularis* (Koelz)

Mesia argentauris aureigularis Koelz, 1953, Jour. Zool. Soc. India 4: 153
(Tura Mountain, Garo Hills, Assam)
Baker, FBI No. 376 (part), Vol. 1: 354

LOCAL NAMES. None recorded.
SIZE. Sparrow; length *c*. 15 cm (6 in.).
FIELD CHARACTERS. As in 1333, q.v.
STATUS, DISTRIBUTION and HABITAT. Common resident in the Garo and Khasi hills of Meghalaya, and Chittagong Hill Tracts of Bangladesh; from *c*. 900 to 1500 m. Habitat as in 1333.
GENERAL HABITS, FOOD and VOICE. As in 1333.
BREEDING. As in 1333.
MUSEUM DIAGNOSIS. More richly coloured than *argentauris*; orange on nape deeper and more extensive, likewise that colour on forehead and underparts. In *vernayi* (1334) the orange does not descend so far on breast and back.
MEASUREMENTS and COLOURS OF BARE PARTS. As in 1333.

LEIOTHRIX LUTEA (Scopoli): REDBILLED LEIOTHRIX

Key to the Subspecies

Inner primaries conspicuously edged with red . . . . *L. l. calipyga*
Red on edge of primaries reduced or absent . . . . *L. l. kumaiensis*

1335. *Leiothrix lutea kumaiensis* Whistler

Leiothrix lutea kumaiensis Whistler, 1943, Bull. Brit. Orn. Cl. 63: 62
(Dehra Dun, United Provinces)
Baker, FBI No. 351 (part), Vol. 1: 328

LOCAL NAMES. *Nanachura* (Dehra Dun); *Peking Robin* (aviculturists' name).
SIZE. Sparrow —; length *c*. 13 cm (5 in.).
FIELD CHARACTERS. As in 1336 but lacks crimson on wing. See Museum Diagnosis.
STATUS, DISTRIBUTION and HABITAT. Uncommon resident, subject to slight vertical or erratic movements in winter. The western Himalayas from Pakistan (Murree foothills, rare winter visitor—T. J. Roberts), Kashmir (1 specimen in Brit. Mus. collected by Biddulph; 2 specimens in the Indian Museum, by Stoliczka) east to Kumaon. Breeds from 1000 to 2400 m. Recorded in winter from 1000 to 2100 m but more frequent at the lower levels. Affects scrub and secondary growth on outskirts of cultivation.
GENERAL HABITS, FOOD and VOICE. As in 1336.
BREEDING. As in 1336.

BABBLERS

MUSEUM DIAGNOSIS. Distinguished from *calipyga* (1336) by the greyer tinge of the green on upperparts, the yellowish wash on crown greener and more restricted, and red on outer edge of inner primaries reduced or absent.

MEASUREMENTS and COLOURS OF BARE PARTS. As in. 1336.

1336, 1337. *Leiothrix lutea calipyga* (Hodgson)

Bahila calipyga Hodgson, 1837, Ind. Rev. 2(2): 88 (Nepal)
Leiothrix lutea luteola Koelz, 1952, Jour. Zool. Soc. India 4: 39 (Mawryngkneng, Khasi Hills)
Baker, FBI No. 351 (part), Vol. 1: 328
Plate 80, fig. 2

LOCAL NAMES. *Jhărjhări* (Pahari); *Rapchil-pho* (Lepcha); *Daotisha-buku-gajao* (Cachari).

SIZE. Sparrow —; length *c.* 13 cm (5 in.).

FIELD CHARACTERS. A sprightly greyish olive bird with bright yellow throat and breast, pale lores and eye-ring, and scarlet bill. Wing black with yellow-and-crimson edges, and a small orange patch. Tail forked, the outer rectrices slightly curved outwards — black above olive below, except tip and edges which are black. The long olive upper tail-coverts have a pale narrow terminal bar; under tail-coverts pale yellow.

F e m a l e differs in having the crimson of wing replaced by yellow.

The olive crown, scarlet bill, and absence of whitish ear-coverts distinguishes this species from the Silvereared Mesia (1333).

STATUS, DISTRIBUTION and HABITAT. Resident, subject to slight vertical or erratic movements in winter, locally common. From western Nepal east through Sikkim, Darjeeling, Bhutan, through Arunachal Pradesh to Burma border; Meghalaya, Nagaland, Manipur and Assam in the Cachar Hills. Breeds from 1500 to 2400 m, locally down to 1000 m or up to 2700 m. Winters between 600 and 2100 m. Affects undergrowth in forest wooded ravines, secondary growth on abandoned cultivation clearings, and tea plantations.

Extralimital. Chin Hills and Arakan Yomas of Burma. The species extends east to northern Vietnam and north to Szechuan and Anhwei. Introduced into Hawaiian Islands.

GENERAL HABITS. Keep in pairs in the breeding season, otherwise in small parties of four to six individuals, usually in association with *Stachyris* or other babblers. Habits similar to those of Silvereared Mesia with which it also frequently consorts. Usually seen scuttling through undergrowth, sometimes hopping on the ground foraging and turning over dead leaves in the manner of a laughing thrush. Pairs form in April.

FOOD. Insects, seeds and berries.

VOICE and CALLS. Alarm, a series of harsh, hissing notes and a somewhat explosive *k'd'k-cha-jö-jü* followed by a call sounding like *pile-pile-pile* (the second note higher); while feeding, keeps up a clear and rather wistful but unvaried piping *pe-pe-pe-pä* (rising-low), or a rapid *pü-pü-pü-pü-pü*; also a soft, muttering *che-che-che* and a loud rustling call (Lister). The song is a loud, cheerful warbling reminiscent of the Redwhiskered Bulbul's, rather more prolonged and more musical (SA). Song period February to August, mainly April to June (proud).

BREEDING. *Season*, April to October, chiefly May and June. *Nest*, cup-shaped, made of fine bents, dead leaves, moss, lichen or other materials, usually lined with rootlets. Placed low down in bushes, less often up to a couple of metres. *Eggs*, 3 or 4, rarely 5, pale blue, rarely white, usually marked with bold blotches of dark red-brown or umber-brown, mostly at the large end. Nest as well as eggs indistinguishable from those of Silvereared Mesia. Average size of 200 eggs 21·9 × 16·1 mm (Baker). The female seems to do most of the nest-building, the male bringing her the material. Incubation shared by both sexes, though female appears to do the greater part. Nest location is soon given away by the fussy demonstrations of the owners. Brood-parasitization by *Cuculus canorus* reported by Baker.

MUSEUM DIAGNOSIS. Differs from *kumaiensis* (1335) in having the back greener, and conspicuous crimson outer edgings on inner primaries. For details of plumage see Baker, loc. cit. Postnuptial moult completed in October.

Young: upperparts browner, with no contrast between crown and mantle. Underparts much paler. Upper tail-coverts narrower and white tips merely indicated. Male rather brighter than female. Postjuvenal moult takes place in October.

MEASUREMENTS

	Wing	Bill (from skull)	Tarsus	Tail	
♂♂	65–72	14–16	25–28	55–59	mm
♀♀	65–70	14–16	24–26	53–58	mm

(BB, SA, Rand & Fleming)

Weight 3 ♂ ♂ 20–23; 1 ♀ 20 g (Diesselhorst); 12 ♂ ♀ 21–25 g (SA).

COLOURS OF BARE PARTS. Iris brown or reddish brown. Bill coral red, black at base of both mandibles. Legs, feet and claws pale horny brown.

Genus MYZORNIS Blyth

Myzornis Blyth, 1843, Jour. Asiat. Soc. Bengal 12: 984.

Type, by monotypy, *Myzornis pyrrhoura* Blyth

Bill slender, nearly as long as the head, notched; culmen gently curved. Nostrils covered by a membrane. Tail about two-thirds the length of wing. Tarsus long and slender.

1338. **Firetailed Myzornis.** *Myzornis pyrrhoura* Blyth

Myzornis pyrrhoura Blyth, 1843, Jour. Asiat. Soc. Bengal 12: 984 (Nepal)
Baker, FBI No. 367, Vol. 1: 345

LOCAL NAME. *Lho-sagvit-pho* (Lepcha).

SIZE. Sparrow −; length *c*. 12 cm (4½ in.).

FIELD CHARACTERS. Male. A brilliant little dark green bird with red and green tail. Feathers of crown green with black centres, giving a scalloped effect. A black stripe from lores through eye. Wing black with a reddish streak, white tips and white inner edge. Female only slightly different (see Museum Diagnosis).

× *c*. 1

STATUS, DISTRIBUTION and HABITAT. Resident, locally distributed, subject to some vertical movements. From central Nepal (Proud, JBNHS 58: 804, breeding; Desfayes, sight record 1964), east through Darjeeling, Sikkim, Bhutan and Arunachal Pradesh (Pachakshiri, Ludlow, *Ibis* 1944: 86) and extends to Burma border (Noa Dihing Watershed, 2600 m, March 1988—Saha & Beehler) at suitable altitudes, from 1600 to 3950 m. Recorded at 2700 m in winter. Breeding zone not determined but probably above 2700 m. Affects rhododendron, juniper and other bushes, also heavy jungle and bamboo thickets, preferably on sunny hillsides.

Extralimital. Southeastern Tibet, northeastern and northwestern Burma.

GENERAL HABITS. Usually keeps solitary, in parties of three or four or small flocks frequently in association with sunbirds, warblers and the smaller babblers, searching for food amongst bushes and shrubs, sometimes up in trees. Habitually visits flowering shrubs and trees, especially rhododendron, to probe into the blossoms for nectar.[1] Presumably assists in their cross-pollination since shot specimens often show pollen adhering to the forehead and breast-feathers. Also takes insects and spiders from the flowers and foliage, and from moss-covered tree-trunks, running up these in the manner of a tree creeper or hovering at sprigs on rapidly beating wings, like a sunbird or flowerpecker. Occasionally captures insects by flycatcher-like aerial sallies. Observed switching up tail like a robin while hopping amongst bushes (J. Panday).

FOOD. Insects and spiders, flower nectar (*Rhododendron, Berberis*, etc.), tree-sap and berries.

VOICE and CALLS. Normally silent. A high-pitched *tsi-tsit* (Desfayes). Song, if any, not recorded.

BREEDING. The only authentic nest so far known was in a deep mossy juniper forest at *c.* 3700 m in Nepal. It was about 6 metres from the ground, close against the trunk of a large juniper and so imbedded in moss and lichen that details could not be seen. Both parents were feeding young at the time—late May (Proud, JBNHS 58: 804 and *in epist.*). One egg sent to Hume (but authenticity not certain) was white without markings. It measured about 17 × 13 mm.

MUSEUM DIAGNOSIS. For details of plumage see Baker, loc. cit.

Female, like male but red tinge of underparts duller, red on wing and tail less bright. The primary-coverts are sometimes pale blue, not always white as given in FBI.

Young, male and female, duller than their respective adults. Juvenal timaline wing and tail characters (i.e. soft blunt first primary; narrower rectrices) not present.

MEASUREMENTS

	Wing	Bill (from skull)	Tarsus	Tail	
5 ♂♂	57–63	15–16	23–24	45–48	mm
8 ♀♀	56–62	13–18	23–24	40–47	mm
				(not 70–75 as in FBI)	
				(SA, MD, Stresemann)	

Weight 3 ♂ ♂ 11–13; 3 ♀ ♀ 11–12.5 g (SA, SDR).

COLOURS OF BARE PARTS. Iris dark brown. Bill black. Legs and feet yellowish brown.

---

[1] The tongue possesses terminal bristles, an adaptation for nectar-feeding as in the Honey-eaters of the Australasian Region, or in the Indo-Malayan genus *Chloropsis*.

TIMALIINAE

Genus Cutia Hodgson

Cutia Hodgson, 1836, Jour. Asiat. Soc. Bengal 5: 773.
Type, by original designation, *Cutia nipalensis* Hodgson

Bill somewhat curved, notched, pointed, slightly longer than half length of head; rictal bristles very short; nostrils longitudinal, covered by a membrane; frontal bristles short and firm. Tail slightly rounded, about two-thirds length of wing; upper tail-coverts very long, reaching nearly to tip of tail.

1339. Nepal Cutia. *Cutia nipalensis nipalensis* Hodgson

Cutia nipalensis Hodgson, 1836, Jour. Asiat. Soc. Bengal 5: 774
(Nepal)
Cutia nipalensis nagaensis Koelz, 1954, Contrib. Inst.
Regional Exploration, No. 1: 9 (Kohima, Naga Hills)
Baker, FBI No. 353, Vol. 1: 329
Plate 80, fig. 3

LOCAL NAMES. *Khatya* (Nepal); *Motum-pho, Rapnūn-pho* (Lepcha).
SIZE. Bulbul; length *c.* 20 cm (8 in.).
FIELD CHARACTERS. Male. A handsome but rather dumpy, arboreal bird with white underparts and bold black rib-like markings on flanks.

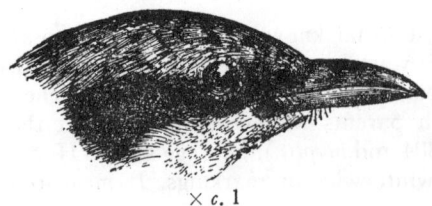

× *c.* 1

Crown slaty blue. A broad black band from lores to nape. Wing black and slaty blue. Back and upper tail-coverts a striking rufous, the latter covering all but the tip of the tail which is black. Under tail-coverts buff. Under surface of wing white and black.

Female duller than male. Head-band chocolate-brown instead of black. Back dull rufous-brown with drop-like oval black spots.

In flight, this species looks like a short, squat woodpecker with rufous upper tail-coverts.

STATUS, DISTRIBUTION and HABITAT. Resident, locally distributed. From Kumaon (specimens in Field Mus., Chicago) and Nepal (mostly between 2100 and 2300 m with some altitudinal movements; regularly met in Phulchowky and the Mail Valley), east through Darjeeling, Sikkim, Bhutan and Arunachal Pradesh; also Nagaland, Manipur, Assam in the Cachar Hills and in Mizoram; from 1350 to 2500 m. Recorded in winter up to 2285 m in Nepal (Inskipp & Inskipp), up to 2100 in Sikkim. and at *c.* 900 m in Cachar. Breeding zone not determined but probably above 1500 m. Affects heavy oak and mossy evergreen forest.

Extralimital. The Chin Hills of Burma. The species extends south through the Malay Peninsula and east to Vietnam.

GENERAL HABITS. Keeps in small parties, commonly in the mixed itinerant associations, usually frequenting the foliage canopy of forest trees, running swiftly along the branches or hopping up mossy trunks with rapidity; may occasionally be seen on the ground, collecting grit. Usually rather silent.

FOOD. Beetles and other insects, larvae, pupae, insect eggs, gastropods, seeds and berries (*Viburnum, Michelia, Magnolia*, etc.).

VOICE and CALLS. A loud, distinctive double *chirp* and a shrill *chip* (Standford). A loud, monotonous *chichip-chip-chichip* [*piou-piou-piou* ... Smythies] repeated 6 to 160 times or more—SA.

BREEDING. Unknown. Fully fledged young noted in June on Mt Victoria (Heinrich).

MUSEUM DIAGNOSIS. For details of plumage see Baker, loc. cit.

Young male, like adult but upperparts paler. Young female, like adult female but crown darker; lores, ear-coverts and sides of neck nearly black; spots on back rounder. Barring on flanks in both sexes replaced by twin spots on the feathers. Timaline primary and tail characters not present (HW).

MEASUREMENTS

	Wing	Bill (from skull)	Tarsus	Tail	
♂♂	88–101	20–23	30–33	53–66	mm
♀♀	83–93	19–21	27–30	56–60	mm

(SA, BB, Stresemann)

Weight 9 ♂ ♂ 48–56; 3 ♀ ♀ 40–46 g (GD, SDR, SA).

COLOURS OF BARE PARTS. Iris dull crimson-brown. Bill black, basal half of lower mandible bluish. Legs and feet bright (gamboge) yellow.

GENUS PTERUTHIUS SWAINSON

Pteruthius Swainson, 1832, Fauna Boreali-Americana, Bds.: 491.
  Type, by original designation, *Lanius erythropterus* Vigors
Hilarocichla Oates, 1889, Fauna Brit. India, Bds. I: 243.
  Type, by monotypy, *Pteruthius rufiventer* Blyth

Bill about half the length of the head, notched and strongly hooked at the tip. Rictal bristles weak. Tarsus strong. Sexes dissimilar.

Key to the Species

		Page
A	Size large: wing over 70 mm	
1	Inner secondaries chestnut	
a	Crown black .... *P. flaviscapis* ♂	73
b	Crown grey .... *P. flaviscapis* ♀	73
2	Secondaries chestnut at tip only	
c	Back chestnut .... *P. rufiventer* ♂	72
d	Back green .... *P. rufiventer* ♀	72
B	Size small: wing under 70 mm	
3	Forehead chestnut (♂) or rufous (♀) .... *P. aenobarbus*	79
4	Forehead grey .... *P. xanthochlorus*	76
5	Forehead and crown green	
e	Tips of wing-coverts white .... *P. melanotis* ♂	77
f	Tips of wing-coverts ochraceous buff .... *P. melanotis* ♀	77

1340. Rufousbellied Shrike-Babbler. *Pteruthius rufiventer* Blyth

Pteruthius rufiventer Blyth, 1842, Jour. Asiat. Soc. Bengal 11: 183
(no locality = Darjeeling)
Baker, FBI No. 360, Vol. 1: 337
Plate 80, fig. 4

LOCAL NAMES. None recorded.

SIZE. Bulbul; length *c.* 17 cm (7 in.).

FIELD CHARACTERS. Male. *Above*, head, wings and tail black, the last and the secondaries narrowly tipped with chestnut. Back and rump chestnut. *Below*, throat and breast ashy, divided from black of head by a white line. A yellow patch on sides of breast. Belly vinous-brown darker on flanks.

Female. *Above*, head grey, back of crown and ear-patch black. Back and folded wings bright greenish yellow, the secondaries tipped with chestnut. Lower rump and upper tail-coverts chestnut. Central rectrices green, outer mostly black, all narrowly tipped with chestnut. *Below*, as in male.

STATUS, DISTRIBUTION and HABITAT. Sparse resident. From central and eastern Nepal, east through Darjeeling, Sikkim, Bhutan and Arunachal Pradesh to the Mishmi Hills, Nagaland and Manipur; between *c.* 1500 and 2500 m (recorded at 2700 m in winter in Burma). Seasonal movements, if any, unknown. Affects dense, moss-covered oak and evergreen forest, and occasionally secondary scrub.

Extralimital. The Chin Hills and northern Burma into Yunnan; another race in northern Vietnam.

GENERAL HABITS. Keeps in small parties in company with tits and babblers. 'A fearless, rather lethargic bird feeding near the ground as well as in the tops of lofty trees, hopping from twig to twig.

FOOD. Prefers larger insects, including beetles *Longicornis* spp., caterpillars, mantids and stick insects (*Phasmida* sp.) as well as berries and seeds (T. J. Roberts).

VOICE and CALLS. Not described; usually very silent. 'A curious shrill *whirr-i-oh* heard from a flock which contained some of these birds was possibly uttered by them' (Stanford).

BREEDING. Unknown.

MUSEUM DIAGNOSIS. See Field Characters.

Young (Sex?), very like adult female with dull black head and green back; lesser, median and greater coverts tipped with the greenish colour of the mantle; wings as in adult male; underparts grey shading into white on belly (HW).

MEASUREMENTS

	Wing	Bill (from skull)	Tarsus	Tail
♂♂	85–90	20–22	29–31	(75) 80–84 mm
♀♀	86–89	20	32 (1)	79–85 mm

(Mayr, MD, SA)

Weight 7 ♂♂ 44–48; 2 ♀♀ 42, 44 g (SDR, SA).

COLOURS OF BARE PARTS. Variable. Iris grey to blue-grey (brown in one ♀); orbital skin grey. Bill: upper mandible black, lower pale grey to blue-grey, pale flesh to grey at tip. Legs and feet pale pinkish brown to chocolate-brown or fawn colour; claws brown to chocolate or grey-brown; soles yellow (one grey) (Stanford).

1341. Redwinged Shrike-Babbler. *Pteruthius flaviscapis validirostris* Koelz

Lanius erythropterus Vigors, 1831, Proc. Zool. Soc. London: 22 (Himalaya Mountains = Murree, Punjab, *vide* Baker, 1922, 'Fauna' 1: 331), *nec Lanius erythropterus* Shaw, 1809, preoccupied

Pteruthius erythropterus validirostris Koelz, 1951, Jour. Zool. Soc. India 3: 28 (Kohima Naga Hills, Assam)

Pteruthius erythropterus nocrecus Koelz, 1952, Jour. Zool. Soc. India 4: 40 (Tura Mountain, Garo Hills)

Pteruthius erythropterus glauconotus Koelz, 1954, Contrib. Inst. Regional Exploration, No. 1: 9 (Sangau, Lushai Hills)

Pteruthius validirostris ripleyi Biswas, 1960, Bull. Brit. Orn. Cl. 80: 106. New name for *Lanius erythropterus* Vigors, preoccupied

Baker, FBI No. 354, Vol. 1: 331
Plate 80, fig. 8

LOCAL NAME. *Dao-kranji* (Cachari).
SIZE. Myna —; length *c.* 16 cm (6½ in.).
FIELD CHARACTERS. Male. A stocky short-tailed black and white arboreal bird with chestnut in wings.

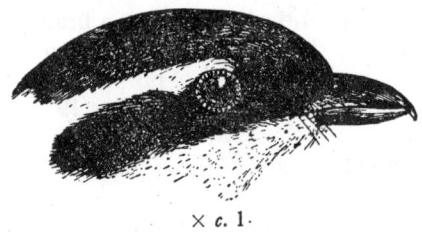
× *c.* 1.

Above, head black with a white post-ocular stripe; back ashy grey. Wings black tipped with white, the inner secondaries chestnut. Inner edge of primaries white, conspicuous in flight. Tail black, short. *Below*, pale ashy, nearly white, the lower flanks vinous brown.

Female, *Above*, head grey, back brownish grey. Outer edge of folded wing yellowish green tipped white, the inner secondaries chestnut. Central rectrices green, outer blackish tipped yellow. Underparts pale buff.

The chestnut inner secondaries distinguish both sexes. In flight white inner edges to the primaries flash, combining with the underside to give the bird a very white appearance.

STATUS, DISTRIBUTION and HABITAT. Resident, fairly common, subject to some vertical and erratic movements in winter. The Himalayas from Murree (Pakistan) and Kashmir (one record, Dickinson, JBNHS 63: 200), east to Sikkim, Bhutan and Arunachal Pradesh in the Mishmi Hills, Meghalaya, the hills of Assam and south to the Mizo Hills; Nagaland and Manipur. Bangladesh hills (?). Breeds between 1500 and 2500 m; winters mostly between 1200 and 2100 m sometimes higher, seldom descending as low as 300 m (Nepal, Rand & Fleming). Affects heavy broad-leaved forest of oak, rhododendron, etc.; in the western Himalayas also pine, deodar and spruce.

Extralimital. Western and northwestern Burma. The species extends north to Szechuan, east to Vietnam, Guangxi and Fujian south to Sumatra, Java and northern Borneo.

GENERAL HABITS. Strictly arboreal. Keeps in pairs in the breeding season, otherwise in parties of six to ten birds or singly in the mixed associations of tits, minivets, drongos and other species. Feeds mostly in the higher canopy, running along the boughs and hopping from branch to branch, exploring leaves and crevices and under the moss; or it works its way slowly up a tree, often hopping sideways along a branch or clinging on the trunk sideways like a nuthatch; 'having arrived at the top it may spend half an hour or more calling persistently before resuming its feeding on another tree' (Smythies). Often seen sitting and peering about in a curious short-sighted manner. A remarkably fearless bird, but sluggish in its movements and easily overlooked. Flight jerky, dipping with weak and hurried wing-beats.

FOOD. Beetles, caterpillars and other insects, seeds and berries.

VOICE and CALLS. A harsh, grating call like the 'churring' of a shrike uttered now and again 10 or more times in quick succession while foraging, varied occasionally by a loud quick-repeated *kewkew-kewkew*, 3 or 4 times. Each is usually prefaced by a subdued *kik*, audible only at close range. In timbre and cadence this call is strangely reminiscent of the 'wailing' of a domestic chicken that has got separated from its mother hen! Song loud and far-carrying, rendered as *cha-chew, cha-cha-chip* and *chu-wip-chip-chip* (Smythies). Song period in Nepal: February to June (Proud).

BREEDING. *Season*, April to June. *Nest*, a loosely made cup of rootlets, fibres and a few twigs, coated on the outside with moss and lichen, lavishly plastered with cobwebs and lined with rootlets. Suspended hammock-wise (like an oriole's nest) in a horizontal fork toward the extremity of a branch, very high up near the top of trees. Extremely difficult to find. *Eggs*, 2 to 4, pinkish white, profusely marked with tiny flecks and spots of purple-brown, forming a ring around the large end. Variable in size between *c*. 24 × 19 and 21 × 16 mm (Baker).

MUSEUM DIAGNOSIS. See Field Characters and Key to the Species. For details of plumage see Baker, loc. cit.

Young male: head grey tinged fulvous; post-ocular streak as in adult; lores, under eye and ear-coverts dark grey. Mantle olive-brown; upper tail-coverts greenish olive. Wings and tail as in adult but lesser, median and greater coverts edged yellowish white, tips of tail yellow. Female differs from adult female in having the head grey-brown like the mantle; ear-coverts dark slate-grey.

Postjuvenal moult involves body, lesser, median and inner half of greater coverts. First-winter birds recognized by remains of edging on the outer greater coverts.

MEASUREMENTS

	Wing	Bill (from skull)	Tarsus	Tail	
♂♂	78–86	18–20	27–30	59–64	mm
♀♀	78–87	17–20	27–29	57–66	mm

(BB, SA, Heinrich, Kinnear)

Weight 3 ♂ ♂ 36–38; 6 ♀ ♀ 34–44 g (SDR, GD, SA).

COLOURS OF BARE PARTS. Iris variable: greenish grey, olive-green, grey, bluish grey, deep blue or brown. Bill: upper mandible black with bluish slate on edges and sides of basal third; lower mandible bluish slate, paler on tip. Legs and feet dark flesh to brownish flesh; claws horny brown; soles fleshy white.

Pteruthius xanthochlorus Gray: Green Shrike-Babbler

Key to the Subspecies

		Page
A	A white eye-ring .... *P. x. hybrida*	77
B	No white eye-ring	
	1 Crown of male blackish .... *P. x. xanthochlorus*	76
	2 Crown of male ashy grey .... *P. x. occidentalis*	75

1342. ***Pteruthius xanthochlorus occidentalis*** Harington

Pterythius [sic] *xanthochloris* [sic] *occidentalis*
Harington, 1913, Bull. Brit.
Orn. Cl. 33: 82 (Dehra Dun)
Baker, FBI No. 359, Vol. 1: 336

LOCAL NAMES. None recorded.

SIZE. Sparrow −; length *c*. 13 cm. (5 in.).

FIELD CHARACTERS. Male. *Above*, head grey. Back olive-green. Closed wing greenish with blackish shoulder-patch and a faint, pale wing-bar. Tail with narrow white tip. *Below*, throat and breast pale ashy; belly yellow.
Female slightly duller than male.

STATUS, DISTRIBUTION and HABITAT. Uncommon resident, subject to some vertical movements. The Himalayas from Murree (Pakistan)[1] to western Nepal (Doti district, Fleming, JBNHS 65 (2): 331, 1968). Breeds between 1800 and 3000 m, optimum zone 2100–2700 m. Descends in winter to 1200 m but is however regularly seen at this season up to at least 2400 m. Affects forests of oak, spruce, hemlock (*Tsuga*) and deodar.

GENERAL HABITS. Strictly arboreal. A quiet and inconspicuous little bird usually found in groups of two to four in the mixed hunting parties of tits, leaf warblers and tit-babblers, among whom it is easily overlooked.

Often looks deceptively like some leaf-warbler in such a mixed gathering but does not droop or nervously flick wings in the characteristic phylloscopine manner. Distinguished also by comparatively sluggish movements and stiff, upright carriage. Frequents the tops of trees where it hops from twig to twig; also descends to the undergrowth. Often perches along a branch instead of across it, and may sometimes be seen clinging to tree-trunks or creeping along a branch like a nuthatch.

FOOD. Insects (ants, beetles, etc.), berries and seeds.

VOICE and CALLS. Usually very silent. Call, a quick-repeated single note *whit*. For description of song, see 1344.

BREEDING. *Season*, April to August. *Nest*, a deep purse-shaped structure, flimsy and fragile-looking, made of fine fibres and hair-like lichen bound together with cobweb, lined with rootlets and exteriorly decorated with a few flakes of lichen. Suspended hammock-wise in the fork of a small branch by means of cobweb, from about half a metre to 5 m above the ground. *Eggs*, 2 to

---

[1] Murree is being included (*fide* H. Whistler, *Ibis* 1930: 81) where a nest with three fresh eggs was taken by General Buchanan on 13. vii. 1900 near Changle Gali. No record has come in since except a tentative sighting by P. Jones in Nathiagali in April 1975 (JBNHS 76: 48).

4, cream-coloured with blotches of red-brown mostly in a zone at the large end. Average size of 34 eggs 18·4 × 13·7 mm (Baker).

MUSEUM DIAGNOSIS. Like nominate race (1343) but whole crown and nape ashy grey and underparts paler.

Young, as adult but upperparts duller, more olive-brown; grey of head and face not differentiated from upperparts. Underparts paler. Greater and median coverts tipped yellow. Postjuvenal moult involves body, lesser, median and inner greater coverts. First-winter birds distinguished by pale tips of unmoulted greater coverts.

MEASUREMENTS. As in 1343.

COLOURS OF BARE PARTS. Iris grey-brown. Bill blue-grey, pale at tip. Legs and feet fleshy grey.

1343. *Pteruthius xanthochlorus xanthochlorus* Gray

Pteruthius xanthochlorus Gray, 1846, Cat. Mamms. Bds. Nepal: 95, 155 (Nepal)

Baker, FBI No. 358, Vol. 1: 335

Plate 80, fig. 5

LOCAL NAMES. None recorded.

SIZE. Sparrow −; length *c.* 13 cm (5 in.).

FIELD CHARACTERS. As in 1342 but crown blackish in male, grey in female.

STATUS, DISTRIBUTION and HABITAT. Resident, locally common, subject to some vertical movements. The eastern Himalayas from central Nepal east through Darjeeling, Sikkim and Bhutan into Arunachal Pradesh (Tawang, and probably east to the Burma border). Affects deciduous or coniferous forest, from 2100 to 3000 m at all seasons, probably descending lower in winter. Recorded at Tawang [27°33′N., 91°48′E.] 3600 m (Whistler MS.; date not given but probably in summer).

GENERAL HABITS and FOOD. As in 1342, q.v.

VOICE and CALLS. As in 1344.

BREEDING. *Season*, chiefly May and June. *Nest*, as in 1342 but apparently somewhat higher—three to eight metres above the ground. *Eggs*, 3 or 4, variable in colour though typical of the genus in character. For details see Baker (*Nidification* 1: 313). Average size of 15 eggs 18·8 × 14·7 mm.

MUSEUM DIAGNOSIS. See Key to the Subspecies. For details of plumage see Baker, loc. cit.

MEASUREMENTS

	Wing	Bill (from skull)	Tarsus	Tail	
♂♂	60–64	12–13	20–23	45–49	mm
♀♀	58–64	12 (1)	22 (1)	45 (1)	mm

(Rand & Fleming, Kinnear, SA, Stresemann)

Weight 2 ♂ ♂ 14, 14; 1 ♀ 15 g (SA).

COLOURS OF BARE PARTS. As in 1342.

1344. *Pteruthius xanthochlorus hybrida* Harington

Pterythius [sic] *pallidus hybrida* Harington, 1913, Bull. Brit. Orn. Cl.
33: 82 (Mt Victoria, Chin Hills)
Not in Baker, FBI

LOCAL NAMES. None recorded.

SIZE. Sparrow —; length c. 13 cm (5 in.).

FIELD CHARACTERS. As in 1343, but male with a conspicuous white eye-ring.

STATUS, DISTRIBUTION and HABITAT. Scarce resident. Nagaland and Mizoram (specimens in the Koelz coll). Affects primeval deciduous forest with dense undergrowth.

Extralimital. The Chin Hills of Burma. The species extends to western Szechuan, southeastern Yunnan and northwestern Fujian.

GENERAL HABITS. Described by Heinrich as a very restless bird found mostly in the undergrowth of forest.

FOOD. As in 1342.

VOICE and CALLS. The song is a rapid, monotonous repetition of a single note reminding one very much of the song of *Stachyris chrysaea* (1212) although not so full and melodious, without any pause between the first and the following notes (Heinrich). Call-note, a single *whit* (Stanford).

BREEDING. Unrecorded.

MUSEUM DIAGNOSIS. Differs from the nominate race (1343) in having a white ring round the eye; head not so dark; sides of neck and flanks paler yellow.

MEASUREMENTS

	Wing	Bill (from skull)	Tarsus	Tail	
♂♂	59–63	11	20–21	45–48	mm
♀♀	59–64	11	19	44–48	mm

(HW, Heinrich)

COLOURS OF BARE PARTS. Iris red-brown. Bill black, lower mandible slaty grey. Legs and feet brown to smoke-brown.

1345. Chestnut-throated Shrike-Babbler. *Pteruthius melanotis melanotis* Hodgson

Pteruthius melanotis Hodgson, 1847, Jour. Asiat. Soc. Bengal 16: 448
(Terai, Southeast Himalayas = Nepal *fide* Gadow, 1883,
Cat. Bds. Brit. Mus. 8: 118)
Pteruthius melanotis melanops Koelz, 1952, Jour. Zool. Soc. India 4: 40
(Kohima, Naga Hills)
Baker, FBI No. 356, Vol. 1: 333
Plate 80, fig. 7

LOCAL NAME. *Ku-er-pho* (Lepcha).

SIZE. Sparrow —; length c. 11 cm (4½ in.).

FIELD CHARACTERS. Male. *Above,* olive-green with yellow forehead and a grey nuchal collar. A conspicuous white eye-ring and a greyish white supercilium. Lores and round the eye black. A black crescentic line behind yellow ear-coverts. Wing grey with a broad black bar between two narrow white bars. Tail greenish and black with white outer rectrices and tips. *Below*, throat and upper breast chestnut, rest yellow.

Female, like male but head markings less distinct, black area around eye replaced by grey. Eye-ring conspicuous. Throat mostly buffish with cinnamon sides or moustache. Wing greenish, the broad black bar between two narrow salmon-rufous ones instead of white.

P. aenobarbus (1346) male has a chestnut forehead and lacks the black upright crescent on sides of neck. The female has a rufous forehead (*v.* yellow).

STATUS, DISTRIBUTION and HABITAT. Uncommon resident, subject to vertical movements. From central Nepal east through Darjeeling, Sikkim, Bhutan and Arunachal Pradesh to the Burma border; from *c.* 1800 to 2700 m, optimum zone 2100–2400 m, in winter between 700 and 2000 m. Also, more scarcely, Nagaland, Manipur, Assam in the Cachar and Meghalaya (Khasi Hills; not found in the Garo Hills where Koelz collected *aenobarbus*), south to the Chittagong Hill Tracts of Bangladesh, from 1200 m up to at least 2100 m. Affects humid but cool, deep evergreen forest in the more open parts (glades, streamsides, etc.).

Extralimital. Extends east through Burma to northern Vietnam; another subspecies in the Malay Peninsula.

GENERAL HABITS. Entirely arboreal; frequents mostly the canopy of trees. In the non-breeding season found either singly, in pairs or in small groups usually in the large mixed roving parties of tits, leaf warblers and other small babblers, sivas, minlas and flycatchers such as *Rhipidura hypoxantha* and *Culicicapa*. Tends to be overlooked amongst these restless gatherings, but its comparatively sluggish movements, dumpier build and upright carriage unmask its identity.

FOOD. Insects (large green grasshoppers, etc.).

VOICE and CALLS. Usually silent. Call-note, a pleasant *too-weet, too-weet* not often uttered unless the birds are separated (Baker).

BREEDING. *Season*, April to June. *Nest*, a flimsy but strong cup of ferns, roots, twigs, bits of tendrils and lichen, strengthened with cobweb, the outside more or less covered with moss and lichen; lining, usually scanty, is of rootlets. The nest is suspended between forking horizontal twigs between *c.* 2 and 5 metres from the ground. *Eggs*, 4 or 5, occasionally 2 or 6, normally pale lilac-white or pale cream stippled with purplish brown or light rufous, especially around the large end. Average size of 40 eggs 17·9 × 13·5 mm (Baker). Incubation by both sexes; period undetermined.

MUSEUM DIAGNOSIS. See Field Characters.

Young, ♂ and ♀ like adult ♀ but browner on upperparts; underparts paler. Postjuvenal moult of body and wing-coverts which then show the sexual distinction (HW).

MEASUREMENTS

	Wing	Bill (from skull)	Tarsus	Tail	
♂♂	55–63	11–12	20–22	39–45	mm
♀♀	56–59	11–13	20–22	38–42	mm

(SA, Rand & Fleming, BB, SDR)

Weight 4 ♂ ♂ 10–15; 1 ♀ 14 g (SDR, SA).

COLOURS OF BARE PARTS. Iris brown (red-brown, grey in a spring bird—Stanford). Bill: upper mandible dark slate-grey, lower grey to pale flesh. Legs whitish brown to pale fleshy brown.

1346. Chestnutfronted Shrike-Babbler. *Pteruthius aenobarbus aenobarbulus* Koelz

Pteruthius aenobarbus aenobarbulus Koelz, 1954, Contrib.
Inst. Regional Exploration, No. 1: 9 (Nokrek, Garo Hills)
Baker, FBI No. 357, Vol. 1: 335 (= *intermedius*)
Plate 80, fig. 6

LOCAL NAMES. None recorded.

SIZE. Sparrow −; length *c.* 11 cm (4½ in.).

FIELD CHARACTERS. M a l e. *Above,* forehead chestnut; forecrown yellow; rest of upper plumage olive-green. A conspicuous white eye-ring surrounded by an incomplete black ring; a greyish white supercilium behind which a patch of grey (not forming a nuchal collar); ear-coverts yellow. Wing green with white outer edge and white tips to secondaries, and a large black shoulder-patch across which are two broad white bars. Outer rectrices black tipped white, the outermost white with a black streak near tip of outer web. *Below,* throat deep chestnut; rest of underparts yellow.

Distinguished from *melanotis* mainly by the chestnut forehead (*v.* yellow) and absence of the crescentic black patch on side of neck.

F e m a l e, forehead rufous-chestnut; no black round white eye-ring; shoulders olive-green with two salmon-buff bars. Throat pale ochraceous; rest of underparts greyish white with a tinge of yellow. Distinguished from female *melanotis* by same characters as male.

STATUS, DISTRIBUTION and HABITAT. Known only from the type specimen collected by Koelz in the Garo Hills, Meghalaya. Altitude not given. Affects fairly open forest and edges of evergreen.

Extralimital. The species extends east to Vietnam and Guangxi, and south through the Malay Peninsula and Java.

GENERAL HABITS. Similar to those of *melanotis*; seems to frequent lower trees and bushes.

FOOD. Insects.

VOICE and CALLS. Unrecorded.

BREEDING. *Season,* probably April and May; Koelz reports his specimen of 7 March as 'breeding male'. *Nest* and *eggs* not recorded in detail but appear to be similar to those of *melanotis*. Average size of 9 eggs 18·5 × 13·3 mm (Baker).

MUSEUM DIAGNOSIS. See Field Characters. Differs from *intermedius* of Burma in having the brown of throat much restricted, not descending over the breast, but sharply defined as a throat-patch. Yellow of forehead deeper; upperparts more yellow, less green (Koelz).

MEASUREMENTS and COLOURS OF BARE PARTS. Unrecorded.

TIMALIINAE

Genus GAMPSORHYNCHUS Blyth

Gampsorhynchus Blyth, 1844, Jour. Asiat. Soc. Bengal 13 (1): 370.
Type, by monotypy, *G. rufulus* Blyth

Tail longer than wing, much graduated. Bill about half the length of head; upper mandible curved at tip. Rictal bristles very long and stiff.

1347. Whiteheaded Shrike-Babbler. *Gampsorhynchus rufulus rufulus* Blyth

Gampsorhynchus rufulus Blyth, 1844, Jour. Asiat. Soc. Bengal 13 (1): 371
(Darjeeling)
Gampsorhynchus rufulus ahomensis Koelz, 1954, Contrib. Inst.
Regional Exploration, No. 1: 4 (Nichuguard, Naga Hills)
Baker, FBI No. 232, Vol. 1: 231
Plate 79, fig. 1

LOCAL NAMES. *Chongto-phep-pho* (Lepcha); *Daophlantu-tiba* (Cachari).
SIZE. Bulbul +; length *c.* 23 cm (9 in.).
FIELD CHARACTERS. Whole head, nape, a shoulder-patch, throat, breast and belly white, the latter washed with buff. Under tail-coverts buff. Rest of plumage olive-brown, wing with buff inner edge. Tail strongly graduated and tipped with buff. Sexes alike.

Young (immature) has light chestnut head and breast.

× *c.* 1

STATUS, DISTRIBUTION and HABITAT. Resident, generally common. The base of the hills from eastern Nepal (specimens in Brit. Mus.) through Darjeeling, Sikkim, N. Bengal (Jalpaiguri dist.), Bhutan, and Arunachal Pradesh into Burma; Nagaland, Manipur and the hills of Assam, Meghalaya, south to the Chittagong region of Bangladesh; from the edge of the plains to *c.* 1200 m. Affects secondary growth, bamboo, bush and grass jungle and undergrowth in evergreen forest in broken foothills country.

Extralimital. Burma. The species extends east to Vietnam and south through the Malay Peninsula.

GENERAL HABITS. Arboreal, gregarious and noisy. Keeps in parties up to fifteen, often in association with wood shrikes, drongos, scimitar babblers and others in thick shrubbery and clumps of bamboo. Usually tame and inquisitive. More active on the wing than most babblers. General appearance, flight and behaviour reminiscent of the bulbuls.

FOOD. Insects; berries (?).

VOICE and CALLS. A weird, grating *kaw-ka-yawk* (Smythies), and a variety of calls incessantly uttered.

BREEDING. *Season*, end of April to August, chiefly May. *Nest*, a shallow flimsy saucer of dead leaves, twigs, roots and lichen bound together with roots and tendrils, plastered with cobweb, and lined with fine grass and rootlets. Placed in bushes a couple of metres from the ground. The species appears to move into deeper evergreen forest for breeding purposes. *Eggs*, 3 or 4, of two types: pale green with blotches of dark brown and secondary markings of dull grey scattered over the whole surface but more numerous at the larger end; or pale dull reddish with primary markings of reddish brown. Except for size, they resemble the eggs of *Pellorneum albiventre* (1164). Average size of 20 eggs 29·9 × 17·6 mm (Baker).

MUSEUM DIAGNOSIS. See Field Characters. For details of plumages see Baker, loc. cit.

MEASUREMENTS

	Wing	Bill (from feathers)	Tarsus	Tail
♂♀	90–100	20–21	26–31	110–120 mm (Baker, SA)

Weight 2 ♂♂ 32, 37 g; 1♀ 34 g (SDR).

COLOURS OF BARE PARTS. Iris yellow ('orange-straw'—SA). Bill plumbeous, paler on lower mandible. Legs, feet and claws flesh; soles yellowish.

Genus ACTINODURA Gould

Actinodura Gould, 1836, Proc. Zool. Soc. London: 17.
Type, by original designation, *Actinodura Egertoni* Gould
Sibia Hodgson, 1836, Asiat. Res. 19: 145. Type, by monotypy, *Sibia? Nipalensis* = *Cinclosoma? Nipalensis* Hodgson
Ixops 'Hodgson' = Blyth, 1843, Jour. Asiat. Soc. Bengal 12: 929, 953.
Type, by monotypy, *Cinclosoma Nipalense* Hodgson = *Sibia Nipalensis* Hodgson

Bill about half the length of head. Rictal bristles long; chin feathers with hair-like tips. Wing feathers narrowly cross-barred. Tail graduated, considerably longer than wing.

Key to the Species

			Page
A	Forehead rufous	*A. egertoni*	81
B	Forehead brown, concolorous with crown	*A. nipalensis*	84
	1 Feathers of crown with pale shaft-streaks	*A. nipalensis*	86

ACTINODURA EGERTONI GOULD: SPECKLED BARWING

Key to the Subspecies

			Page
A	Crown-feathers edged with black	*A. e. lewisi*	83
B	No black edging to crown-feathers		
	1 Crown ashy brown or dark grey contrasting with mantle		
	a Back rufous-brown	*A. e. egertoni*	81
	b Back olive-brown	*A. e. ripponi*	84
	2 Crown browner, less ashy, less contrasting with mantle	*A. e. khasiana*	83

1348. *Actinodura egertoni egertoni* Gould

Actinodura Egertoni Gould, 1836, Proc. Zool. Soc. London: 18 (Nepal)
Baker, FBI No. 321, Vol. 1: 303

Plate 79, fig. 4

LOCAL NAME. *Ramnio-pho* (Lepcha).

SIZE. Bulbul +; length *c.* 23 cm (9 in.).

FIELD CHARACTERS. *Above*, forehead rufous; loose, mop-like ashy brown crest, and ear-coverts paler. Back and rump rufous-brown. Wing narrowly cross-barred, a black patch within a large rufous patch, and a pale outer edge. Tail strongly graduated, rufous-brown, narrowly cross-barred, each feather tipped white. *Below*, chin rufous, throat and breast pinkish brown; rest of underparts tawny olive, centre of belly white. Under surface of tail dark grey with white spots. Sexes alike.

May be mistaken for *Garrulax lineatus* but distinguished by barred wings. *Actinodura nipalensis* and *A. waldeni* have a darker crown and black tail.

STATUS, DISTRIBUTION and HABITAT. Common resident. The Himalayas from central Nepal (Biswas, JBNHS 59: 220; Stevens *apud* Baker, *Nidification* 1: 281), east through Darjeeling, Sikkim, Bhutan and Arunachal Pradesh to the Miri Hills. Breeds from 1200 to at least 2000 m (2400 m, Osmaston *apud* Baker); recorded in winter from 1000 to 1800 m and around Buxa duars (W. Bengal). Affects dense secondary growth and mixed trees-and-scrub in evergreen forest.

GENERAL HABITS. Keeps in pairs during the breeding season, otherwise in small slow-travelling parties of six to twelve, usually not associating with other species though sometimes found in company of sibias and laughing thrushes whose habits and behaviour are very similar. Frequents dense thickets, bushes and low trees, occasionally the high canopy, clambering about and poking into holes and crannies, and foraging about amongst clumps of orchids and other epiphytes, at times clinging upside-down like a tit or fluttering in front of a sprig. Dives into undergrowth when alarmed. Pairs form in the latter half of April.

FOOD. Mainly insects (grasshoppers, ants, etc.); also berries, figs and seeds.

VOICE and CALLS. A feeble conversational *cheep* constantly uttered while feeding (SA). For description of song see 1351.

BREEDING. *Season*, April to July. *Nest*, cup-shaped, made of fern, bamboo or other dead leaves held together with fibrous roots, coated exteriorly with green moss and lined with rootlets; amount and proportion of each material varies in different nests. Placed in bushes or small trees from one to six metres above the ground. *Eggs*, 3, or sometimes 4, pale blue, marked all over with dark brown blotches, spots, loops, whorls and fine lines. Many eggs very similar to those of *Garrulax phoeniceus* (1332). Average size of 25 eggs 22·9 × 17·7 mm (Baker).

MUSEUM DIAGNOSIS. See Key to the Subspecies. For details of plumage see Baker, loc. cit.

Young, like adult but crown brown instead of ashy brown; underparts more uniformly ochraceous brown, no white on belly.

MEASUREMENTS

	Wing	Bill (from skull)	Tarsus	Tail
♂♀	79–90	17–19	29–33	93–115 mm (SA, Stresemann)

Weight 2 ♂ ♂ 33, 37; 2 o? 33, 36 g (SA).
COLOURS OF BARE PARTS. As in 1349.

1349. *Actinodura egertoni lewisi* Ripley

Actinodura egertoni lewisi Ripley, 1948, Proc. Biol. Soc. Washington 61: 105
(Dreyi, Mishmi Hills)
Not in Baker, FBI

LOCAL NAMES. None recorded.
SIZE. Bulbul +; length *c.* 23 cm (9 in.).
FIELD CHARACTERS. As in 1348, q.v.
STATUS, DISTRIBUTION and HABITAT. Common resident. Arunachal Pradesh in the Mishmi Hills (Ali & Ripley, JBNHS 48: 28; Baker, *Records Ind. Mus.* 8: 265), probably intergrading with *egertoni* in the Abor Hills. Affects scrub jungle and dense shrubbery in evergreen forest.
GENERAL HABITS, FOOD and VOICE. As in 1348.
BREEDING. Unrecorded; probably as in 1348.
MUSEUM DIAGNOSIS. Differs from *egertoni* (1348) in having pronounced dark edgings on a very grey head, in having more dark grey wash on the neck, a darker back, and a tendency to broader striping on the tertiaries. Differs from *khasiana* in being darker and not having the pronounced barring on the central rectrices. Differs from *ripponi* in being much darker, more rufous brownish on the back and more rufous on tail.

MEASUREMENTS

	Wing	Bill (from skull)	Tarsus	Tail	
4 ♂♂	81–90	20	29–31	104–116	mm
8 ♀♀	81–91	19–20	29–31	101–111	mm
				(MD)	

COLOURS OF BARE PARTS. Iris brownish grey. Bill: upper mandible brown, lower mandible yellow at base, fleshy at tip. Legs and feet brownish flesh.

1350. *Actinodura egertoni khasiana* Godwin-Austen

Actinodura khasiana Godwin-Austen, 1876, Jour. Asiat. Soc. Bengal 45: 76
(Khasia Hills)
Actinodura egertoni montivaga Koelz, 1954, Contrib. Inst.
Regional Exploration, No. 1: 7 (Kohima, Naga Hills)
Baker, FBI No. 322, Vol. 1: 304

LOCAL NAMES. *Nya-si, Ko-yu* (Naga).
SIZE. Bulbul +; length *c.* 23 cm (9 in.).
FIELD CHARACTERS. As in 1348, q.v.
STATUS, DISTRIBUTION and HABITAT. Common resident; Khasi Hills (Meghalaya), Cachar Hills (Assam), Arunachal Pradesh (Tirap and Changlang dists.), Nagaland and Manipur. Breeds between 800 and 2200 m. Inhabits dense evergreen forest, scrub jungle, and secondary growth.
GENERAL HABITS, FOOD and VOICE. As in 1348.
BREEDING. As in 1348. Average size of 100 eggs 23·4 × 17·7 mm (Baker).
MUSEUM DIAGNOSIS. Differs from *lewisi* (1349) in being pale and having more pronounced barring on central rectrices. Differs from *egertoni* (1348) in having the crown browner, less ashy and not so contrasting with mantle, the rufous of forehead

paler and not extending to the crown; back, rump and upper tail-coverts more ochraceous; central rectrices more distinctly barred.

MEASUREMENTS. As in 1349.

Weight 6 ♂ ♂ 28.5, 27.5, 25.5, 38(3); 1♀ 35 g (SDR).

COLOURS OF BARE PARTS. As in 1349.

1351. *Actinodura egertoni ripponi* Ogilvie-Grant

Actinodura ripponi Ogilvie-Grant, 1907, Ibis: 186
(Mt Victoria, Chin Hills, 6000–7000 feet)
Baker, FBI No. 323, Vol. 1: 305

LOCAL NAMES. None recorded.

SIZE. Bulbul +; length *c*. 23 cm (9 in.).

FIELD CHARACTERS. As in 1348, q.v.

STATUS, DISTRIBUTION and HABITAT. Common resident, subject to some vertical movements. Hills of Mizoram. Altitudinal distribution undetermined here, but in the Chin Hills breeds above 1500 m and on Mt Victoria between 2000 and 2600 m; descends in winter to 1200–1800 m, some straying down to the foothills. Affects high bushes and low trees in dense, shady, primeval forest.

Extralimital. Southwestern Burma.

GENERAL HABITS and FOOD. As in 1348.

VOICE and CALLS. The song consists of a three-noted whistle reminding one of the notes of the Grey Tit but louder and sharper, rendered as *ti-ti-tā*, the first note accentuated, the last lower (Heinrich).

BREEDING. Mostly April and May. *Nest* and *eggs* as in 1348. Average size of 30 eggs 22·6 × 17·4 mm (Baker).

MUSEUM DIAGNOSIS. Like *khasiana* (1350) but crown dark grey; back, rump and upper tail-coverts olive-brown.

MEASUREMENTS

Wing 10 ♂ ♂ 86–93; 3 ♀ ♀ 83–90 mm (Heinrich)
Tail ♂ ♀ 113–122 mm (Ticehurst)

COLOURS OF BARE PARTS. Iris olive-yellow; orbital skin grey-green. Bill brown, yellow at gape. Legs and feet grey-brown.

ACTINODURA NIPALENSIS (Hodgson): HOARY BARWING

Key to the Subspecies

Black band on tail under 20 mm wide .... *A. n. nipalensis*
Throat and breast grey, streaked
 with rufous-brown .... *A. n. daflaensis*
Throat and breast tawny brown with fulvous streaks .*A. n. waldeni*
Dark and dark-streaked on crown and mantle; throat
 dark and breast dark streaked .... *A.n. poliotis*

1352, 1353. *Actinodura nipalensis nipalensis* (Hodgson)

Cinclosoma? Nipalensis Hodgson, 1836, Asiat. Res. 19: 145
(Nepal, restricted to slopes of Kathmandu Valley, central Nepal, by Ripley)
Actinodura nipalensis vinctura Ripley, 1950, Proc. Biol. Soc. Washington 63: 104
(Mangalbare, Dhankuta Dist, east Nepal)
Baker, FBI No. 326 (part), Vol. 1: 307
Plate 79, fig. 2

LOCAL NAMES. None recorded.

SIZE. Bulbul; length *c.* 20 cm (8 in.).

FIELD CHARACTERS. *Above*, forehead and loose mop-like crest dark brown with pale shaft-streaks. A pale eye-ring. Ear-coverts grey. Moustachial stripe black. Back and rump rufous-brown. Wing rufous, narrowly barred with black; a black shoulder-patch (primary-coverts) and a grey patch above it (tips to secondary-coverts). Tail graduated; central rectrices narrowly barred, terminal third and outer rectrices black, the latter tipped with white. Throat and breast grey, lower belly and under tail-coverts rufous-brown. Sexes alike.

× *c.* 1

A. egertoni is distinguished by the longer tail, rufous breast, white centre of belly, lack of grey patch on wing and lack of black on upper surface of tail.

STATUS, DISTRIBUTION and HABITAT. Nepal from 1800 (even in winter) to 3300 m; probably does not breed at lower levels; east through Darjeeling, Sikkim and Bhutan, from 2100 to at least 3000 m at all seasons. Affects the upper oak forest and mixed oak, conifer and rhododendron with plenty of undergrowth.

GENERAL HABITS. Arboreal. Keeps in small parties of three to ten in company with yuhinas, sivas and sibias. Frequents mostly the upper branches of middle-sized trees, feeding chiefly on insects concealed in the mossy growth adhering to trunks and branches. May be seen hopping swiftly up a tree-trunk or feeding while clinging to it. Occasionally descends to or near the ground.

FOOD. Mostly insects (beetles, caterpillars, etc.); also gastropods, berries, seeds, nectar, flower-buds and moss.

VOICE and CALLS. A whistle *tui whee-er* very like that of the Streaked Laughing Thrush (Proud). Alarm, a loud, rapid *je-je* . . . repeated eight or ten times (Fleming).

BREEDING. Little known. *Season*, May and June. *Nest*, a neat, compact cup, rather small for the size of the bird, made of fine grass bents with some lichen and moss on the exterior, lined with rootlets. One nest was reported to be placed on a sapling. *Eggs*, apparently 2, pale pinky white marked with bold blotches of reddish brown and secondary blotches of inky grey, forming a well-defined ring at the large end. Two eggs measured 27·1 × 18·7 and 25·2 × 18·8 mm (Baker).

MUSEUM DIAGNOSIS. For details of plumage see Baker, loc. cit.

Young, as adult. Primary and tail characters of juvenal timalines not present.

TIMALIINAE

Actinodura nipalensis

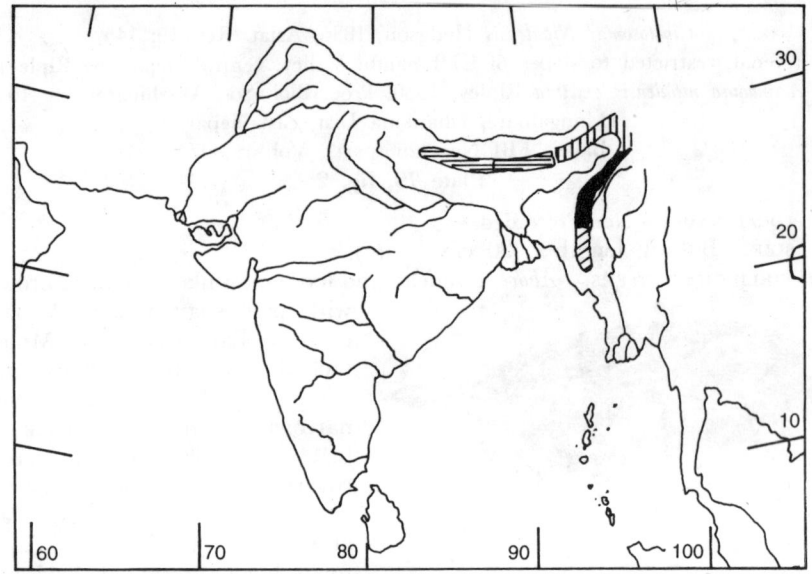

Distributional ranges

≡ *A. n. nipalensis* (1352, 1353).   ⦀⦀⦀ *A. n. daflaensis* (1354).

■ *A. n. waldeni* (1355).   ▨ *A. n. poliotis* (1356).

MEASUREMENTS

	Wing	Bill (from skull)	Tarsus	Tail	
♂♂	90–96	21–22	c. 31	82–84	mm
♀♀	83–97	21–22	—	77–80	mm

(BB, Rand & Fleming)

Weight 5 ♂ ♂ 44–48; 5 ♀ ♀ 39–45 g (GD). 2 ♂ ♂ 44, 46; 3 ♀ ♀ 39–44 g (SA).

COLOURS OF BARE PARTS. Iris brown. Bill dark brown. Legs and feet fleshy to greyish brown or brown.

1354. *Actinodura nipalensis daflaensis* Godwin-Austen

Actinodura daflaensis Godwin-Austen, 1875, Ann. Mag. Nat. Hist. 16: 340
(Dafla Hills, NE. Bengal)
Baker, FBI No. 329, Vol. 1: 309

LOCAL NAMES. None recorded.

SIZE. Bulbul; length *c.* 20 cm (8 in.).

FIELD CHARACTERS. *Above*, very similar to *A. nipalensis* but crown-feathers a darker brown with pale edges, giving the crown a scalloped aspect. Rest of upperparts as in *nipalensis*. Moustachial stripe and ear-coverts paler, streaked with whitish. *Below*, grey with rufous-brown streaks on throat and breast (*A. nipalensis* has plain grey underparts). Sexes alike.

Distinguished from *A. egertoni* by the shorter tail, black on its upper surface terminal half, and by the lack of white on belly.

STATUS, DISTRIBUTION and HABITAT. Common resident, subject to some vertical movements. Arunachal Pradesh from the Dafla Hills to the Mishmi Hills; from *c.* 2400 to 3300 m in summer, descending to *c.* 1500 m in winter. Affects mossy evergreen and mixed forest.

Extralimital. Southeastern Tibet in the Pachakshiri area.

GENERAL HABITS and FOOD. As in 1355.

VOICE and CALLS. As in 1355.

BREEDING. Unknown; presumably as in *nipalensis* (1352).

MUSEUM DIAGNOSIS. See Field Characters.

MEASUREMENTS and COLOURS OF BARE PARTS. As in 1355.

1355. *Actinodura nipalensis waldeni* Godwin-Austen

Actinodura waldeni Godwin-Austen, 1874, Proc. Zool. Soc. London: 46, pl. 12
(Japoo Peak, Naga Hills)
Baker, FBI No. 327, Vol. 1: 308
Plate 79, fig. 5

LOCAL NAMES. None recorded.

SIZE. Bulbul; length *c.* 20 cm (8 in.).

FIELD CHARACTERS. As in 1354, q.v.

STATUS, DISTRIBUTION and HABITAT. Common resident, probably subject to some vertical movements. Manipur, Nagaland and adjacent north Cachar [Assam] (Baker, JBNHS 8: 203); from 2200 m up in winter. Affects shady moss forest in steep valleys.

Extralimital. The subspecies *saturatior* is found in northeastern Burma and northwestern Yunnan; the subspecies *poliotis* (1356) with deep rufous or maroon-brown upperparts and dark grey ear-coverts is found in the Chin Hills of Burma and *may* occur in the Mizo Hills.[1]

GENERAL HABITS do not differ from those of *A. nipalensis* (1352). Usually seen clambering about mossy trunks, pulling the moss to pieces and feeding on the insects it harbours. Generally keeps to medium heights in trees but does not shun low bushes. Actions somewhat phlegmatic; remarkably tame and unperturbed by the presence of man.

FOOD. As in 1352.

VOICE and CALLS. Call-note, a soft *chup, chup*; also a mewing note and a *churr* (Smythies).

BREEDING. Unknown; presumably as in *A. nipalensis* (1352).

MUSEUM DIAGNOSIS. Differs from *daflaensis* (1354) in being paler above with the chestnut brighter; underparts entirely tawny brown with paler, fulvous streaks on throat and breast.

MEASUREMENTS

	Wing	Bill (from skull)	Tarsus	Tail	
♂♂	89–97	21	31	79	mm
♀♀	86–99	22	32	75	mm

(Wing SDR, Bill, tarsus, tail 1 ♂, 1 ♀, MD)

Weight 10 ♂♂ 39–56; 8 ♀♀ 41–53 g (SDR).

---

[1] Omitted here for want of any definite record within Indian limits. It is rare on Mt Victoria.

COLOURS OF BARE PARTS. Iris brownish grey. Bill dark brown. Legs and feet brown.

1356. *Actinodura nipalensis poliotis* (Rippon)

Ixops poliotis Rippon, 1905, Bull. Brit. Ornith. Cl. 15: 97 (Mt Victoria)

LOCAL NAMES. None recorded.
SIZE. Bulbul; length 20 cm (8 in.).
FIELD CHARACTERS. As in 1355 but darker dorsally. Crown dark, with crest feathers blackish; mantle and back dark maroon brown. Ear-coverts are darker grey with broader brown centres.
STATUS, DISTRIBUTION and HABITAT. Known only from the type locality, Mount Victoria, Chin Hills, western Burma, but might be expected in India territory in the mountains of eastern Manipur and Mizoram. Collected between 2100 and 2400 m in forest.
GENERAL HABITS. Unknown, but presumably as for 1355.
FOOD. Unknown.
VOICE and CALLS. Unknown.
MUSEUM DIAGNOSIS. For details of plumage see Baker, FBI no. 328, Vo. 1: 309.

Genus MINLA Hodgson

Minla Hodgson, 1837 (Apr. 13), Ind. Rev. 2 (1): 32, 44. Type,
by original designation, *Minla ignotincta* Hodgson
Siva Hodgson, 1837 (May 13), Ind. Rev. 2 (2): 88. Type,
by original designation, *Siva cyanouroptera* Hodgson
Staphida Swinhoe, 1871, Proc. Zool. Soc. London: 373. Type,
by original designation, *Siva torqueola* Swinhoe

Bill slender, pointed and slightly notched. Rictal bristles well developed. Outer edge of wing and tail brightly coloured.

Key to the Species

		Page
A	Edge of wing blue .... *M. cyanouroptera*	94
B	Edge of wing yellow or red	
1	Throat barred .... *M. strigula*	90
2	Throat not barred .... *M. ignotincta*	88

1357. **Redtailed Minla.** *Minla ignotincta ignotincta* Hodgson

Minla ignotincta Hodgson, 1837, Ind. Rev. 2 (1): 32, 44
(Central and northern regions of the hills, Nepal)
Baker, FBI No. 377, Vol. 1: 355
Plate 80, fig. 12

LOCAL NAMES. *Minla* (Nepal); *Megblim-ayene, Megblim-adum* (Lepcha); *Pobhum dasin* (Dafla).
SIZE. Sparrow; length *c.* 14 cm (5½ in.).

FIELD CHARACTERS. Male. *Above*, head black with a long, white supercilium from lores to nape. Back chocolate-brown. Wings black with white tips and crimson outer edge. Tail black with crimson outer edge and tip; a white patch at base. *Below*, chin and throat whitish, rest pale yellow.

Female, like male but back olive-brown, red on wings and tail and yellow of underparts paler. When flitting high up in tall trees, the tail looks comparatively long and narrow, reminiscent of Redheaded Tit (1819).

STATUS, DISTRIBUTION and HABITAT. Common resident, subject to marked vertical movements. From central Nepal east through Darjeeling, Sikkim, Bhutan and Arunachal Pradesh east to Burma border; also Assam in Cachar hills, Nagaland, Meghalaya and Manipur, south to the Chittagong region in Bangladesh. From 1800 to 3100 m, locally up to 3400 m (Nepal—Diesselhorst) and perhaps as low as 1350 m (Sikkim—Stevens). Winters mostly below 1800 m (but recorded (?) at 2400 m) and commonly reaches the foothills and duars (e.g. Buxa—Inglis, JBNHS 26: 990). Affects humid, dense forest, mixed, deciduous or evergreen, in the breeding season being most common in oak and arborescent rhododendron forest.

Extralimital. The species extends to southeastern Sichuan, Guangxi and northern Vietnam.

GENERAL HABITS. Arboreal, sociable, and very similar to those of its congeners. Usually found in parties, sometimes of considerable size, with yuhinas, shrike-babblers, tits, sunbirds, etc. Frequents tree-tops and higher branches, clinging upside down and sideways to the sprigs and working through the foliage in the manner of tits, but slower in its movements. Sometimes makes short hopping or sidling spurts along the moss-covered trunks and branches like a creeper or nuthatch, searching methodically for insects under the moss and lichen.

FOOD. Chiefly insects and their larvae; also seeds.

VOICE and CALLS. A peculiar, rattling, shrill *trri-krö . . . tschitsesa. . . trri-krrö* (Heinrich). A high-pitched *wi-wi-wi*; a loud *chik* repeated seven or eight times; a frequent, high-pitched *tsi . . . tsi*; a high-pitched *chitititit*; a tit-like *whi-whi-te-sik-sik*; song, some loud, ringing notes: *twiyi-twiyuwi. . .* (Lister; for more details see JBNHS 52: 32).

BREEDING. Little known. *Season*, May and June. *Nest*, a rather deep cup of moss and rootlets, lined with hair and rootlets, and placed in the fork of 'some bushy tree at no great height from the ground' (Hodgson). *Eggs*, 2 to 4, pale blue, marked with specks and tiny spots of black or reddish brown, generally forming a ring around the large end. Average size of 12 eggs 19·4 × 14·4 mm (Baker).

MUSEUM DIAGNOSIS. See Field Characters. For details of plumages see Baker, loc. cit.

Young, like adult female but mantle slightly darker and white parts of plumage not so pure.

MEASUREMENTS

	Wing	Bill (from skull)	Tarsus	Tail	
♂♂	63–70	12–14	20–22	52–58	mm
♀♀	59–66	12–14	19–21	51–58	mm

(SA, Rand & Fleming, Kinnear, Mayr)

Weight ♂ ♂ 12–16; ♀ ♀ 11–16 g (SDR, GD, SA). 5 ♂ ♀ 13–18 g (SA).

COLOURS OF BARE PARTS. Iris brown to pale yellow. Bill blackish brown, plumbeous on lower mandible. Legs, feet and claws olive-brown; soles yellow.

MINLA STRIGULA (Hodgson): BARTHROATED SIVA

Key to the Subspecies

		Page
A Chestnut on tail restricted to basal half		
1 Crown paler	*M. s. simlaensis*	90
2 Crown darker	*M. s. strigula*	91
B Chestnut on tail extending to *c.* 20 mm or less from tip		
3 Eye-ring pale yellow	*M. s. yunnanensis*	93
4 Eye-ring whitish	*M. s. cinereigenae*	93

1358. *Minla strigula simlaensis* (Meinertzhagen)

Siva strigula simlaensis Meinertzhagen, 1926, Bull. Brit. Orn. Cl. 46: 128 (Simla)
Baker, FBI No. 333 (part), Vol. 1: 313

LOCAL NAMES. None recorded.
SIZE. Sparrow; length *c.* 14 cm (5½ in.).
FIELD CHARACTERS. As in 1359, q.v.

Minla strigula

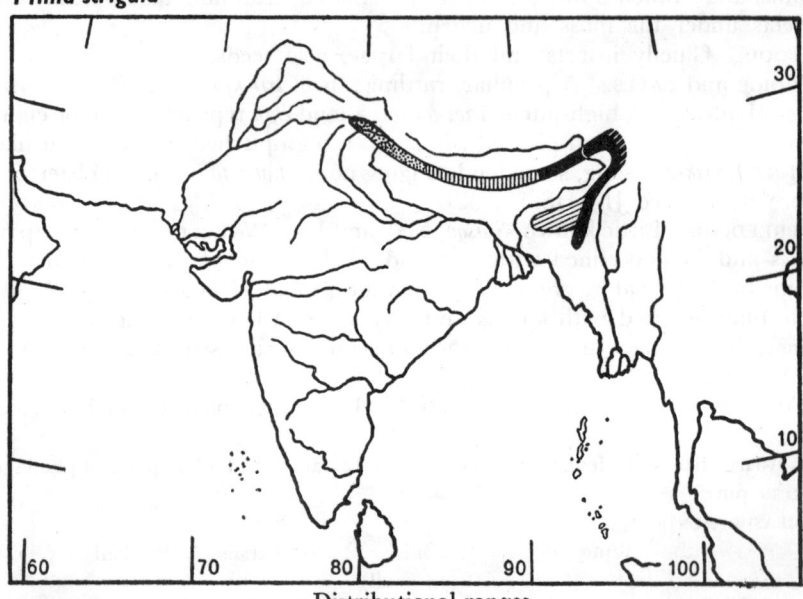

Distributional ranges

simlaensis (1358).   strigula (1359).
yunnanensis (1360).   cinereigenae (1361).

BABBLERS

STATUS, DISTRIBUTION and HABITAT. Fairly common resident, subject to vertical movements. The western Himalayas from Kangra to western Nepal (Fleming, *Fieldiana* 53: 175). Breeds from 2100 m (Kangra) and 2400 m (Garhwal) up to 3600 m. Winters between 1300 and 2250 m. Affects open forest of birch, willow, barberry, etc., oak and rhododendron scrub or other low bushes; in winter frequents bushes in heavy jungle.

GENERAL HABITS and FOOD. As in 1359.

VOICE and CALLS. A distinctive, varied combination of *pip* and *peep* (HW). In the breeding season, a melancholic whistle of three notes uttered somewhat slowly and deliberately at equal intervals, the last note lower (Osmaston); see also No. 1359.

BREEDING. As in 1359. Average size of 12 eggs 20·4 × 15.3 mm (Baker).

MUSEUM DIAGNOSIS. Like *strigula* (1359) but crown paler; back with variable amount of grey; chestnut on tail confined to basal half and rather paler.

MEASUREMENTS and COLOURS OF BARE PARTS. As in 1359.

1359. *Minla strigula strigula* (Hodgson)

Siva strigula Hodgson, 1837, Ind. Rev. 2 (2): 89 (Nepal)
Baker, FBI No. 333 (part), Vol. 1: 313
Plate 80, fig. 9

LOCAL NAME. *Megblim* (Lepcha).
SIZE. Sparrow; length *c.* 14 cm (5½ in.).
FIELD CHARACTERS. *Above,* crown orange-brown, slightly tufted. A pale yellow eye-ring and post-ocular stripe. Ear-coverts grey. Back greyish olive. Closed wing with a bright orange outer edge, a black shoulder-patch; secondaries pale ashy and black, tipped with white. Tail black, partly chestnut on the basal half, edged and tipped with bright yellow. *Below,* chin orange; throat whitish narrowly barred with black; a black malar stripe. Rest of underparts including under surface of tail yellow. Sexes alike.

× *c.* 1

STATUS, DISTRIBUTION and HABITAT. Common resident, subject to vertical movements. From west-central Nepal east through Darjeeling, Sikkim and Bhutan. Breeds between 1800 and 3750 m, chiefly in the subalpine forest at timber-line. Winters mostly between 1300 and 2700 m [recorded as low as 800 m (Meinertzhagen) and as high as 3700 m (Schäfer) in Sikkim during the cold season]. In the breeding season, affects mostly subalpine oak and rhododendron forest. In winter it is met with in pine, mixed forest, bamboo and scrub jungle.

GENERAL HABITS. Arboreal. Keeps in restless parties of six to twenty individuals, usually in the mixed itinerant bird associations, often with Redtailed Minlas and Bluewinged Sivas, hunting actively in the higher bushes and crowns of medium height trees, in the same niche as *Yuhina gularis*. The birds run or sidle up branches and hanging creepers with the agility of a mouse, and 'flow' in waves from tree to tree. Sometimes they may be seen resting side by side on a twig, snuggled up in the manner of munias.

Display. 'The birds collect into flocks of fifteen to fifty birds, and these break up into pairs which all remain together in the same tree. The two birds of each pair sit very close together, tails on opposite sides of the branch and separated by at least six inches (15 cm) from the next pair. They then alternately bow and stretch their heads up to the fullest extent, the feathers of the head and neck fluffed out, all the time keeping up a continual churring note and a sweet whistle. They frequently reverse their positions on the branch, and both birds always do this at the same instant as if at a signal, so that they never both have their tails the same side of a branch even for a second. The whole performance is like an elaborate dance and will continue for an hour or more without a break. Each pair normally takes no notice of any other birds in the tree, but I have occasionally seen one bird leave its partner and commence to bow to the nearest bird of the next pair, this leading to much scuffling and disturbance. The three-noted song is not heard during this display which appears to be performed only collectively' (Proud, JBNHS 53: 59).

FOOD. In summer almost exclusively insects (caterpillars, beetles, etc.); berries and seeds are also taken in other seasons. Often visits rhododendron blossoms for nectar and insects.

VOICE and CALLS. Song, a loud, rather hoarse-sounding strophe, most often of four syllables, rendered as *tsi-tsä-ti-si* (Diesselhorst) or *too-sweet-sweet*, the second note highest (Heinrich, Smythies). Another song is described as a jumble of sweet notes mingled with harsh squeaks and churrs uttered continuously for a minute or more. Song period in Nepal: March to July with a resumption in October (Proud). Call-note, a mellow *peera-tzip* and a louder *pe-eo*.

BREEDING. *Season*, end of April to July. *Nest*, a cup of grass, bamboo leaves, moss and birch bark, held together with thread-like *Usnea* lichen, and lined with hair, rootlets, pine needles, or fern stalks. Placed in bushes, usually within three metres from the ground. *Eggs*, normally 3, deep blue or blue-green (fading quickly if exposed to light), lightly spotted or freckled with black or pale red at the larger end. Average size of 10 eggs, 20·4 × 15·3 mm (Baker).

MUSEUM DIAGNOSIS. See Key to the Subspecies. For details of plumage, see Baker, loc. cit. Skins fade rapidly; some old skins lose all trace of yellow. Fresh specimens are much richer yellow.

Young, like adult but crown and mantle paler. Underparts paler, throat-bars less distinct.

MEASUREMENTS

	Wing	Bill (from skull)	Tarsus	Tail	
♂♂	65–76	12–16	25–28	66–75	mm
♀♀	63–69	13–15	26–28	67–68	mm

(SDR, BB, Rand & Fleming, SA).

Weight 13 ♂ ♂ 15–21; 5 ♀ ♀ 18–24 g (SA, SDR, GD). 22 ♂ ♀ 17–20 g (SA).

COLOURS OF BARE PARTS. Iris brown. Bill: upper mandible greyish brown, lower greyish white. Legs and feet brownish grey.

1360. *Minla strigula yunnanensis* (Rothschild)

Siva strigula yunnanensis Rothschild, 1921, Novit. Zool. 28: 40
(Lichiang Range, NW. Yunnan)
Siva strigula victoriae Meinertzhagen, 1926, Bull. Brit. Orn. Cl. 26: 128
(Mount Victoria, Chin Hills)
Baker, FBI No. 333 (part) and 334 (part), Vol. 1: 313–14

LOCAL NAMES. None recorded.

SIZE. Sparrow; length *c*. 14 cm (5½ in.).

FIELD CHARACTERS. As in 1359, q,v.

STATUS, DISTRIBUTION and HABITAT. Common resident, subject to vertical movements. Arunachal Pradesh from the Dafla to the Mishmi hills, Nagaland and Manipur? (Whistler MS.); from 1300 (winter) to 3600 m (summer). Affects forests of oak and rhododendron, bamboo and bushes.

Extralimital. Southeastern Tibet, western and northern Burma, Yunnan and northern Vietnam; other subspecies in northwestern Thailand and the Malay Peninsula.

GENERAL HABITS, FOOD and VOICE. As in 1359.

BREEDING. As in 1359.

MUSEUM DIAGNOSIS. Darker and more slaty on upperparts than other races. Chestnut of tail extending to *c*. 20 mm from tip.

MEASUREMENTS

	Wing	Tail	
♂♂	64–73	74 (2)	mm
♀♀	66–70	67–71	mm

(Heinrich, SDR, Mayr)

Weight 1 ♂ 21; 2 ♀ ♀ 17, 21 g (SDR).

COLOURS OF BARE PARTS. As in 1359.

1361. *Minla strigula cinereigenae* (Ripley)

Siva strigula cinereigenae Ripley, 1952, J. Bombay Nat. Hist. Soc. 50: 500
(Mt Japvo, western Naga Hills, Assam)
Baker, FBI No. 333 (part), Vol. 1: 313

LOCAL NAMES. None recorded.

SIZE. Sparrow; length *c*. 14 cm (5½ in.).

FIELD CHARACTERS. As in 1359, q.v.

STATUS, DISTRIBUTION and HABITAT. Resident, uncommon or rare. Meghalaya (Khasi Hills) to western Nagaland (Mt Japvo) and presumably the adjacent region of northwestern Manipur, above *c*. 1800 m (Godwin-Austen). Affects oak and rhododendron forest, and bamboo-and-scrub jungle.

GENERAL HABITS, FOOD and VOICE. As in 1359.

BREEDING. Not recorded; probably as in 1359.

MUSEUM DIAGNOSIS. Differs from *yunnanensis* in having a whitish eye-ring and whitish supercilium; sides of cheeks grey, mottled with whitish and dusky. Differs from *strigula* in being darker, more brownish orange on the crown and more olive brownish on back and having a greater area of chestnut on tail; eye-ring, supercilium and cheeks greyish rather than suffused with yellowish; chin colour less bright.

MEASUREMENTS

	Wing	Bill (from feathers)	Tail	
5 ♂♂	66–71	12–13	64–73	mm
7 ♀♀	63–67	12–13	44–69 (SDR)	mm

Weight ♂ ♂ 18–21; ♀ ♀ 17–19 g (SDR).
COLOURS OF BARE PARTS. As in 1359.

1362. **Bluewinged Siva.** *Minla cyanouroptera cyanouroptera* (Hodgson)

Siva cyanouroptera Hodgson, 1837, Ind. Rev. 2 (2): 88 (Nepal)
Leiothrix lepida Horsfield, 1839, Proc. Zool. Soc. London: 162: *ex* McClelland MS. (Assam = Naga Hills, *fide* Koelz, 1954, op. cit., below: 8)
Siva cyanouroptera aglaë Deignan, 1942, Notulae Naturae, Philadelphia, No. 100: 2 (Mt Victoria, 2600 metres, Chin Hills, near Pakokku, Burma)
Siva cyanouroptera thalia Koelz, 1954, Contrib. Inst. Regional Exploration, No. 1: 8 (Mawphlang, Khasi Hills)
Siva cyanouroptera rama Koelz, 1954, Contrib. Inst. Regional Exploration, No. 1: 8 (near Nokrek, Garo Hills)

Baker, FBI No. 335, Vol. 1: 314
Plate 80, fig. 10

LOCAL NAME. *Megblim adum* (Lepcha).
SIZE. Sparrow; length *c.* 15 cm (6 in.).
FIELD CHARACTERS. *Above*, slightly tufted crown dark blue striped with whitish; supercilium and eye-ring white. Back fulvous, paler on rump. Folded wing blue with a small white spot and white tips. Upper surface of tail dark grey with blue edges, narrowly tipped with white; spread tail shows white outer rectrices edged with black. *Below*, entire underparts and ear-coverts pale vinous grey, centre of belly whitish. Under surface of tail white edged with black. Sexes alike.

Blue of head, wing and tail not noticeable except at very close range. The head appears grey, contrasting with the fulvous back. The bird appears long and thin with a flat head and very clear-cut square-ended tail—a curious, distinctive shape.

STATUS, DISTRIBUTION and HABITAT. Resident, locally common, subject to vertical movements. The Himalayas from Naini Tal east through Nepal, Darjeeling, Sikkim, Bhutan and Arunachal Pradesh; also the hills of Meghalaya, Assam, Nagaland and Manipur south to the Chittagong Hill Tracts of Bangladesh. Altitudinal distribution in summer not satisfactorily determined: Smythies (JBNHS 49: 514) records it as a resident above 2400 m in central Nepal while Stevens (JBNHS 29: 736) found it up to 1800 m in summer in Sikkim; its breeding zone thus probably lies between 1500 and 2500 m; most other Himalayan data are winter records. In this season it is found mostly between 1200 and 2200 m and reaches the foothills in Sikkim, Bhutan and Arunachal Pradesh, even entering a short distance the plains of the Brahmaputra. In Manipur it occurs above 1400 m (Hume) and on Mt Victoria between 1400 and 2600 m (Heinrich). Affects bushes in evergreen forest or mixed deciduous and evergreen secondary growth; also cultivations, pine and bamboo.

Extralimital. Western Burma. The species extends to Sichuan, Guangxi, Vietnam and the Malay Peninsula.

GENERAL HABITS. Usually met with in parties of five to fifteen; a regular constituent of the large roving associations of yuhinas, mesias, shrike-babblers, etc. Moves through the tops of bushes and trees much like mesias. Pairs form in April.

FOOD. Mostly insects.

VOICE and CALLS. Call-note, a chick-like *cheep* or *cree-cree*. Song, a three noted whistle, the first lowest, the second highest, the third between these two.

BREEDING. *Season*, May and June. *Nest*, cup-shaped, made of bamboo leaves, rootlets and moss, tightly held together by tendrils and creeper stems; the exterior of the nest is often completely covered with moss; lining is usually of rootlets, sometimes hair. Well hidden in bushes, with a preference for stream banks, within a couple of metres from the ground, generally under one metre. *Eggs*, normally 3 or 4, deep blue, marked with a few small black spots at the larger end. Average size of 24 eggs, 18·4 × 14·1 mm (Baker).

MUSEUM DIAGNOSIS. See Field Characters. For details of plumage see Baker, loc. cit.

Young, like adult but upperparts paler. Tertials edged grey-buff. Underparts more buffish. Primary character of juvenal timaline wing absent. Postjuvenal moult partial. Postnuptial moult completed in October.

MEASUREMENTS

	Wing	Bill (from skull)	Tarsus	Tail	
♂♂	60–69	14–16	22–24	60–70	mm
♀♀	60–67	14–16	22–24 (once 26)	65–68	mm

(BB, SA, Rand & Fleming, Heinrich, Stresemann)

Bill ♂ 17, ♀ ♀ 16–17 mm (BB)

Weight ♂♂ 17–20; 2 ♀ ♀ 16, 17 g (SDR, GD, SA). 16 ♂ ♀ 15–20 g (SA).

COLOURS OF BARE PARTS. Iris greyish brown. Bill yellowish flesh, brownish on culmen for about one-quarter length at tip. Legs and feet brownish flesh; claws horny brown.

Genus YUHINA Hodgson

Yuhina Hodgson, 1836, Asiat. Res. 19: 165. Type, by subsequent designation (Gray, 1841), *Yuhina gularis* Hodgson

Ixulus Hodgson, 1844, *in* J. E. Gray, Zool. Misc., No. 3: 82. Type, by monotypy, *Yuhina? flavicollis* Hodgson

Erpornis Hodgson, 1844, *in* Blyth, Journ. Asiat. Soc. Bengal 13: 379, footnote. Type, by original designation and monotypy, *Erpornis zantholeuca* Blyth

Bill about two-thirds or as long as the head; upper mandible well curved at tip. An erectile crest. Tail rather short and square.

TIMALIINAE

Key to the Species

		Page
A Upperparts yellowish green .... *Y. zantholeuca*		106
B Tips of outer rectrices white .... *Y. castaniceps*		96
C Upperparts olive-brown		
1 Chin and lores black .... *Y. nigrimenta*		105
2 Chin not black		
a Outer webs of secondaries bright rufous .... *Y. gularis*		103
b No rufous on wing		
i A white occipital patch .... *Y. bakeri*		98
ii A rufous occipital patch .... *Y. occipitalis*		104
iii Throat whitish, crown chocolate-brown .... *Y. flavicollis*		100

YUHINA CASTANICEPS (Moore): WHITEBROWED YUHINA

Key to the Subspecies

A Crown grey		
1 A rufous patch above posterior part of supercilium .... *Y. c. rufigenis*		96
2 No rufous patch above supercilium .... *Y. c. plumbeiceps*		97
B Crown mostly chestnut-brown .... *Y. c. castaniceps*		98

1363. *Yuhina castaniceps rufigenis* (Hume)

Ixulus rufigenis Hume, 1877, Stray Feathers 5: 108
(Himalayas = Darjeeling)
Baker, FBI No. 332 (part), Vol. 1: 311

LOCAL NAMES. None recorded.

SIZE. Sparrow −; length *c*. 13 cm (5 in.).

FIELD CHARACTERS. *Above*, head crested, grey scalloped with paler grey; ear-coverts rufous-brown; a narrow white supercilium. Back and wings grey-brown. Tail dark brown, rounded when spread and showing white tips of outer rectrices. *Below*, greyish white. Sexes alike.

The only yuhina with white in tail.

STATUS, DISTRIBUTION and HABITAT. Resident, fairly common. The Himalayan foothills from Darjeeling and Sikkim east through Bhutan and Arunachal Pradesh to the Subansiri river (Stevens, JBNHS 23: 243), from *c*. 600 to 1500 m. Affects secondary forest with scrubby undergrowth.

GENERAL HABITS. Keeps in parties of as many as twenty or thirty birds, often in the mixed itinerant associations, hunting feverishly among the foliage of higher bushes or lower trees, and rapidly sweeping on to the next tree in disorderly 'follow-my-leader' fashion. Movements reminiscent of tits, as they cling to the leaves and branches.

Display (?). 'Has a curious habit of soaring 20 or 30 feet into the air and then sinking down with outstretched wings to the lower bushes' (Baker).

FOOD. Chiefly insects; also seeds. Visits flowers of various trees for nectar. Partial to aphid-infested trees.

VOICE and CALLS. Flocks keep up a comparatively loud cheeping or twittering rendered as *chir-chit . . . chir-chit*.

BREEDING. *Season*, April to July, mostly April and May. *Nest*, a compact cup of very fine and soft shreds of grass and some silk-like vegetable material, surrounded by moss, dead leaves, roots and fibres. Placed in bee-eaters' or kingfishers' burrows, or any natural hollow in steep banks, usually in roadside cuttings. *Eggs*, 3 or 4, white, marked with small, brown or reddish brown blotches, more numerous at the large end. Average size of 100 eggs 16.6 × 13.3 mm (Baker). Incubation by both sexes; period *c*. 12 days.

MUSEUM DIAGNOSIS. Differs from *plumbeiceps* by the presence of a rufous patch above the posterior part of the supercilium.

MEASUREMENTS

	Wing	Bill (from skull)	Tarsus	Tail	
4 ♂ ♀	60–62	11–12	16–18	54–55	mm (SA)

Weight 2 ♂ ♂ 12, 12 g (SA).

COLOURS OF BARE PARTS. Iris reddish brown. Bill horny brown with a pinkish tinge. Legs and feet brownish yellow; claws horny brown.

1364. *Yuhina castaniceps plumbeiceps* (Godwin-Austen)

Staphida plumbeiceps Godwin-Austen, 1877, Ann. Mag. Nat. Hist. 20: 519 (near Sadya and Brahmakhend, Eastern Assam)

Staphida castaneiceps conjuncta Mayr, 1941, Ibis: 86 (Chipwi-Laukkaung Road, Myitkyina district, Burma)

Baker, FBI No. 332 (part), Vol. 1: 311

LOCAL NAMES. None recorded.

SIZE. Sparrow −; length *c*. 13 cm (5 in.).

FIELD CHARACTERS. As in 1363, q.v.

STATUS, DISTRIBUTION and HABITAT. Resident. Arunachal Pradesh from the Dikrang Valley (Godwin-Austen, SF 7: 144) east to the Mishmi Hills and south to eastern Nagaland; from the foothills to *c*. 1500 m. Affects scrub and under-growth in evergreen or light deciduous forest.

Extralimital. Northern Burma and western Yunnan. The species extends to Vietnam, Guangdong and Fujian; also Borneo.

GENERAL HABITS, FOOD and VOICE. As in 1363.

BREEDING. As in 1363.

MUSEUM DIAGNOSIS. Differs from *rufigenis* by the absence of rufous patch above the posterior part of the supercilium where only a few scattered rufous feathers are found. Differs from *castaniceps* by the grey, not rufous crown.

MEASUREMENTS

	Wing	Bill (from skull)	Tarsus	Tail
♂♂	58–62	11	16	53–57 mm
♀♀	58–62	—	—	50–57 mm (SDR, Mayr)

Weight 4 ♂ ♂ 11 (3), 12 g (SDR).

COLOURS OF BARE PARTS. As in 1363.

1365. *Yuhina castaniceps castaniceps* (Moore)

Ixulus castaniceps Moore, 1854, *in* Horsfield & Moore,
Cat. Bds. Mus. E. I. Co. 1: 411 (Afghanistan, *errore* = Cachar)
Baker, FBI No. 330, Vol. 1: 310
Plate 79, fig. 13

LOCAL NAME. *Daotisha-magini* (Cachari).
SIZE. Sparrow −; length *c.* 13 cm (5 in.).
FIELD CHARACTERS. As in 1363, but crown rufous-brown scalloped with pale grey on forehead.
STATUS, DISTRIBUTION and HABITAT. Common resident. Western Nagaland, Manipur and the hills of Assam and Meghalaya, south to the Chittagong Hill Tracts of Bangladesh from 600 to 1500 m. Affects secondary scrub and evergreen forest.
Extralimital. Chin Hills and Arakan Yomas.
GENERAL HABITS, FOOD and VOICE. As in 1363.
BREEDING. As in 1363.
MUSEUM DIAGNOSIS. Differs from other races by its rufous crown. For details of plumage see Baker, loc. cit.
Young, as adult but upperparts tinged brown. Edges of secondaries and coverts browner; occipital feathers not so long.
MEASUREMENTS. As in 1364.
COLOURS OF BARE PARTS. Iris pale hazel. Bill light reddish horny, gape and base of both mandibles purplish. Legs and feet dull reddish or flesh colour.

1366. Whitenaped Yuhina. *Yuhina bakeri* Rothschild

Siva occipitalis Blyth, 1844, Jour. Asiat. Soc. Bengal 13: 937 (Darjeeling)
Yuhina bakeri Rothschild, 1926, Novit. Zool. 33: 276.
New name for *Siva occipitalis* Blyth, 1844, preoccupied.
Yuhina occipitalis atrovinacea Koelz, 1954, Contrib. Inst.
Regional Exploration, No. 1: 8 (Laikul, Cachar)
Baker, FBI No. 344, Vol. 1: 321
Plate 79, fig. 19

LOCAL NAMES. *Temgyeng-pho, Turringing-pho* (Lepcha).
SIZE. Sparrow −; length *c.* 13 cm (5 in.).
FIELD CHARACTERS. A perky hair-brown tit-like bird with chestnut head and upstanding crest, conspicuous blackish lores and a white nape-patch.
Above, head, upper back and sides of neck rusty brown; crest erect, posteriorly white; ear-coverts streaked with white; lores blackish. Rest of upperparts olive-brown with inconspicuous white shaft-streaks. *Below*, throat white; breast vinaceous with fine dark streaks; belly olivaceous; under tail-coverts ferruginous. Sexes alike.

× *c.* 1

STATUS, DISTRIBUTION and HABITAT. Common resident, subject to vertical movements. From eastern Nepal[1] (Inskipp & Inskipp) east through the Himalayas to Bhutan and Arunachal Pradesh to the Burma border, Nagaland, Manipur, Cachar Hills (Assam), Khasi Hills (Meghalaya), and Bangladesh in the northeastern highlands (winter). There are no records for Garhwal nor for the Chittagong Hill Tracts as given in FBI. Altitudinal distribution not satisfactorily known, especially in summer. Recorded in winter from the foothills to *c.* 2000 m. Gammie (*apud* Hume & Oates's *Nests and Eggs*) found a nest at about 3000 feet (900 m), this being the only Himalayan breeding record. In Assam, according to Baker (*Nidification* 1: 300) it is common between 900 and 1500 m and 'although doubtless it often breeds much higher, I never found its nest below 2500 feet' (750 m). Affects secondary jungle and evergreen forest.

Extralimital. Northern Burma.

GENERAL HABITS. Arboreal. In the non-breeding season, keeps in parties, commonly in association with other small insectivorous species, moving about in tree-tops and bushes.

FOOD. Mainly insects; also berries.

VOICE and CALLS. A shrill *chip* and a soft chatter.

BREEDING. *Season*, April to July, chiefly May and June. *Nest*, cup-shaped or domed according to the site selected: domed when unsheltered, a deep cup when built in a hollow in a bank, low down in bushes, against a moss-covered trunk or other well-protected position. Generally made of moss more or less mixed with dead leaves, roots and bits of bracken, lined with fine rootlets. Comparatively open forest is chosen for the purpose, such as along paths, streams, clearings or forest edges. *Eggs*, normally 4, occasionally 3, white, marked with blotches varying from reddish brown to deep umber-brown, more numerous at the large end. Average size of 60 eggs 19·3 × 14·2 mm (Baker). Building of nest and incubation by both sexes.

MUSEUM DIAGNOSIS. See Field Characters.

Young, like adult but mantle browner, underparts paler, streaks less sharply marked, ferruginous paler. Primary and tail characters of juvenal timalines present, i.e. soft blunt first primary and narrow, pointed rectrices.

MEASUREMENTS

	Wing	Bill (from skull)	Tarsus	Tail	
♂♂	62–71	12–16	20–22	50–53	mm
♀♀	65–72	11–13	20–23	46–50	mm
				(SA, SDR)	

Weight 4 ♂ ♀ 14–21 g (SA).

COLOURS OF BARE PARTS. Iris brown, yellowish brown or reddish brown. Bill dark brown. Legs, feet and claws brown; soles yellow.

---

[1] Rand & Fleming's records of 'Chestnut-headed Ixulus' in west-central Nepal (*Fieldiana* 41: 139) do belong to *Y.o. occipitalis* Hodgson, not to *Y. bakeri* as reported by Biswas, JBNHS 59: 222 (confirmed by Rand, *in epist.*).

Biswas's record for Khumbu 3960 m (JBNHS 59: 222) was reported as *bakeri* because of a nomenclatural confusion. It also pertains to *Y. o. occipitalis* Hodgson (confirmed by Biswas, *in epist.*).

TIMALIINAE

Yuhina flavicollis Hodgson: Yellownaped Yuhina

Key to the Subspecies

		Page
A	Nuchal collar narrow, yellowish rust colour .... *Y. f. albicollis*	100
B	Nuchal collar broader, more rufous	
	a Generally paler; sides of breast more olive............ *Y.f. flavicollis*	100
	b Generally darker; sides of breast more rufous .... *Y.f. rouxi*	102

1367. *Yuhina flavicollis albicollis* (Ticehurst & Whistler

Ixulus flavicollis albicollis Ticehurst & Whistler, 1924, Bull. Brit. Orn. Cl. 44: 71
(Dharmsala, 400 feet, Kangra)
Baker, FBI No. 345 (part), Vol. 1: 322

LOCAL NAMES. None recorded.

SIZE. Sparrow; length *c.* 13 cm (5 in.).

FIELD CHARACTERS. As in 1368 but nuchal collar less conspicuous and almost reduced to patches on the sides of neck.

STATUS, DISTRIBUTION and HABITAT. Resident, subject to vertical movements, locally common. The western Himalayas from Chamba to west central Nepal; from *c.* 1700 to 3000 m in summer and from 1200 m (occasionally lower) to at least 2300 m in winter. Affects oak forest and open jungle.

GENERAL HABITS, FOOD and VOICE. As in 1368.

BREEDING. As in 1368.

MUSEUM DIAGNOSIS. Like nominate *flavicollis* but nuchal collar narrower and much paler. Colour of back slightly less olive.

MEASUREMENTS and COLOURS OF BARE PARTS. As in 1368.

1368, 1369. *Yuhina flavicollis flavicollis* Hodgson

Yuhina? flavicollis Hodgson, 1836, Asiat. Res. 19: 167
(Nepal, restricted to central Nepal by Ripley, Synopsis: 408)
Yuhina flavicollis baileyi Baker, 1914, Bull. Brit. Orn. Cl. 35: 17
(Tembang, Drang Valley, 7000 ft)[1]
Baker, FBI No. 345 (part), Vol. 1: 322
Plate 79, fig. 15

LOCAL NAME. *Srip-chong-pho* (Lepcha).

SIZE. Sparrow; length *c.* 13 cm (5 in.).

FIELD CHARACTERS. *Above*, crested head chocolate-brown, ear-coverts and nape paler; a white eye-ring, a dark brown malar stripe and a rusty yellow nuchal collar. Rest of upperparts olive-brown. *Below*, chin and throat white,

---

[1] As already pointed out by Bailey, JBNHS 24: 75, the type locality is not in the Mishmi Hills as given by Baker (loc. cit.), Ripley (*Synopsis:* 408) and Deignan (Peters's *Check-list* 10: 423). According to Bailey's itinerary, Tembang is located between Tulang La and Dirang Dzong in western Arunachal Pradesh. In his manuscript, Whistler states that this race does not seem to be separable from typical *flavicollis*, while Ripley (JBNHS 58: 282) would include the Mishmi Hills population with *rouxi* of Assam and northern Burma. It thus appears unnecessary to retain a name for this intermediate population of a very variable species.

the latter with fine dark streaks. Centre of breast and abdomen washed with yellow; flanks olivaceous with a few white streaks on sides of breast. Vent and under tail-coverts buff. Sexes alike.

The erectile crest, white eye-ring and rusty yellow nape are conspicuous characters. May be confused with *occipitalis* (1373), q.v.

STATUS, DISTRIBUTION and HABITAT. Common resident, subject to vertical movements. From central Nepal east through northern Bengal (Darjeeling and Jalpaiguri districts), Sikkim, Bhutan and Arunachal Pradesh to the Abor-Miri hills. Intergrades with *albicollis* in west-central Nepal and with *rouxi* in Arunachal Pradesh. From 1800 to 3000 m in summer and from the edge of the plains to 2700 m in winter. Affects deciduous forest and secondary jungle.

GENERAL HABITS. Arboreal; similar to other yuhinas etc. Keeps in small parties, associating with tits, warblers, nuthatches, sibias and other small babblers. Frequents bushes and lower branches of trees. Actions very tit-like. Sometimes also makes vertical aerial sallies from a bush-top after winged insects, doubling back to the perch after the capture. In pairs during the breeding season.

FOOD. Mainly insects and their larvae; also flower-nectar, berries (*Rubus, Galium, Trema*, etc.) and small seeds (*Michelia cathcarti* and others). Stanford records small snails taken on the ground.

VOICE and CALLS. Flocks keep up a continual murmuring twitter punctuated by a harsh *chi-chi-chiu* (SA). Song rendered as *twe-tyurwi-tyawi-tyawa* (Lister). Other notes described as a metallic *tse-kling* (Jones) and a querulous screech followed by a pleasant warble.

BREEDING. *Season*, May and June. *Nest*, either domed or cup-shaped, made mostly of moss with some rootlets and thickly lined with the latter, sometimes with wool. Suspended to some twigs or among the moss of a branch at heights up to two, rarely four metres, or in a bank among roots or tufts of grass. *Eggs*, 3 or 4 similar to those of *Y. bakeri*. Average size of 28 eggs 19·8 × 14·2 mm (Baker). Building of nest and incubation by both parents; period 13 or 14 days.

MUSEUM DIAGNOSIS. For details of plumage see Baker, loc. cit. For distinction from *albicollis* and *rouxi* see 1367 and 1370 respectively. Young very similar to adult.

MEASUREMENTS

	Wing	Bill (from skull)	Tarsus	Tail	
♂♂	58–66	13–17	19–22	45–53	mm
♀♀	58–66	12–15	19–22	45–54	mm

(SA, BB, Kinner)

Weight 2 ♂ ♂ 16, 18; 3 ♀ ♀ 14–16; 30 ♂ ♀ 13–22 g (SA).

COLOURS OF BARE PARTS. Iris brown. Bill: upper mandible dark brown, lower light brown. Legs and feet yellowish brown or yellowish flesh.

1370. *Yuhina flavicollis rouxi* (Oustalet)

Ixulus rouxi Oustalet, 1896, Bull. Mus. d'Hist. Nat. Paris 2: 184, 186
(Ly-Sien-Kiang or Black River, Yunnan)
Ixulus flavicollis harterti Harington, 1913, Bull. Brit. Orn. Cl. 33: 62
(Sinlum, Bhamo)
Baker, FBI No. 346 and 347, Vol. 1: 323

LOCAL NAMES. None recorded.
SIZE. Sparrow; length *c*. 13 cm (5 in.).
FIELD CHARACTERS. As in 1368, q.v.
STATUS, DISTRIBUTION and HABITAT. Common resident, subject to vertical movements. Arunachal Pradesh in the Mishmi and Patkai hills, south through Nagaland, Manipur, the hills of Assam, Meghalaya and the Chittagong Hill Tracts of Bangladesh. Breeding zone in Arunachal Pradesh probably as in 1368; in winter, from the edge of the plains to 2600 m. On Mt Victoria, breeds mostly between 2000 and 2600 m while in Assam it breeds down to *c*. 1000 m. Affects broad-leaved forest.
GENERAL HABITS, FOOD and VOICE. As in 1368.
BREEDING. As in 1368.
MUSEUM DIAGNOSIS. Somewhat more richly coloured than *flavicollis* (1368), particularly in the colour of the nuchal collar and of the back; yellow wash of underparts slightly richer; crown darker; lower flanks and thighs more suffused with olive-brown.
Note. The yellow of underparts fades rapidly in museum specimens while the upperparts tend to become darker.
MEASUREMENTS and COLOURS OF BARE PARTS. As in 1368.
Weight 2 ♂ ♂ 15, 19; 1 ♀ 16 g (SDR).

YUHINA GULARIS Hodgson: STRIPETHROATED YUHINA
Key to the Subspecies

Paler . . . . *Y. g. vivax*
Darker . . . . *Y. g. gularis*

1371. *Yuhina gularis vivax* Koelz

Yuhina gularis vivax Koelz, 1954, Contrib. Inst. Regional Exploration,
No. 1: 8 (above Luni, Tehri, United Provinces, 10,000 feet)
Not in Baker, FBI

LOCAL NAMES. None recorded.
SIZE. Sparrow; length *c*. 14 cm (5½ in.).
FIELD CHARACTERS. As in 1372, q.v.
STATUS, DISTRIBUTION and HABITAT. Apparently uncommon. Recorded only from Garhwal, 2250 m (Osmaston, JBNHS 28: 143), from Tehri Garhwal, 3000 m (Koelz, loc. cit.) and Mussooree 1800 m (Rand & Fleming, *Fieldiana* 41: 139). Altitudinal distribution and habitat probably as in 1372.
GENERAL HABITS, FOOD and VOICE. As in 1372.
BREEDING. Unrecorded, probably as in 1372.

BABBLERS

MUSEUM DIAGNOSIS. Differs from *gularis* (1372) by being paler generally, especially on throat and breast.

MEASUREMENTS and COLOURS OF BARE PARTS. As in 1372.

1372. *Yuhina gularis gularis* Hodgson

Yuhina gularis Hodgson, 1836, Asiat. Res. 19: 166 (Nepal)
Yuhina yangpiensis Sharpe, 1902, Bull. Brit. Orn. Cl. 13: 12
(Yang-pi, Tali-fu road, Yunnan)
Yuhina gularis griseotincta Rothschild, 1921, Novit. Zool. 28: 42
(Shweli-Salwin Divide, Yunnan)
Baker, FBI No. 339 and 340, Vol. 1: 317, 318
Plate 79, fig. 18

LOCAL NAME. *Fugi-pho* (Lepcha).

SIZE. Sparrow; length *c.* 14 cm (5½ in.).

FIELD CHARACTERS. A plump active little brown bird with a well-marked erectile crest, striped throat, and orange-fulvous longitudinal bar on blackish wing. Crest brown; rest of upperparts olive-brown. *Below*, throat streaked with dark brown. Breast vinaceous brown; rest of underparts tawny olive-brown. Sexes alike.

STATUS, DISTRIBUTION and HABITAT. Common resident, subject to vertical movements. From western Nepal east through Darjeeling, Jalpaiguri district (winter only), Sikkim, Bhutan and Arunachal Pradesh to the Mishmi Hills, then south through Nagaland, Manipur and the Chin Hills of Burma. There are no records from Assam west of Nagaland and Manipur. Occurrence in the Chittagong Hill Tracts (*fide* Rashid, 1967) needs confirming. Found in summer between 2400 and 3700 m, optimum zone 2700–3300 m, and in winter mostly between 1700 and 3300 m, rarely descending to the foothills (Jalpaiguri duars). Affects forest of oak, birch, rhododendron or mixed conifers and rhododendron, occasionally low scrub or bamboo.

Extralimital. Extends to western and northern Burma, and southwestern Szechuan; also northern Vietnam.

GENERAL HABITS. Arboreal. Usually found in small parties, by themselves or mixed with other small babblers, working through the higher bushes or lower branches of trees, sometimes in low scrub or bamboo. Movements somewhat tit-like but slower.

FOOD. Insects (beetles, wasps, etc.); also berries; flower-nectar, and seeds (*Prunus, Magnolia*). Regularly visits rhododendron blossoms for nectar, the forehead feathers of the birds becoming thickly coated with pollen and often giving them a startlingly new look!

VOICE and CALLS. A quiet, rustling *shr . . . shr . . .* continually uttered and a curious and characteristic, long-drawn-out *kweeeee* (Proud). A note described as 'a rather mournful tinkling call' probably refers to the same. A call-note regularly uttered is rendered as *zäi zäi* (Diesselhorst). Alarm, a sharp *cheep*.

BREEDING. Little known. *Season,* apparently May and June. *Nest* and *eggs* not well authenticated. One nest is described as a cradle of roots well interlaced, lined with finer roots, and attached to the pendent roots of plants sticking through an overhanging bank. The four eggs were dingy grey-green, speckled with dark reddish brown forming a ring or cap at the large end, rather like those of *Y. nigrimenta* (1374). Size, between 17 × 12·3 and 17·5 × 12·8 mm (Baker). Other nests are described by Hodgson as large, globular structures of moss placed in a fork of a branch or between ledges of rocks.

MUSEUM DIAGNOSIS. See Field Characters; for details of plumage Baker, loc. cit.

Young, as adult but rather darker rufous-brown above, especially on upper tail-coverts. Crest shorter. Primary and tail characters not present.

MEASUREMENTS

	Wing	Bill (from skull)	Tarsus	Tail	
♂♂	70–79	17–19	21–25	56–63	mm
♀♀	68–79	16–18	21–23	53–58	mm
				(BB, SA)	

Weight 14 ♂ ♂ 19–24; 4 ♀ ♀ 18–22 g (GD, SDR, SA).

COLOURS OF BARE PARTS. Iris clay- or reddish brown. Bill: upper mandible dark brown, lower basally brownish yellow, distally dark brown. Legs and feet yellow, brownish yellow or orange-brown; claws dark brown.

1373. **Slatyheaded** or **Rufousvented Yuhina.** *Yuhina occipitalis occipitalis* Hodgson

Yuhina occipitalis Hodgson, 1836, Asiat. Res. 19: 166 (Nepal)
Baker, FBI No. 342, Vol. 1: 319
Plate 79, fig. 17

LOCAL NAME. *Turringing-pho* (Lepcha).
SIZE. Sparrow; length *c*. 13 cm (5 in.).
FIELD CHARACTERS. *Above*, head, ear-coverts and neck grey; the erectile crest grey in front, bright rufous posteriorly. A conspicuous pale eye-ring and a black malar stripe. *Below*, throat and breast vinaceous; belly and under tail-coverts pale rufous. Sexes alike.

The somewhat similar *Y. flavicollis* (1368) has a brown crest and a rusty yellow neck.

× *c*. 1   STATUS, DISTRIBUTION and HABITAT. Common resident. From western Nepal (Inskipp & Inskipp) east through Darjeeling, Sikkim, Bhutan and Arunachal Pradesh; from 2400 to 3900 m in summer and between 1500 and at least 2700 m in winter. Affects evergreen forest, especially rhododendron and oak.

Extralimital. Southeastern Tibet. The species extends to northern Burma and northwestern Yunnan.

GENERAL HABITS. In the non-breeding season keeps in parties up to fifteen individuals in company with tits, Nepal Babblers, Bluewinged Sivas, and other small babblers. Hunts in the high foliage, on moss-covered trunks and branches, and in bushes; seems to keep more to the canopy than *Y. gularis* with which it often associates.

FOOD. Mostly insects in summer. Very partial to rhododendron blossoms, probing into the flowers in quest of nectar and probably also insects. Berries are also taken in winter.

VOICE and CALLS. A deep churring conversational note; a harsh, grating series of alarm-notes; 'a gay little song'.

BREEDING. *Season*, April to June. The only nest known is described as a cup of moss and leaves, built into a large lump of moss which formed a dome over the nest. It was in the fork of a small tree about three feet (90 cm) from the ground, and well masked since similar large lumps of moss were found on practically every bush and tree in this damp area. It was lined with fine roots and contained two young, sparsely covered with long black down (Proud).

MUSEUM DIAGNOSIS. See Field Characters.

Young, like adult but crest shorter; rufous of nape paler; vinaceous of breast less marked.

MEASUREMENTS

	Wing	Bill (from skull)	Tarsus	Tail	
♂♀	62–66	15–17	18–19	51–53	mm

(Rand & Fleming, Mayr, Stresemann, SA)

Weight 3 ♂♂ 12–16; 2 ♀ ♀ 12, 12 g (GD, SA).

COLOURS OF BARE PARTS. Iris brown. Bill pinkish or reddish brown. Legs and feet yellow-brown to orange-brown.

1374. Blackchinned Yuhina. *Yuhina nigrimenta nigrimenta* Hodgson

Yuhina nigrimenta Hodgson, 1845, *in* Blyth, Jour. Asiat. Soc. Bengal 14: 562 (Nepal)

Yuhina nigrimentum titania Koelz, 1954, Contrib. Inst. Regional Exploration, No. 1: 9 (Karong, Manipur)

Baker, FBI No. 343, Vol. 1: 320

Plate 79, fig. 16

LOCAL NAME. *Turringing-pho* (Lepcha).

SIZE. Sparrow −; length *c.* 11 cm (4½ in.).

FIELD CHARACTERS. A small yuhina with erectile black crest, black lores and chin and black-and-red bill. *Above*, crest black with scale-like grey edgings. Lores black. Nape and sides of head grey. Rest of upperparts olive-brown. *Below*, chin black; throat white. Rest of underparts pale fulvous. Sexes alike.

STATUS, DISTRIBUTION and HABITAT. Resident, uncommon to rare west of Sikkim; common eastward. The Himalayan foothills from Garhwal (Hume collection) and Kumaon (Whymper, JBNHS 14: 607) east through the Nepal duns (Biswas, JBNHS 59: 224; Fleming & Traylor, *Fieldiana* 53: 176), Darjeeling, Sikkim, Bhutan and Arunachal Pradesh; Nagaland, Manipur, the hills of Assam, Meghalaya (rare in the Khasi Hills), and the Chittagong Hill Tracts of Bangladesh. From *c.* 300 to 1800 m; these altitudes are winter records; breeding zone within these extremes not satisfactorily determined in the Himalayas. In Assam the breeding zone is 1000 to 1800 m (Baker). Affects evergreen forest and secondary jungle, particularly in overgrown cultivation clearings.

Extralimital. The species extends to southern Sichuan and northern Vietnam; also Fukien.

GENERAL HABITS. Very gregarious, active, restless and noisy; usually seen in flocks of 15 to 20 or in mixed parties with other 'tinies', busily hunting in the canopy of lofty trees as well as in low shrubs, clinging sideways or upside-down to the sprigs to peer under the leaves for insects. Sometimes in the tall grass which grows under trees.

FOOD. Chiefly insects; also berries, seeds and flower-nectar.

VOICE and CALLS. Birds of a party keep up a lively chorus of low cheeping twitters occasionally breaking out into louder, shriller calls.

BREEDING. *Season*, March to July. *Nest*, a compact cup of moss and moss roots lined with very fine grass stems or rootlets. It is either suspended in the lichen hanging from branches of trees within a couple of metres from the ground, or fastened to the exposed, hanging roots on banks from which earth has fallen away. *Eggs*, normally 4, either a pale sea-green profusely spotted all over with very pale brown, small blotches forming fairly well-defined broad rings at the large end, or a pale clay or clay-green with the spots numerous but smaller. Average size of 12 eggs 16·2 × 12·3 mm (Baker).

MUSEUM DIAGNOSIS. See Field Characters.

Young, like adult but upperparts, edges of wing and tail browner. Crest shorter, pale edges to the feathers less clear. Primary and tail characters present; postjuvenal moult complete.

MEASUREMENTS

	Wing	Bill (from skull)	Tarsus	Tail	
♂♂	52–59	13–14	16–18	38–39	mm
♀♀	53–59	13–14	c. 16	36–39	mm
			(BB, SA, Kinnear)		

Weight 1♂ 8.9 g (SDR); 4 ♂♂ 9–10; 2 ♀♀ 8, 9 g (SDR, SA). 3 ♂♀ 10–11 g (SA).

COLOURS OF BARE PARTS. Iris hazel-brown. Bill: upper mandible horny brown to black, lower and gape orange-red, coral-red or pink with brownish tip; mouth bright orange-red. Legs and feet yellowish brown or orange-flesh; claws pale horny brown.

1375. Whitebellied Yuhina. *Yuhina zantholeuca zantholeuca* (Hodgson)

Erp. [*ornis*] *zantholeuca* Hodgson *in* Blyth, 1844, Jour. Asiat. Soc. Bengal 13: 380 (central region of Nepal)
Baker, FBI No. 350, Vol. 1: 325
Plate 79, fig. 12

LOCAL NAME. *Dung-pu-pho* (Lepcha).

SIZE. Sparrow −; length *c.* 11 cm (4 in.).

FIELD CHARACTERS. *Above*, entirely olive-green, edges of tail yellow; head tufted. *Below*, and ear-coverts greyish white; under tail-coverts yellow. Sexes alike.

STATUS, DISTRIBUTION and HABITAT. Resident, locally common, subject to vertical or erratic movements. From extreme western Nepal (Fleming & Traylor, *Fieldiana* 53: 176), east through Sikkim, Bhutan and Arunachal Pradesh; Nagaland, Manipur, the hills of Meghalaya, Assam

× c. 1

(scarce or rare) and the Chittagong Hill Tracts of Bangladesh. Breeding zone not satisfactorily determined. Appears to have a wide altitudinal range, both summer and winter, remarkably so in the latter season. Biswas found it in the duns at *c.* 600 m in May which is the breeding season, while Fleming recorded it at 2250 m and 900 m in winter. According to Proud it is most common around 1500 m in autumn and winter. In Sikkim it is found below 1000 m *fide* Stevens while Sálim Ali noted it from 360 to 2250 m in winter. Farther east it is said to be common in the duars and to extend well into the plains of north Lakhimpur in winter (Stevens) and up to 2600 m in northern Burma in the same season. In Assam it is found from the foothills to 900 m, breeding mostly between 300 and 600 m, once only as high as 1200 m (Baker). Affects rhododendron trees, secondary growth and evergreen forest, especially in open spaces such as glades, stream sides or forest edges, and light deciduous forest.

Extralimital. Extends to Burma and western Thailand. The species ranges north to northwestern Yunnan, east to Fujian, Guangdong and Vietnam, south through the Malay Peninsula to Sumatra and Borneo.

GENERAL HABITS. Seems to be less gregarious than other yuhinas; often found solitary, in pairs or in small parties by themselves although they also consort with tits, minlas and other small babblers, the flocks 'flowing' rapidly from tree to tree. Frequents mostly the lower canopy and higher bushes, clinging to sprigs and searching the foliage like a tit. Its sprightly and restless disposition reminds one also of *Phylloscopus* or *Zosterops.*

FOOD. Mostly insects and their larvae, particularly small caterpillars; also berries and flower-nectar.

VOICE and CALLS. Unrecorded; a very silent species.

BREEDING. *Season,* from the end of March through April and May. *Nest,* a little cup of fine fibres, moss and rootlets, lined with fine, dark-coloured leaf-stems. Suspended between two horizontal twigs from about half a metre to 2 m above the ground, usually around one metre. Banks of streams, glades and forest edges are favourite nesting-sites. *Eggs,* 2 or 3, white, thinly marked with speckles and small blotches of pale pinkish red, sometimes forming an ill-defined ring at the large end. Average size of 20 eggs 16·7 × 12·7 mm (Baker).

MUSEUM DIAGNOSIS. See Field Characters.

Young, a dull edition of the adult; upperparts with a brownish cast. Primary and tail characters present; postjuvenal moult complete.

MEASUREMENTS

	Wing	Bill (from skull)	Tarsus	Tail	
♂♂	66–72	14–15	16–19	44–49	mm
♀♀	60–66	14–15	16–18	42–47	mm

(HW, SA, Heinrich)

Weight 4 ♂♂ 12–17; 3 ♀♀ 10.5 11, 12 g (GD, SDR). 2 oo? 8, 11 g (SA).

COLOURS OF BARE PARTS. Iris brown. Bill: upper mandible brown, lower whitish grey. Legs and feet whitish grey.

TIMALIINAE

Genus ALCIPPE Blyth

Alcippe Blyth, 1844, Jour. Asiat. Soc. Bengal 13: 370, 384. Type, by monotypy, *T. poioicephala* Jerdon

Proparus Blyth, 1844, Jour. Asiat. Soc. Bengal 13: 938. Type, by original designation, *Pr. vinipectus, nec Proparus* Hodgson, 1841 (= *Minla*)

Schoeniparus Hume, 1874, Stray Feathers 2: 449. Type, by subsequent designation (Sharpe, 1883, Cat. Birds Brit. Mus. 7: 606), *Minla rufogularis* Mandelli

Fulvetta David & Oustalet, 1877, Ois. Chine, text: 220. Type, by subsequent designation (Sharpe, 1883), *Siva cinereiceps* Verreaux

Lioparus Oates, 1889, Fauna Brit. Ind., Bds. 1: 174. Type, by original designation, *Proparus? chrysaeus* = *chrysotis* Blyth

Pseudominla Oates, 1894, Ibis: 480; new name for *Sittiparus* Oates. Type, by original designation, *Minla cinerea* Blyth

Alcippornis Oberholser, 1922, Smith. Misc. Coll. 74: 1. New name for *Alcippe* Blyth (based on type, *Alcippe cinerea* Blyth *nec* Eyton)

Cf. Riley, J. H., 1933, *Auk* 50: 363–4 (use of *Alcippe* and genotype).

Bill stout, slightly curved. Nostrils covered by a membrane. Rictal bristles well developed in most species. Wing short and rounded.

Key to the Species

				Page
A	A well-marked supercilium			
	1	A yellow supercilium	*A. cinerea*	110
	2	A black supercilium, no white	*A. nipalensis*	124
	3	A white supercilium		
		a A chestnut band across throat	*A. rufogularis*	118
		b Throat streaked with brown	*A. vinipectus*	112
		c Not as a or b		
		i A black shoulder-patch	*A. castaneceps*	111
		ii No black shoulder-patch	*A. brunnea*	120
B	Supercilium absent or obscure			
	4	Underparts yellow	*A. chrysotis*	109
	5	Underparts not yellow		
		d Wing edged with contrasting pale grey		
		iii Dark streaking on nape	*A. striaticollis*	117
		iv Cheek, crown, nape red-brown	*A. ludlowi*	116
		v Crown dusky grey with obscure dark supercilium	*A. cinereiceps*	116
		e Wing without contrasting outer edge	*A. poioicephala*	121

ALCIPEE CHRYSOTIS (Blyth): GOLDENBREASTED TIT-BABBLER

Key to the Subspecies

A white streak through crown .... *A. c. albilineata*
No streak through crown .... *A. c. chrysotis*

1376. *Alcippe chrysotis chrysotis* (Blyth)

Pr.(oparus) chrysotis Blyth, 1844, Jour. Asiat. Soc. Bengal 13: 938
(Himalaya = Nepal)
Baker, FBI No. 309 (part), Vol. 1: 293

Plate 80, fig. 11

LOCAL NAME. *Prong-samyer-pho* (Lepcha).

SIZE. Sparrow −; length *c.* 11 cm (4½ in.).

FIELD CHARACTERS. *Above*, crown blackish; ear-coverts pale silver-grey. Back olive. Wing blackish with orange-yellow outer edge and an orange longitudinal patch; inner edge white; secondaries tipped with white. Tail brown, the basal two-thirds edged with orange-yellow. *Below*, throat grey with silvery tips. Rest of underparts yellow. Sexes alike.

STATUS, DISTRIBUTION and HABITAT. A rather scarce resident, subject to some vertical movements. The Himalayas from central Nepal (Inskipp & Inskipp, p. 318) east through Darjeeling, Sikkim, Bhutan and Arunachal Pradesh; from *c.* 2400 to 3000 m in summer and between 2000 and at least 2600 m in winter. Affects dense growth on steep hillsides, particularly bamboo jungle.

Extralimital. The species extends to northern Burma, southwestern Sichuan and northern Vietnam.

GENERAL HABITS. A confiding species met with in large parties of up to fifty birds in the non-breeding season, sometimes associated with parrotbills (*Paradoxornis nipalensis*) or other small babblers. Forages low down in thickets, 'flowing' on from bush to bush. Movements very tit-like; often seen hanging upside-down on a sprig peering under the leaves for insects.

FOOD. Insects, small berries and seeds.

VOICE and CALLS. Utters a continual low twitter as it flits from twig to twig. Not intelligibly described.

BREEDING. Not recorded since Hodgson. *Season*, May and June. *Nest*, egg-shaped, fixed with its longest diameter perpendicular to the ground in a bamboo clump, between the small lateral shoots, a few feet from the ground; entrance on the side. Made of bamboo leaves and broad blades of grass, lined with a little grass and rootlets. *Eggs*, 3 or 4, pinkish white, thinly speckled and spotted with brownish red, tending to form a cap or zone at the large end. Measurements, *c.* 12·7 × 17·7 mm (Hodgson *in* Hume & Oates's *Nests and Eggs*, Vol. 1: 120).

MUSEUM DIAGNOSIS. Differs from *albilineata* in lacking the white coronal stripe. For details of plumage, see Baker loc. cit.

MEASUREMENTS

	Wing	Bill (from skull)	Tarsus	Tail	
♂♂	52–54	10 (5)	20–22	46–49	mm
♀♀	50–55				
				(Kinnear, SA)	

Weight 1 ♂ 5.5 g (SA).

COLOURS OF BARE PARTS. As in 1377.

1377. *Alcippe chrysotis albilineatus* (Koelz)

Lioparus chrysotis albilineatus Koelz, 1954, Contrib. Inst.
Regional Exploration, No. 1: 7 [Karong (Sungtun), Manipur]
Baker, FBI No. 309 (part), Vol. 1: 293

LOCAL NAMES. None recorded.

SIZE. Sparrow −; length *c.* 11 cm (4½ in.).

FIELD CHARACTERS. Like 1376 but with a white coronal stripe from forehead to nape.

STATUS, DISTRIBUTION and HABITAT. Scarce resident. Assam in the Cachar Hills, Nagaland and Manipur, from 1800 to 2700 m. Same biotope as 1376.

GENERAL HABITS, FOOD and VOICE. As in 1376.

BREEDING. Unrecorded; probably as in 1376.

MUSEUM DIAGNOSIS. Differs from *chrysotis* (1376) in having a conspicuous white coronal stripe and a richer colour overall.

MEASUREMENTS
10 ♂ ♀ Wing 49–55 mm (Koelz)

COLOURS OF BARE PARTS. Iris hazel. Bill: upper mandible yellow on distal half, orange on basal half, bluish on the commissure; lower mandible blue, paler at tip. Legs and feet purple-flesh; soles tinged yellow (Koelz).

1378. Dusky Green or Yellowthroated Tit-Babbler. *Alcippe cinerea* (Blyth)

Minla cinerea Blyth, 1847, Jour. Asiat. Soc. Bengal 16: 449 (Darjeeling)
Alcippe delacouri Yen, 1936, L'Oiseau 6: 449. New name for
Minla cinerea Blyth, not preoccupied
Baker, FBI No. 301, Vol. 1: 287

Plate 79, fig. 7

LOCAL NAME. *Dao-péré kashiba* (Cachari).

SIZE. Sparrow −; length *c.* 10 cm (4 in.).

FIELD CHARACTERS. *Above*, crown and nape yellowish green, the feathers edged with black; a black stripe on sides of crown, a conspicuous yellow supercilium from lores to nape and another black stripe from lores through eye. Rest of upperparts greyish olive. *Below*, yellow, olivaceous on sides and lower belly. Sexes alike.

× *c.* 1

STATUS, DISTRIBUTION and HABITAT. Resident, locally distributed. The Himalayan foothills from central Nepal (Proud, JBNHS 48; 700) east through Darjeeling, Sikkim, Bhutan and Arunachal Pradesh; Meghalaya (rare) in the Khasi and Assam in the Cachar hills, Nagaland, Manipur and the Chittagong Hill Tracts of Bangladesh; from *c.* 1000 to 2100 m at all seasons. Affects deep evergreen forest, mainly in glades or breaks such as made by streams, jungle tracks, etc.; also bamboo clumps and cut-over scrub.

Extralimital. Extends east to northern Laos.

GENERAL HABITS. Usually found in large flocks in the roving associations of small babblers, darting in and out of the undergrowth and bustling to and fro in incessant movement.

FOOD. Unrecorded. Doubtless as in other tit-babblers.

VOICE and CALLS. A low *chip-chip* (Stanford) and a soft twittering while feeding.

BREEDING. *Season*, April to July, chiefly May and June. *Nest* a deep cup, sometimes domed or semi-domed, made of bamboo leaves and fern fronds, and lined with fine shreds of grass or rootlets, the materials blending well with the surroundings. Most nests are placed on the ground, some between boulders, others low down in bamboo clumps. *Eggs*, normally 4, uncommonly 3, pale buff to warm buff, with reddish brown stipplings, coalescing to form a well-defined ring around the large end. Average size of 60 eggs 18·3 × 14·3 mm (Baker).

MUSEUM DIAGNOSIS. See Field Characters. Yellow of feathers in fresh specimens fades to white in older skins.

MEASUREMENTS

	Wing	Bill (from skull)	Tarsus	Tail
♂♀	51–58	11–13	20–23	38–44 mm (Baker, SA)

Weight 2 ♀ ♀ 11, 11 g (SA).

COLOURS OF BARE PARTS. Iris brown (SA) or pink. Bill: upper mandible blackish, lower bone-colour. Legs, feet and claws dull yellow (Stanford).

1379. Chestnut-headed Tit-Babbler. *Alcippe castaneceps castaneceps* (Hodgson)

Minla Castaneceps Hodgson, 1837, Ind. Rev. 2 (1): 33 (Nepal, restricted to Chandragiri Pass, central Nepal, by Ripley, 1950, JBNHS 49: 397)
Minla brunneicauda Sharpe, 1883, Cat. Bds. Brit. Mus. 7: 609 (Shillong)[1]
Pseudominla castaneiceps garoensis Koelz, 1951, Jour. Zool. Soc. India 3: 29 (Tura Mt., Garo Hills)[2]
Alcippe castaneiceps wagstaffei Wynne, 1954, North Western Naturalist, Key-List of Palaearc, and Oriental Pass. Bds., pt. 3: 397. New name for *A. brunneicauda* Sharpe, preoccupied
Baker, FBI No. 302 and 303, Vol. 1: 288–9
Plate 79, fig. 6

LOCAL NAMES. None recorded.

SIZE. Sparrow −; length *c*. 10 cm (4 in.).

FIELD CHARACTERS. *Above*, crown chestnut streaked with white on fore-head and with rufous on crown and nape. A broad white supercilium; a blackish post-ocular stripe; ear-coverts mostly white; a narrow dark malar stripe. Wing with a noticeable black shoulder-patch behind which a rufous patch, and pale outer edge. Rest of upperparts olive. *Below*, whitish with olive-rufous sides.

[1, 2] In view of Biswas's remarks (JBNHS 59 (1): 225) and a subsequent re-examination, we would admit the validity of *brunneicauda* (Khasi Hills) on the basis of paler head colour, *Garoensis* and *wagstaffei*, however, seem to us synonymous with nominate *castaneceps*.

STATUS, DISTRIBUTION and HABITAT. Common resident, subject to vertical movements. The Himalayas from central Nepal east through Darjeeling, Sikkim, Bhutan and Arunachal Pradesh; the hills of Meghalaya, Assam, Nagaland, Manipur, and the Chittagong Hill Tracts of Bangladesh. Breeds between 1500 and 3000 m, mostly above 1800 m. Noted in winter between 700 and 2745 m. Breeds somewhat lower in Assam where recorded in winter as low as 300 m. Affects heavy evergreen undergrowth at edge of forest and on abandoned clearings.

Extralimital. Extends to central Tenasserim and northwestern Thailand. The species ranges east to Vietnam and south through the Malay Peninsula.

GENERAL HABITS. In the non-breeding season keeps in large flocks, by themselves or mixed with minlas, yuhinas, tits, leaf warblers, etc., the composition of flocks varying according to altitude and season. Behaves much like the smaller parrotbills, hunting feverishly in the foliage of dense bushes and undergrowth, the flocks 'flowing' from tree to tree. Frequently climbs up vertical trunks, clinging to the bark or moss, searching crannies and crevices.

FOOD. Insects and, on occasion, tree sap.

VOICE and CALLS. A distinctive call of three notes *tu-twee-twee* in crescendo (Smythies). A tit-like *cheep* and a rather distinctive churring *purr* which can be very soft or loud and harsh.

BREEDING. *Season*, April to July, chiefly May and June. *Nest*, usually domed, sometimes a deep cup; made mostly of green moss with some bamboo leaves, and lined with fine grass or rootlets. Placed in a tangle of creepers climbing up trees or bushes within three metres or so from the ground, among the moss covering tree-trunks or on a sloping moss- and fern-covered bank, snugly hidden in clumps of ferns or orchids. *Eggs* 4, occasionally 3, chalky white, marked with blotches of inky black, forming a well-marked ring around the large end, with secondary markings of pale inky lavender. Average size of 28 eggs 17·7 × 13·4 mm (Baker). Incubation by both sexes; period undetermined.

MUSEUM DIAGNOSIS. See Field Characters. Young like adult.

MEASUREMENTS

	Wing	Bill (from skull)	Tarsus	Tail	
♂♂	56–61	12–13	20–21	41–46	mm
♀♀	50–56	10–13	20–21	40–46	mm
		(BB, Rand & Fleming, SA)			

Weight 19 ♂ ♀ 8–12 (av. 11) g–SDR, SA.

COLOURS OF BARE PARTS. Iris brown. Bill horny brown, yellowish at base of lower mandible. Legs, feet and claws olive-brown; soles yellow.

ALCIPPE VINIPECTUS (Hodgson): WHITEBROWED TIT-BABBLER.

Key to the Subspecies

Page

A Throat white.
 1 Head brighter, ear-coverts darker .... *A. v. kangrae*   113
 2 Head duller, ear-coverts paler .... *A. v. vinipectus* 113

B Throat streaked
   3 Coronal stripes dark brown .... *A. v. chumbiensis*   114
   4 Coronal stripes reddish brown .... *A.v. austeni*   115
   5 Coronal stripes black, mask black .... *A. v. perstriata*   115

1380. *Alcippe vinipectus kangrae* (Ticehurst & Whistler)

Fulvetta vinipecta kangrae Ticehurst & Whistler, 1924, Bull. Brit. Orn. Cl.
44: 71 (Palumpur, 6000 ft, Kangra)
Baker, FBI No. 304 (part), Vol. 1: 290

LOCAL NAMES. None recorded.
SIZE. Sparrow −; length *c.* 11 cm (4½ in.).
FIELD CHARACTERS. As in 1381, q.v.
STATUS, DISTRIBUTION and HABITAT. Resident, locally common, subject to vertical movements. The western Himalayas from Dharmsala to Garhwal; from *c.* 2700 to over 3300 m in summer and between 1500 and at least 2700 m in winter. Affects low scrub, dwarf willow and ringal bamboo in summer, and scrub-covered hillsides and bushy nullahs in winter.
GENERAL HABITS, FOOD and VOICE. As in 1381.
BREEDING. As in 1381.
MUSEUM DIAGNOSIS. Differs from *vinipectus* (1381) by the brighter head, darker ear-coverts, more rusty lower back, rump, and edges of the primaries and their coverts, and the slightly browner abdomen (Kinnear, *Ibis* 1939: 750).
MEASUREMENTS and COLOURS OF BARE PARTS. As in 1381.

1381. *Alcippe vinipectus vinipectus* (Hodgson)

Siva Vinipectus Hodgson, 1837, Ind. Rev. 2 (2): 89
(Nepal, restricted to central Nepal by Ripley, Synopsis: 412)
Baker, FBI No. 304 (part), Vol. 1: 290
Plate 79, fig. 14

LOCAL NAMES. None recorded.
SIZE. Sparrow −; length *c.* 11 cm (4½ in.).
FIELD CHARACTERS. *Above*, crown, ear-coverts and back brown; a broad white supercilium from eye to nape, above which a darker brown stripe. Rump and wings rusty, the latter (when closed) with a black line and pale outer edge. *Below*, throat and breast white, lower belly olive-brown. Sexes alike.
STATUS, DISTRIBUTION and HABITAT. A high-elevation babbler, fairly common and resident, subject to vertical movements. The Himalayas in western and central Nepal, from *c.* 2400 to 4200 m in summer, mostly above 3000 m, and from 1500 to 3000 m in winter. Affects rhododendron and juniper scrub, light forest of pine, birch or spruce, and forest edges with clearings and plenty of undergrowth and ringal bamboo.
GENERAL HABITS. In the non-breeding season keeps in flocks of up to twenty individuals, by themselves or in mixed company, hunting in bushes, undergrowth and low trees—even when heavily shrouded in snow in

winter—in acrobatic tit-like manner though somewhat slower in its movements. Confiding and inquisitive: will often approach within arm's reach of an observer to fussily investigate. Behaviour also reminiscent of *Chrysomma sinense* and *Dumetia hyperythra*. Flocks break up during the breeding season though small groups may still be observed, probably family parties.

FOOD. In the breeding season, almost exclusively insects—caterpillars being a favourite food; at other times also takes small seeds and berries.

VOICE and CALLS. A soft, high-pitched and incessant *chip, chip* (Smythies) or a clear *tsuid, tsuid* (Schäfer). A *churr* of alarm. Song, a faint *chit-it-it-or-key* given while flicking its tail up and its head forward (Fleming).

BREEDING. *Season*, mid April to July. *Nest*, a deep cup exteriorly made of moss and grass, followed by a layer of fibres, birch bark, bamboo leaves, then a thick layer of rootlets, finally a lining of hair. Placed in a fork in a bush within a couple of metres from the ground, or suspended among the fine branchlets of a bamboo clump. *Eggs*, normally 3, greenish with olive fleckings (Diesselhorst), grey-blue with a few black blotches (Baker), pale green with brown blotches (Osmaston), with a few purple freckles, all markings chiefly around the large end. Unlike any other timaline eggs in appearance. Average size of 49 eggs 18·8 × 13·7 mm (Baker).

MUSEUM DIAGNOSIS. See Field Characters. Differs from *chumbiensis* in having a white, unstreaked throat (intergrades are found in eastern Nepal). For distinction from *kangrae*, see 1380 under Museum Diagnosis. Colour of crown varies in depth.

Young, like adult but upperparts more rusty; dark coronal stripes less distinct; no streaks on throat; flanks paler (Whistler).

MEASUREMENTS

	Wing	Bill (from skull)	Tarsus	Tail	
♂♂	54–62	10–11	23–24	48–55	mm
♀♀	51–60	10–11	23–24	46–54	mm
			(BB, Rand & Fleming)		

Weight 6 ♂ ♂ 11–13; 5 ♀ ♀ 11–13 g (GD).

COLOURS OF BARE PARTS. Iris variable: creamy white, biscuit colour, brown or pale greenish yellow. Bill horny brown, flesh-colour at base of lower mandible. Legs, feet and claws horny brown, somewhat lighter than bill.

1382. ***Alcippe vinipectus chumbiensis*** (Kinnear)

Fulvetta vinipectus chumbiensis Kinnear, 1939, Ibis: 751
(Yatung, Chumbi Valley)
Baker, FBI No. 304 (part), Vol. 1: 290

LOCAL NAMES. None recorded.

SIZE. Sparrow −; length *c.* 11 cm (4½ in.).

FIELD CHARACTERS. As in 1381 but throat streaked with brown.

STATUS, DISTRIBUTION and HABITAT. Common resident, subject to vertical movements. Eastern Nepal, from the Okhaldunga district east through Darjeeling, Sikkim and Bhutan to the Rudo La, east of which it appears to be replaced by *A. cinereiceps*. Collected in Noa Dihing Watershed, Arunachal Pradesh, at 2600 m, March 1988 (Saha & Beehler) and it probably occurs throughout Arunachal Pradesh at suitable altitudes as the species is found again in northern Burma. Both this species and *cinereiceps* occur in the same ranges south of the Brahmaputra. Altitudinal distribution and biotope as in 1381.

GENERAL HABITS, FOOD and VOICE. As in 1381.

BREEDING. As in 1381.

MUSEUM DIAGNOSIS. Differs from *vinipectus* in being generally darker and with dusky streaks on throat and upper breast.

MEASUREMENTS

	Wing	Bill (from skull)	Tarsus	Tail	
4 ♂♂	56–62	10–11	23–24	52–55	mm
6 ♀♀	58–62	10–11	23–25	52–55	mm (SA)

Weight 6 ♂ ♀ 11–13 g (SA).

COLOURS OF BARE PARTS. As in 1381.

1382a. *Alcippe vinipectus perstriata* (Mayr)

Fulvetta vinipectus perstriata Mayr, 1941, Ibis, p. 79
(Chawngmawhka, Burma-Yunnan border)

LOCAL NAMES. None reported.

SIZE. Sparrow −; length *c.* 11 cm (4½ in.).

FIELD CHARACTERS. As in 1381, q.v.

STATUS, DISTRIBUTION and HABITAT. Resident, eastern Arunachal Pradesh. At present known only from the mountains of the upper Noa Dihing; collected at 2500 m in winter (S. Saha). Has been collected at this single locality along with *A. cinereiceps manipurensis* and *A. ludlowi*.

GENERAL HABITS, FOOD and VOICE. As in 1381.

MUSEUM DIAGNOSIS. Differs from all other Indian populations in the combination of heavily streaked throat and coal-black mask and upper supercilium.

MEASUREMENTS

1 o: Wing 60, bill (from skull) 11, tarsus 22.5, tail 49.5 (SDR).

Weight 1 o 8.6 g (SDR).

1383. *Alcippe vinipectus austeni* (Ogilvie-Grant)

Proparus austeni Ogilvie-Grant, 1895, Bull. Brit. Orn. Cl. 5: 3
(Manipur and the Naga Hills)
Baker, FBI No. 305, Vol. 1: 291

LOCAL NAME. *Dao-péré-gajao* (Cachari).

SIZE. Sparrow −; length *c.* 11 cm (4½ in.).

FIELD CHARACTERS. As in 1381, q.v.

STATUS, DISTRIBUTION and HABITAT. Resident, Nagaland, Manipur and Assam in the Barail range, from 1500 to at least 2700 m in winter, probably higher in summer (since the adjoining extralimital subspecies *ripponi* is found from 2500 m to the highest peak on Mt Victoria). Affects stunted rhododendron and oak forests, dense reed-bamboo and scrub at the edge of forest.

Extralimital. The species extends to southeastern Szechuan and northern Vietnam.

GENERAL HABITS, FOOD and VOICE. As in 1381.

BREEDING. Unrecorded; probably as in 1381.

MUSEUM DIAGNOSIS. Differs from *vinipectus* in having the head duller brown, the coronal stripes more reddish brown, and the throat streaked with reddish brown. For distinction from *perstriata* of Burma, see JBNHS 50: 502.

MEASUREMENTS and COLOURS OF BARE PARTS. As in 1381.

Weight 2 ♂ ♂ 11, 12 g (SDR).

1384. Brownheaded Tit-Babbler. *Alcippe ludlowi* (Kinnear)

Fulvetta ludlowi Kinnear, 1935, Bull. Brit. Orn. Cl. 55: 134
(Sakden, eastern Bhutan, 9000 ft)
Not in Baker, FBI

LOCAL NAME. None recorded.

SIZE. Sparrow −; length *c.* 11 cm (4½ in.).

FIELD CHARACTERS. *Above,* head chocolate-brown, sides of head and nape reddish brown. Rest of upperparts and underparts as in *A. vinipectus* (1381). Distinguished from the latter species by the absence of white supercilium and the dark brown line above it. Sexes alike.

STATUS, DISTRIBUTION and HABITAT. Common resident, subject to some vertical movements. Eastern Bhutan and easternmost Arunachal Pradesh (upper Noa Dihing, Saha & Beehler, March 1988). Presumably also in the intervening areas of northern, central, and western Arunachal Pradesh at suitable altitudes (where recorded across the border in southeastern Tibet). From 2200 m (winter) to 3500 m (summer). Affects bamboo and rhododendron forest.

GENERAL HABITS, FOOD and VOICE. As in 1385.

BREEDING. Unknown.

MUSEUM DIAGNOSIS. Differs from all other *Alcippe* in the entirely concolourous crown, nape and cheeks, a deep vinaceous red-brown without any streaking or supercilium.

Young (juvenile) like adult but paler (*Ibis* 1937: 38).

MEASUREMENTS

Wing, ♂ ♂ 59–64; ♀ ♀ 56–60 mm (NBK). Other measurements as in 1385.

Weight 1 o 10.5 g (SDR).

COLOURS OF BARE PARTS. Iris brown. Bill dark horn, fleshy at base of lower mandible. Legs and feet fleshy brown (Ludlow. More details under 1385).

1385. Greyheaded Tit-Babbler. *Alcippe cinereiceps manipurensis* (Ogilvie-Grant)

Proparus manipurensis Ogilvie-Grant, 1906, Bull. Brit. Orn. Cl. 16: 123
(Owenkulno Peak, Manipur Hills)
Baker, FBI No. 307, Vol. 1: 292
Plate 79, fig. 10

LOCAL NAMES. None recorded.

SIZE. Sparrow —; length c. 11 cm (4½ in.).
FIELD CHARACTERS. As in 1384 but ground colour of throat pale smoke-grey instead of white.
STATUS, DISTRIBUTION and HABITAT. Resident, probably subject to some vertical movements. Nagaland, Manipur (Owenkulno Peak) and Assam in the Barail Range, from 1500 m in winter to over 2500 m in summer. Collected in Arunachal Pradesh (Noa Dihing Drainage, 2600 m, March 1988—Saha & Beehler). Appears to occupy a slightly lower zone than *A. vinipectus* and to replace it, the two species having never been recorded in the same area. Affects secondary scrub-jungle, especially bramble, and also dense bamboo.
Extralimital. Extends through Burma (including the Chin Hills) to western Yunnan. The species ranges north to southwestern Kansu and Hopeh, and east to northern Vietnam, Fujian and Taiwan.
GENERAL HABITS. Very similar to those of *A. vinipectus* (1381). In the non-breeding season keeps in parties of six to ten, often in company with other small babblers or leaf warblers. The birds scramble about in shrubs close to the ground, sometimes ascending trees. Usually very confiding.
FOOD. Chiefly insects.
VOICE and CALLS. A rattling song of three or four notes (Stanford). Call, a tit-like *cheep* (SDR).
BREEDING. Unknown. *Season,* apparently beginning in April.
MUSEUM DIAGNOSIS. Differs from *ludlowi* (1384) in having grey-brown crown, and the ground colour of throat pale smoke-grey instead of white. For details of fresh plumage see JBNHS 50: 503.
MEASUREMENTS

	Wing	Bill (from skull)	Tarsus	Tail	
2 ♂♂	55, 56	10, 10	23, 23	52, 53	mm
1 ♀	51	10	—	50	mm
				(SDR)	

Wing ♂ ♀ 51–57; tail 49–54 mm (Mayr)
Weight 2 ♂♂ 10, 10 g (SDR).
COLOURS OF BARE PARTS. Iris variable; pale pinkish yellow or straw yellow (Stanford); ♂ yellowish brown, ♀ pale pinkish brown (SDR). Bill: ♂ black, ♀ dark brown. Legs and feet: ♂ dark brown to greyish brown, ♀ brownish flesh; claws grey or bone colour.

1385a. **Streakthroated Tit-Babbler.** *Alcippe striaticollis* (Verreaux)

Siva striaticollis Verreaux, 1870, Nouv. Arch. Mus. Hist. Nat. (Paris)
6: 38 (Muping)
Not in Baker, FBI

LOCAL NAMES. None recorded.
SIZE. Sparrow —; length c. 11 cm (4½ in.).
FIELD CHARACTERS. *Above,* brown with darker streaks on head, nape and upper back; lores black; wing chestnut with a pale grey outer edge. *Below,* throat and breast white, conspicuously streaked with dark brown. Belly whitish grey. Sexes alike.

Distinguished from both *vinipectus* and *cinereiceps* by the dark brown streaks on upper back and the lack of rusty on rump.

STATUS, DISTRIBUTION and HABITAT. Not recorded within our limits but it may occur in northern Arunachal Pradesh since it has been collected by Ludlow in many localities of southeast Tibet north of the main range, from the Tsari Valley (upper Subansiri) to Tripé (*Ibis* 1944: 81 and 1951: 556). Also obtained at Tripé by Bailey [JBNHS 24: 75, but erroneously identified as *vinipectus* by Baker, *fide* Ludlow (1944)]. Has a wide altitudinal range in summer, from 2800 to 4200 m. Replaces *vinipectus* at higher altitudes when both occur. There is presumably a withdrawal from the upper altitudes in the cold season. Affects shrubbery and is particularly fond of rhododendron and holly-oak.

Extralimital. Extends to eastern Sichuan and southwestern Kansu.

GENERAL HABITS. A gregarious, confiding but inconspicuous species. Very active.

FOOD. Unrecorded.

VOICE and CALLS. A clear *tserrr-tserr*, deceptive as it varies in intensity, so that one may believe the bird to be a stone's throw away while it suddenly appears within a couple of metres (Schäfer).

BREEDING. A nest found by Bailey on 15 July was made of grass, covered with moss and lichen on the outside and lined with hair. It was 'hanging in a branch of bamboo' and contained four eggs.

MUSEUM DIAGNOSIS. See Field Characters.

Young, like adult but darker above and duller below.

MEASUREMENTS

Wing 6 ♂♂ 61–62; 7 ♀♀ 58–60 mm (Kinnear)
Bill from skull *c.* 12; tarsus *c.* 23; tail *c.* 59 mm.

COLOURS OF BARE PARTS. Iris yellowish white to lemon-yellow. Bill: upper mandible horny, lower fleshy pink. Legs and feet horny brown to fleshy brown.

ALCIPPE RUFOGULARIS (Mandelli): REDTHROATED TIT-BABBLER

Key to the Subspecies

Darker on crown; back rufescent brown .... *A. r. collaris*
Crown paler; back brown .... *A. r. rufogularis*

1386. *Alcippe rufogularis rufogularis* (Mandelli)

Minla rufogularis Mandelli, 1873, Stray Feathers 1: 416 (Bhutan Duars)
Baker, FBI No. 300 (part), Vol. 1: 286
Plate 79, fig. 8

LOCAL NAMES. None recorded.

SIZE. Sparrow −; length *c.* 12 cm (4½ in.).

FIELD CHARACTERS. *Above*, crown rufous-brown bordered by a broad black stripe from forehead to nape. Lores, supercilium and eye-ring white. Ear-coverts dark brown. Rest of upperparts brown. *Below*, chin and throat white with a wide chestnut band across the latter. Centre of belly whitish, sides olive-brown; under tail-coverts fulvous. Sexes alike.

The chestnut throat-band easily identifies this species.

STATUS, DISTRIBUTION and HABITAT. Resident, fairly common. The Himalayan foothills from the Jalpaiguri duars east to the Dihang river, from the plains level to *c.* 900 m. Affects undergrowth in evergreen forest.
GENERAL HABITS, FOOD and VOICE. As in 1387.
BREEDING. As in 1387.
MUSEUM DIAGNOSIS. See Field Characters. For distinction from *collaris* see 1387.
MEASUREMENTS and COLOURS OF BARE PARTS. As in 1387.

1387. *Alcippe rufogularis collaris* Walden

Alcippe collaris Walden, 1874, Ann. Mag. Nat. Hist. 14: 156
(Sadiya, upper Assam)
Baker, FBI No. 300 (part), Vol. 1: 286

LOCAL NAMES. None recorded.
SIZE. Sparrow −; length *c.* 12 cm (4½ in.).
FIELD CHARACTERS. As in 1386, q.v.
STATUS, DISTRIBUTION and HABITAT. Common resident. Arunachal Pradesh east of the range of the nominate race (about the Dihang river) and south through the Sadiya Frontier Tract, the Patkai Range, Nagaland, Manipur, Assam in the Cachar hills, and Bangladesh in the northeastern highlands and the Chittagong Hill Tracts; from the plains level to *c.* 900 m. Affects bamboo jungle, scrub, secondary growth and undergrowth in evergreen forest.
Extralimital. The species extends east to Vietnam.
GENERAL HABITS. Creeps about and skulks in dense cover close to the ground, in small restless parties often mixed with *Stachyris* and other small babblers, also feeding on the ground. In pairs in the breeding season.
FOOD. Mainly insects.
VOICE and CALLS. A musical *chip-churr* (Stanford) or cheeping *chree-chree* while on the move (SA).
BREEDING. *Season*, March to June, mostly April. *Nest*, domed, rather loosely made of dead leaves, grass, roots, moss and a few tendrils, lined first with rootlets with an inner layer of dead leaves. Usually placed on the ground or on fallen rubbish, rarely at the bottom of a bush or in tangles of creepers and cane, generally blending well with the surroundings. *Eggs*, normally 3, seldom 4, similar to those of *A. brunnea* (1388) but paler, less brown and more grey. Average size of 100 eggs 19·5 × 14·7 mm (Baker).
MUSEUM DIAGNOSIS. Differs from *rufogularis* (1386) in being darker on the crown and rufescent on the back. Flanks more heavily washed with brown.
MEASUREMENTS

	Wing	Bill (from skull)	Tarsus	Tail	
♂♂	55–60	13–14	21	46–49	mm
♀♀	55–57	13–14	21	46–47	mm
				(SA)	

Weight 3 ♀♀ 10, 12, 13 g (SDR).
COLOURS OF BARE PARTS. Iris brown. Bill black. Legs and feet yellow.

1388. **Rufousheaded Tit-Babbler.** *Alcippe brunnea mandelli*
(Godwin-Austen)

Minla mandelli Godwin-Austen, 1876, Ann. Mag. Nat. Hist. 18: 33
(Naga Hills, Northeast Bengal)
Schoeniparus dubius certus Koelz, 1952, Jour. Zool. Soc. India 4: 39
(Shillong Peak, Khasi Hills)
Baker, FBI No. 298, Vol. 1: 284
Plate 79, fig. 3

LOCAL NAME. *Dao-chitter* (Cachari).
SIZE. Sparrow −; length *c.* 13 cm (5 in.).
FIELD CHARACTERS. *Above,* crown rufous-brown, more rufous on forehead, bordered by black stripes starting above the eye and meeting on the upper back. A broad white supercilium; lores dark brown; ear-coverts umber-brown. Rest of upperparts olive-brown. *Below,* buffish white. Sexes alike.

Very similar to *A. rufogularis* but lacks the chestnut band on throat.

× *c.* 1

STATUS, DISTRIBUTION and HABITAT. Common resident. Meghalaya in the Khasi and Assam in the Cachar hills, Nagaland and Manipur. Two records north of the Brahmaputra: Arunachal Pradesh in the Dafla Hills [Godwin-Austen? (Whistler's MS.)] and Kobo (Baker, 1913, *Rec. Ind. Mus.* 8: 273); from 900 m to 1800 m in Assam and from *c.* 1400 to 2400 m in Nagaland. Affects dense bushes, especially bracken and brambles in forest margins or light forest; in winter may be seen in bamboo jungle, scrub and secondary growth.

Extralimital. The Chin Hills of Burma. The species extends north to Sichuan and Anhui and east to Guangxi, Fujian and Taiwan (whence the nominate race).

GENERAL HABITS. Keeps in small parties often in company with *Stachyris*, hunting in dense undergrowth, close to or on the ground. In spite of being shy and secretive, it seldom escapes notice thanks to its restless vivacity. Keeps in pairs in the breeding season.

FOOD. Mainly insects.

VOICE and CALLS. A constant *chir-r-r-r* alternating with a sharp *chit* while hopping through cover. Also a chattering note of alarm and a distinctive *chee-chee-chee-chee-chee-hpwit* (Smythies).

BREEDING. *Season,* April to June. *Nest,* oval-shaped with entrance near the top sometimes occupying so much of the side that the nest looks semi-domed or even a deep cup in shape. Made of bamboo leaves, grass, bracken leaves and roots, lined with rootlets with an inner layer of dead leaves. Placed on the ground or on fallen leaves and rubbish, concealed among bracken, preferably on sloping ground. *Eggs,* 3 or 4, occasionally 5, clay-white to deep clay-colour marked with dark brown spots and a few short lines and blotches, with secondary smudges of paler brown or grey. Average size of 200 eggs 20·8 × 15·6 mm (Baker). Building and incubation by both sexes; incubation period undetermined.

MUSEUM DIAGNOSIS. See Field Characters.

MEASUREMENTS

	Wing	Bill (from skull)	Tarsus	Tail
10 ♂♂	57–61	c. 14	c. 25	c. 60 mm
7 ♀♀	54–58			

(Stresemann, SDR)

Weight ♂ ♀ 16–19 g (SDR).

COLOURS OF BARE PARTS. Iris yellowish red, pale yellow to slaty pink. Bill dark brown to dull black. Legs and feet fleshy.

ALCIPPE POIOICEPHALA (Jerdon): QUAKER BABBLER

Key to the Subspecies

	Page
Paler and greyer .... *A. p. brucei*	121
Darker and browner; underparts more fulvous .... *A. p. poioicephala*	122
Underparts and ear-coverts more ochraceous .... *A. p. fusca*	123

1389. *Alcippe poioicephala brucei* Hume

Alcippe brucei Hume, 1870, Jour. Asiat. Soc. Bengal 39: 122 (Mahableshwar)

Baker, FBI No. 289, Vol. 1: 278

LOCAL NAME. *Seetimār lalédo* (Gujarati).

SIZE. Sparrow; length *c.* 15 cm (6 in.).

FIELD CHARACTERS. As in 1390, q.v.

STATUS, DISTRIBUTION and HABITAT. Resident, locally very common. The hills of the Indian peninsula [except for the range of the nominate race (1390, q.v.)], south of a line running from Kathiawar and Mt Abu across southern Madhya Pradesh (Pachmarhi, Balaghat) to Parasnath (southern Bihar) and Orissa; from the foothills to the highest elevations. Affects wet forest of teak, mixed deciduous or evergreen as well as bamboo jungle and frequently open scrub, occasionally gardens.

GENERAL HABITS, FOOD and VOICE. As in 1390. Often in small parties (SDR); associate often with *Stachyris* or *Macronous*. Song all year round, most prominent component of dawn chorus in April. Has also another song of similar quality, but with many more notes in a long drawn cadence, accompanied by the chattering of the ♀ (Trevor Price, JBNHS 76: 413).

MOULT. Still going through complete prenuptial moult in May (Eastern Ghats, Trevor Price, JBNHS 76: 413).

BREEDING. *Season*, nests or birds in breeding condition have been recorded from January to November; main period appears to be January to April. *Nest* and *eggs* as in 1390. Average size of 35 eggs 20 × 15 mm (Baker).

MUSEUM DIAGNOSIS. Differs from *poioicephala* (1390) in being paler. Crown and nape a clearer colder grey; rest of upperparts greyer, less brown. Wings and tail not so dark. Lower parts also much paler, washed with greyer brown (*v.* rich fulvous). Perhaps slightly larger. Some specimens are difficult to assign because of individual variation and possible intergradation (See Abdulali, 1983, JBNHS 80: 358).

Alcippe poioicephala

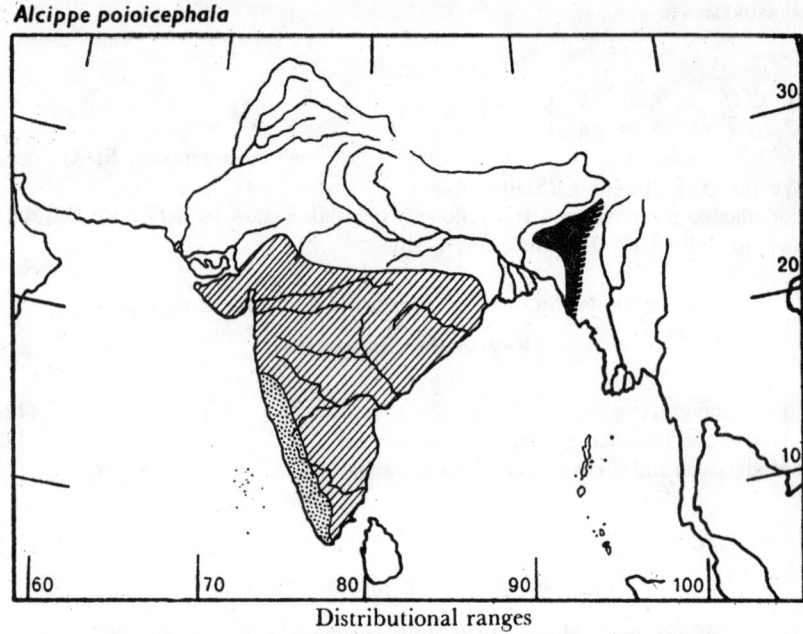

Distributional ranges

▨ brucei (1389).   ⣿ poioicephala (1390).   ■ fusca (1391).

MEASUREMENTS
	Wing	Bill (from skull)	Tarsus	Tail	
♂♂	66–77	15–17	21–24	62–70	mm
♀♀	66–74	15–16	21–23	59–70	mm
				(HW, SA, Koelz)	

Weight 14 ♂ ♀ 18–23 (av. 20·7) g (SA). 2 ♂ ♂ –, 16.5 g, 1 ♀ 17 g, 20 o? 17, 26 g (SDR).

COLOURS OF BARE PARTS. As in 1390.

1390. *Alcippe poioicephala poioicephala* (Jerdon)

Thimalia poioicephala Jerdon, 1844, Madras Jour. Lit. Sci. 13: 169 (Coonoor Ghat, Nilgiris)
Baker, FBI No. 288, Vol. 1: 277
Plate 79, fig. 11

LOCAL NAME. *Kana chilappan* (Malayalam).
SIZE. Sparrow; length *c.* 15 cm (6 in.).
FIELD CHARACTERS. A nondescript olive-brown babbler with grey crown and nape, rusty brown on wings and tail, and pale fulvous underparts. Sexes alike.
STATUS, DISTRIBUTION and HABITAT. Common resident. Western Karnataka along the Ghats south through Kerala and western Tamil Nadu (Nilgiris and Palnis). Intergrades with *brucei* in southwestern Maharashtra,

Goa and the Biligirirangan Hills; from the foothills to 2100 m. Affects evergreen and moist-deciduous forest, sholas, ravines, mixed bamboo jungle and canebrakes.

GENERAL HABITS. Usually seen in parties of six to ten individuals, sometimes up to twenty or more, hopping about among the undergrowth, often ascending to the canopy, flitting from sprig to sprig, clinging upside-down in acrobatic positions, continually calling to each other; often forming the nucleus of the mixed hunting parties of small insectivorous species. Behaviour very similar to that of *Macronous gularis* (1228).

FOOD. Ants and other insects. Very fond of nectar, especially of *Erythrina* spp.

VOICE and CALLS. Members of a party maintain contact by a harsh, rather subdued *chur-r-chur-r*. Song, most often heard in the first half of the year, a quavering trill of four (sometimes up to 7 or 8) sweet whistling notes of Magpie-Robin quality recalling those of the Spotted Babbler—constantly uttered (SA). See also 1391.

BREEDING. *Season*, not well defined; chiefly January to May but nests also found in August and October. *Nest*, a roughly built cup of green moss, rootlets, lichen, leaves and grass, lined with rootlets and placed in a fork or suspended from a branch, usually between one and three metres above the ground, in bushes or saplings. *Eggs*, usually 2, sometimes 3, pale salmon, marked with blotches and smudges of deep purple-brown with secondary markings of pale grey or pinkish grey, and in many eggs, some short lines and hieroglyphic markings. Average size of 42 eggs 20 × 15·1 mm (Baker). Incubation by both sexes; period undetermined.

MUSEUM DIAGNOSIS. More richly coloured than *brucei*; for distinction from it see 1389; for details of plumage Baker, loc. cit.

Young, similar to adult, but showing the juvenal timaline characters—soft blunt first primary, narrow, pointed rectrices.

MEASUREMENTS

	Wing	Bill (from skull)	Tarsus	Tail	
♂♂	68–73	15–16	*c.* 23	60–67	mm
♀♀	65–75	15–16	*c.* 23	61–65	mm
				(SA)	

COLOURS OF BARE PARTS. Adult: Iris greyish brown. Bill dark horny brown; commissure and lower mandible greyish; mouth pale flesh-colour. Legs, feet and claws greyish brown. Young: Iris slaty grey. Gape and mouth bright yellow; paler portions of bill yellowish. Legs, feet and claws as in adult.

1391. *Alcippe poioicephala fusca* Godwin-Austen

Alcippe fusca Godwin-Austen, 1877, Jour. Asiat. Soc. Bengal 45: 197
(Naga Hills)
Baker, FBI No. 290 (part), Vol. 1: 278

LOCAL NAME. *Dao-péré-gadeba* (Cachari).
SIZE. Sparrow; length *c.* 15 cm (6 in.).
FIELD CHARACTERS. As in 1390, q.v.

STATUS, DISTRIBUTION and HABITAT. Common resident. Bangladesh in the northeastern highlands and the Chittagong Hill Tracts. Meghalaya in the Khasi and Assam in the Cachar hills, Nagaland and Manipur (where it intergrades with *phayrei*); in the foothills up to *c.* 1000 m and adjacent plains. Affects deep evergreen forest, bamboo jungle and secondary growth.

Extralimital. Extends to northwestern Burma. The species ranges east to Vietnam. *A. brunneicauda* of the southern Malay Peninsula, Sumatra and Borneo and *A. pyrrhoptera* of Sumatra are considered as subspecies by Delacour; their ranges are complementary.

GENERAL HABITS and FOOD. As in 1390.

VOICE and CALLS. Has a distinctive musical *chewy-chewy-chewy-chewy-chewy* uttered quickly with each *chewy* alternately higher and lower in pitch; may be heard the whole year (Smythies). See also 1390.

BREEDING. *Season*, overall March to September, chiefly April to June. *Nest and eggs* as in 1390. Average size of 150 eggs 19·6 × 15 mm.

MUSEUM DIAGNOSIS. Like *poioicephala* but not so fulvous above; ear-coverts and sides of neck ochraceous; edges of wings and tail fulvous-olive; underparts more ochraceous.

Young, like adult but upperparts, edges of wings and tail browner. Timaline juvenal characters, i.e. soft blunt first primary and narrow, pointed rectrices present. Postjuvenal moult complete.

MEASUREMENTS

	Wing	Bill (from skull)	Tarsus	Tail	
♂♂	67–72	15–17	21–22	63–70	mm
♀♀	63–69	15–17	20–21	62–68	mm
				(HW)	

COLOURS OF BARE PARTS. As in 1390.

ALCIPPE NIPALENSIS (Hodgson): NEPAL QUAKER BABBLER

Key to the Subspecies

		Page
Crown darker	*A. n. nipalensis*	124
Crown paler	*A. n. stanfordi*	126

1392, 1393. *Alcippe nipalensis nipalensis* (Hodgson)

Siva nipalensis Hodgson, 1837, Ind. Rev. 2(2): 89 (Nepal)
Alcippe nipalensis commoda Ripley, 1948, Proc. Biol. Soc. Washington 61: 104 (Dening, Mishmi Hills, NE. Assam)
Alcippe nipalensis turensis Koelz, 1952, Jour. Zool. Soc. India 4: 39 (Tura Mountain, Garo Hills)
Alcippe nipalensis khasiensis Koelz, 1954, Contrib. Inst. Regional Exploration, No. 1: 16 (Cherrapunji, Khasia Hills)
Baker, FBI No. 286 (part), Vol. 1: 275
Plate 79, fig. 9

LOCAL NAMES. None recorded.

SIZE. Sparrow −; length *c.* 12 cm (4½ in.)

FIELD CHARACTERS. *Above*, head grey with a blackish supercilium from eye to nape, and a conspicuous white eye-ring. Rest of upperparts fulvous brown. *Below*, uniformly buff. Sexes alike.

× *c.* 1

STATUS, DISTRIBUTION and HABITAT. Common resident, subject to vertical movements. From central Nepal east through Darjeeling, Sikkim, Bhutan, Arunachal Pradesh, Nagaland, Manipur, hills of Assam and Meghalaya in the Garo and Khasi hills and the adjacent Hill States south to the Lushai and Chittagong Tracts of Bangladesh; from the foothills to *c.* 1800 m in winter, and up to 2400 m (locally) in summer; lower limit of breeding zone not satisfactorily known; mostly above 1500 m in Nepal, but apparently lower down farther east. Affects moist-deciduous or evergreen forest with dense undergrowth.

GENERAL HABITS. Keeps in active, restless flocks, often in company with other small insectivorous species. Feeds mostly in the undergrowth and in the crowns of lower trees, occasionally on the ground. May be seen clinging sideways or upside-down on twigs, racing along some branch or fluttering in front of a sprig in its feverish search for food, the flocks rapidly 'flowing' from tree to tree. Pairs form in March.

FOOD. Insects and berries, the latter forming a large part of its diet in the non-breeding season; also flower-nectar.

VOICE and CALLS. A constant twittering given while foraging— 'a shrill whinnying note' (Stanford). One call is rendered as *p-p-p-p-jet* (Fleming), others as a rapid *dzi-dzi-dzi-dzi-dzi* and a soft, high-pitched *pi-pi-pi-pi-pi* (Lister).

BREEDING. *Season*, March to July. *Nest*, a deep cup of bamboo leaves, grasses, fern fronds, roots, bark fibre or other materials, lined with rootlets. Placed in bushes or bamboo clumps, generally between upright twigs, occasionally semi-pendent in a horizontal fork, between 30 and 150 cm above the ground. *Eggs*, 3 or 4, less often 5, very variable: white to pink with reddish specks, spots or blotches forming a ring or cap around the large end (for more details see Baker, *Nidification* 1: 242). Average size of 200 eggs including those of the Assam race, 18·4 × 14 mm (Baker). Building and incubation by both sexes; period about 12 days.

MUSEUM DIAGNOSIS. See Field Characters. For distinction from *commoda*, see 1393.

Young, like adult but upperparts richer fulvous, flanks and under tail-coverts more fulvous. Juvenal timaline primary and tail characters present.

MEASUREMENTS

	Wing	Bill (from skull)	Tarsus	Tail	
♂♂	57–63	13–15	22–24	57–66	mm
♀♀	58–61	12–15	22–24	57–64	mm
			(BB, SA, Kinnear)		

Weight 7 ♂ ♂ 14–18; 4 ♀ ♀ 14–16 g (GD). 6 ♂ ♀ 10–14 g (SA). ♂ ♀ (Miao, Arunachal Pradesh) 13–16 g (SDR)

COLOURS OF BARE PARTS. Iris greyish brown. Bill greyish horn, brown on basal third of upper mandible. Legs, feet and claws plumbeous horny brown.

TIMALIINAE

1394. *Alcippe nipalensis stanfordi* Ticehurst

Alcippe nepalensis stanfordi Ticehurst, 1930, Bull. Brit. Orn. Cl. 50: 84
(Taungup-Prome Cart Road, Arakan Yoma, 2900 ft.)
Baker, FBI No. 286 (part), Vol. 1: 275

LOCAL NAMES. None recorded.
SIZE. Sparrow —; length *c*. 12 cm (4½ in.).
FIELD CHARACTERS. As in 1392, 1393, q.v.
STATUS, DISTRIBUTION and HABITAT. Common resident. The Chin Hills and Arakan Yomas of Burma, probably intergrading with *nipalensis* in the Mizo Hills and the Chittagong Hill Tracts; from the foothills to *c*. 1800 m. Biotope as in 1392, 1393.
GENERAL HABITS, FOOD and VOICE. As in 1392, 1393.
BREEDING. As in 1392, 1393.
MUSEUM DIAGNOSIS. Differs from *nipalensis* in being paler.
MEASUREMENTS and COLOURS OF BARE PARTS. As in 1392, 1393.

Genus HETEROPHASIA Blyth

Heterophasia Blyth, 1842, Jour. Asiat. Soc. Bengal 11: 186.
Type, by monotypy, *H. cuculopsis* Blyth = *Sibia picaoides* Hodgson
Leioptila Blyth, 1847, Jour. Asiat. Soc. Bengal 16: 449.
Type, by monotypy, *Leioptila annectans* [sic] Blyth

Bill shorter than head, slender and curved; nostrils covered by a membrane. Rictal bristles moderate. Tail long, well graduated.

Key to the Species

		Page
A Crown black		
1 Underparts cinnamon	*H. capistrata*	127
2 Underparts mostly white or greyish white		
a Rump chestnut	*H. annectens*	126
b Rump grey	*H. gracilis*	131
B Crown grey or blue-grey		
3 A white patch on wing	*H. picaoides*	133
4 No white on wing	*H. pulchella*	132

1395. **Chestnutbacked Sibia.** *Heterophasia annectens annectens* (Blyth)

Leioptila annectans [sic] Blyth, 1847, Jour. Asiat. Soc. Bengal 16: 450
(Darjeeling)
Baker, FBI No. 317, Vol. 1: 300
Plate 80, fig. 14

LOCAL NAME. *Rubnun-pho* (Lepcha).
SIZE. Bulbul; length *c*. 18 cm (7 in.).
FIELD CHARACTERS. *Above,* crown, ear-coverts, sides of neck and sides of upper back black; hindneck streaked with white. Centre of back, rump and upper tail-coverts chestnut. Wing black with a chestnut bar, pale ashy edges to primaries and white tips to tertiaries. Tail black, graduated, the rectrices

tipped with white. *Below*, white; flanks, vent and under tail-coverts fulvous. Sexes alike.

STATUS, DISTRIBUTION and HABITAT. A rather scarce resident except in Manipur where it is common; subject to vertical movements. From eastern Nepal, Darjeeling and Sikkim, east through Bhutan and Arunachal Pradesh to the Burma border; Meghalaya in the Khasi, and Assam in the Cachar hills; Mizo Hills, Nagaland and Manipur; from *c.* 1200 m to 2300 m in summer, straying down to the foothills in winter, obtained at *c.* 810 m in Arunachal Pradesh on 18. iii. 1979. Affects dense, humid evergreen forest.

Extralimital. Extends to western Burma and western Yunnan. The species ranges east to Vietnam.

GENERAL HABITS. Arboreal. Small parties keep to the canopy of tall trees. Creeps along the branches or clambers on the trunks like a nuthatch, searching amongst the moss, lichen and crevices of bark for insects.

FOOD. Beetles and other insects; also seeds.

VOICE and CALLS. Song: 'a musical phrase of four notes, the first two on the same pitch, followed by two notes dropping in pitch, the whole sometimes preceded by an introductory grace note' (Smythies). Alarm-note *chirr-r-r* (Baker). A clear single whistle has also been recorded.

BREEDING. *Season*, May and June. *Nest*, cup-shaped, very neatly and compactly built; exteriorly of moss with a few leaves and fine grass, followed by a deep layer of grass, bamboo or other leaves and lined with fine fibres, rootlets or rhizomorphs. Usually placed on a branch of a small tree between two and six metres from the ground, generally in a fork of an outer branch, slender and difficult to reach. *Eggs*, normally 3, sometimes 2, exceptionally 4, very variable: the most common type is a very pale blue-grey or green-grey marked with blotches and smears of reddish brown, a few spots and short wavy lines of the same colour, and secondary blotches of pale lavender and pale brown. For futher details see Baker, *Nidification* 1: 279. Average size of 25 eggs 22 × 15·5 mm (Baker). Incubation by both sexes. When disturbed from the nest, the bird's persistent alarm-notes draw attention to it.

MUSEUM DIAGNOSIS. See Field Characters; for details of plumage, Baker loc. cit.

MEASUREMENTS

	Wing	Bill (from feathers)	Tarsus	Tail	
♂♀	75–84	15–16	*c.* 24	81–87	mm
				(Baker, Mayr)	

Bill from skull *c.* 18 mm.
Weight 1♂ 26·5 g (SDR).

COLOURS OF BARE PARTS. Iris brownish to deep crimson. Bill black, yellow at base of lower mandible. Legs and feet wax yellow; claws brownish.

HETEROPHASIA CAPISTRATA (Vigors): BLACKCAPPED SIBIA

Key to the Subspecies

	Page
Centre of back pale brownish grey .... *H. c. capistrata*	128
Centre of back rufous .... *H. c. nigriceps*	130
Centre of back sooty brown .... *H. c. bayleyi*	130

TIMALIINAE

1396. *Heterophasia capistrata capistrata* (Vigors)

Cinclosoma capistratum Vigors, 1831; Proc. Zool. Soc. London: 56
(Himalayas; restricted to Simla by Ripley, Synopsis)
Malacias capistrata pallida Hartert, 1891, Kat. Vög. Mus. Senckenb.: 21
(northwestern India; restricted to Simla by Baker, 1921, JBNHS 27: 460)
Baker, FBI Nos. 311 (part), 312, Vol. 1: 296, 298
Plate 80, fig. 13

LOCAL NAMES. None recorded.

SIZE. Bulbul +; length *c.* 21 cm (*c.* 8 in.).

FIELD CHARACTERS. *Above*, ear-coverts, crown and erectile crest black, often suffused with reddish brown. Back, rump and part of tail rufous, the middle back tinged with greyish brown. Wings slaty with paler outer edge and a black shoulder-patch; a white patch conspicuous in flight. Tail long, graduated, rufous with black outer edge and subterminal band, tipped with slaty grey. *Below*, entirely cinnamon. Sexes alike.

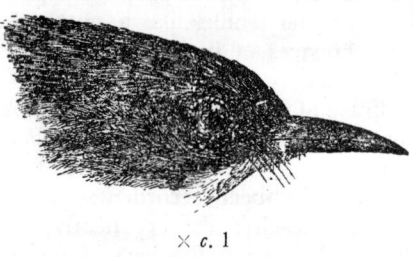

× *c.* 1

STATUS, DISTRIBUTION and HABITAT. Common resident, subject to vertical movements. The western Himalayas from mid Swat (P. Jones, JBNHS 76: 45) to extreme western Nepal (Fleming & Traylor, *Fieldiana* 53: 177). Breeds between 1800 and 2700 m; from October to March it is found mostly between 1200 and 2100 m but may also be seen as high as 2700 m and has strayed down to *c.* 100 m. Affects various types of forest but shows a preference for mixed evergreen: oak, fir, chestnut, etc.

GENERAL HABITS. Keeps in pairs or small parties according to the season. Strictly arboreal; hunts in the canopy and on moss-covered trunks and branches of tall trees, occasionally descending to the undergrowth. Very active and lively; hops swiftly from branch to branch like a laughing thrush, hanging upside-down on sprigs to peer under leaves, occasionally flying out to catch an insect on the wing; often flies up to a trunk and clings to the bark while exploring crevices. Flight rather laboured.

FOOD. Insects and berries. Often visits flowers of Silk Cotton trees and rhododendron for insects and nectar.

VOICE and CALLS. Song, a clear, flute-like, very pleasing and far-carrying *tee-dee-dee-dee-dee-o-lu*, the first five notes on the same tone, the sixth lowest and the last in between, sometimes shortened of the first two or three notes but ending always typical (Desfayes); a very characteristic sound of the forests of its zone. In the western part of its range, the song is simplified, and consists of similar repeated notes without the rising and falling two notes at the end (TJR). Alarm, a harsh, rasping *chrai-chrai-chrai-chrai-chrai* (Lister). Call-note, a loud, rapid *chi-chi* (Fleming); *tee-riri-reeri-reeri*, like the jingle of a silver bell, repeated every half-minute or so (SA).

BREEDING. *Season*, April to August, chiefly June and July. *Nest*, a neat cup of green moss, dry grasses and leaves, firmly interwoven and lined with rootlets, pine needles or other fine materials. Generally built on trees or bushes, often near the extremity of a branch, from two to eighteen metres

Heterophasia capistrata and *H. gracilis*

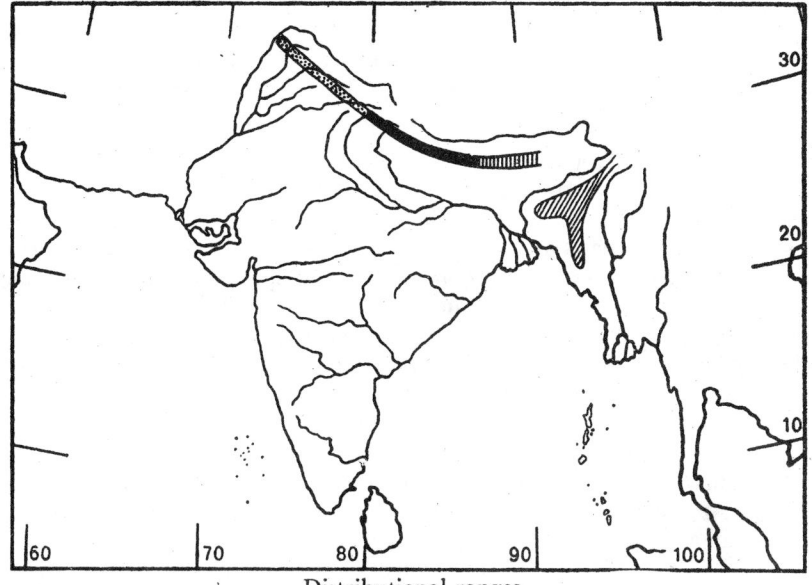

Distributional ranges

▒ *H. c. capistrata* (1396).   ■ *H. c. nigriceps* (1397).

▥ *H. c. bayleyi* (1398).   ▨ *H. gracilis* (1399).

above the ground, on average about seven or eight metres. *Eggs,* usually 3, sometimes 2, very distinctive: pale bluish grey, clouded, spotted and lined with various shades of brown. For further details see Baker, *Nidification* 1: 272–4. Average size of fifty eggs 25·5 × 18·3 mm (Baker). Both sexes share in nest building. Incubation apparently by female alone. The birds desert the nest on the least provocation—merely climbing the tree and peeping into the nest in some cases is enough.

MUSEUM DIAGNOSIS. Differs from *nigriceps* (1397) in being paler and having the centre of the back pale brownish grey. Also a little larger. Postnuptial moult in September-October.

Young (juvenile). Crown and sides of head brownish black with a tendency to pale shaft-lines. Rest of upperparts bright rufous, darker on upper tail-coverts and tinged with brown on mantle. Underparts paler rufous. Wing and tail much as in adult, but white on greater coverts not so extensive and tinged with buff. Outer tail-feathers narrower. Postjuvenal moult of body and probably lesser, median and greater coverts.

MEASUREMENTS

Wing11 ♂ ♂ 100–110; ♀ ♀ 99–110 (Kinnear)

COLOURS OF BARE PARTS. Iris reddish brown. Bill black. Legs and feet flesh-brown.

1397. *Heterophasia capistrata nigriceps*
(Hodgson)

Sibia nigriceps Hodgson, 1839, Jour. Asiat. Soc. Bengal 8: 38
(Nepal, restricted to central Nepal by Ripley, 1950, JBNHS 49: 399)
Baker, FBI No. 311 (part), Vol. 1: 296

LOCAL NAME. *Sibya* (Nepal).
SIZE. Bulbul +; length *c*. 21 cm (*c*. 8 in.).
FIELD CHARACTERS. As in 1396, but back rufous *v*. greyish brown.
STATUS, DISTRIBUTION and HABITAT. Common resident, subject to vertical movements. Nepal, meeting *capistrata* in the western extremity of its range and *bayleyi* in the eastern extremity (Fleming, JBNHS 65: 331–2). Breeds from *c*. 2000 to 3500 m, locally as low as 1600 m. Winters (October to March) between 1000 and 2700 m. Affects evergreen or deciduous forest in wet zone; a characteristic species of oak forest.
GENERAL HABITS, FOOD and VOICE. As in 1396.
BREEDING. *Season*, May to August. *Nest* and *eggs* as in 1396.
MUSEUM DIAGNOSIS. Differs from both *capistrata* and *bayleyi* in having the centre of back rufous, hardly tinged with brown.
MEASUREMENTS

	Wing	Bill (from skull)	Tarsus	Tail	
♂♂	88–102	22–24	*c*. 28	99–106	mm
♀♀	89–93	22–24	*c*. 28	94–99	mm

(BB, Rand & Fleming)

Weight 9 ♂ ♂ 37–47; 4 ♀ ♀ 36–41 g (GD).
COLOURS OF BARE PARTS. As in 1396.

1398. *Heterophasia capistrata bayleyi* (Kinnear)

Leioptila capistrata bayleyi Kinnear, 1939, Ibis: 752
(Taktoo, near Sakdan, E. Bhutan)
Baker, FBI No. 311 (part), Vol. 1: 296

LOCAL NAMES. *Sambriak-pho* (Lepcha); *Sesigona* (Bhutanese).
SIZE. Bulbul +; length *c*. 21 cm (*c*. 8 in.).
FIELD CHARACTERS. As in 1396, q.v. See Museum Diagnosis.
STATUS, DISTRIBUTION and HABITAT. Common resident, subject to vertical movements. Eastern Nepal from the Okhaldunga district east through Darjeeling, Sikkim, the Chumbi Valley (Tibet), Bhutan, and Arunachal Pradesh to the Dafla Hills; from *c*. 1800 to at least 2500 m in summer, descending to the foothills in winter. Affects tall moist-deciduous or evergreen forest.
GENERAL HABITS, FOOD and VOICE. As in 1396.
BREEDING. As in 1397.
MUSEUM DIAGNOSIS. Differs from *nigriceps* in having the back sooty brown tinged with grey.

MEASUREMENTS

	Wing	Bill (from skull)	Tarsus	Tail	
♂♂	88–101	22–23	28–31	102–107	mm
♀♀	83–92	21–23	27–30	90–109	mm
				(SA)	

Weight 25 ♂ ♀ 28–45 (av. of 15, 39·4) g–SA.

COLOURS OF BARE PARTS. Iris brown. Bill black. Legs and feet brownish flesh; claws horny brown.

1399. Grey Sibia. *Heterophasia gracilis* (Horsfield)

Hypsipetes gracilis Horsfield, 1839 (1840), Proc. Zool. Soc. London: 159 (Assam, restricted to Naga Hills by Koelz, 1954, loc. cit. below; but see Biswas, 1963, JBNHS 60: 683)
Leioptila gracilis ardosiaca Koelz, 1954, Contrib. Inst. Regional Exploration, No. 1: 7 (Mawphlang, Khasi Hills)
Leioptila gracilis dorsalis Stresemann, 1940, in Stresemann & Heinrich, Mitt. Zool. Mus. Berlin 24: 153 (Mount Victoria, Upper Burma)
Baker, FBI No. 313, Vol. 1: 298
Plate 80, fig. 17

LOCAL NAME. *Titi* (Angami Naga).
SIZE. Bulbul +; length *c.* 21 cm (8 in.).
FIELD CHARACTERS. *Above*, crown and ear-coverts black. Back and rump grey. Wing black, tertiaries grey. Tail graduated, grey with broad black outer edges and subterminal band; tip grey. *Below*, throat and centre of belly white; sides pale ashy; under tail-coverts buff. Sexes alike.
STATUS, DISTRIBUTION and HABITAT. Common resident. Eastern most Arunachal Pradesh, Assam in Cachar Hills, Meghalaya in Khasi Hills, Nagaland and Manipur, from 1200 m to the highest summits (1400 to 2800 m on Mt Victoria). Wanders down to about 900 m in the cold weather. A geographical representative of *H. capistrata*, differing from it only by the lack of any rufous pigment. Affects deciduous or evergreen primeval forest and, in the Khasi Hills, mostly pine forest.
Extralimital. Western and northern Burma to western Yunnan.
GENERAL HABITS. Arboreal, active and restless. Keeps in pairs or small parties according to the season, often in mixed company. Frequents the tops of trees, hopping along the branches, sometimes momentarily cocking tail like a magpie-robin, dodging in and out of clumps of epiphytes; only occasionally descending to bushes.
FOOD. Insects, berries and seeds; often visits the blossoms of various trees for nectar and insects.
VOICE and CALLS. A melancholy song of four flute-like descending notes, oft-repeated at short intervals; a variation consists of the first and second notes being doubled (Heinrich); this song appears to be very similar, if not identical with that of the Blackheaded Sibia. Alarm-note, a *churr* heard especially when going to roost (Stanford). Also various notes of a conversational character.

BREEDING. *Season*, chiefly May and June, extending into August. *Nest*, a deep, compact cup of moss-roots, leaves and fibres, covered exteriorly with green moss, lined wih rootlets or rhizomorphs. Usually placed in the crown of trees, especially pine, very difficult to find. *Eggs*, 2 or 3, one as often as the other, occasionally 4, pale bluish or greenish grey, lightly freckled with reddish brown or dark brown, more numerous at the large end. Average size of 60 eggs 23·9 × 17·7 mm (Baker). Building of nest, incubation and care of young by both sexes. Incubation period unknown.

MUSEUM DIAGNOSIS. See Field Characters.

Young, like adult but colours not so pure.

MEASUREMENTS

Wing ♂ ♀ 84–95; tail 101–112 mm (Stanford)
Wing 10 ♂ ♂ 89–96; 10 ♀ ♀ 87–95; tail 100–110 mm (Stresemann)
Bill from skull *c.* 24; tarsus *c.* 28 mm
Weight ♂ ♀ 34–42 g (SDR).

COLOURS OF BARE PARTS. Iris red, reddish brown to brown. Bill black. Legs and feet dark brown to black; soles yellow.

1400. **Beautiful Sibia.** *Heterophasia pulchella* (Godwin-Austen)

Sibia pulchella Godwin-Austen, 1874, Ann. Mag. Nat. Hist. 13: 160
(Kunho Peak, eastern Barail Range, Naga Hills)
Lioptila pulchella caeruleotincta Rothschild, 1921, Novit. Zool. 28: 38
(Shweli-Salwin Divide)
Lioptila pulchella nigroaurita Kinnear, 1944, Ibis 86: 83
(Lhalung, Pachakshiri dist., SE. Tibet, 7000 ft)
Baker, FBI No. 320, Vol. 1: 302
Plate 80, fig.15

LOCAL NAMES. None recorded.

SIZE. Bulbul +; length *c.* 22 cm (9 in.).

FIELD CHARACTERS. *Above*, head bluish slate; forehead, lores and line through eye black. Back and rump slaty. Wing with pale bluish outer edge and a conspicuous black shoulder-patch. Centre of tail dark brown with black edges and subterminal band; terminal band slaty. *Below*, uniformly slate-grey, paler than above. Sexes alike.

STATUS, DISTRIBUTION and HABITAT. Resident, subject to vertical movements. Southeast Tibet and Arunachal Pradesh from the Dafla to the Mishmi hills, Nagaland and Assam in the Cachar and Meghalaya in the Khasi hills (winter only ?); from 1200 to 2700 m in winter, breeding mostly above 2100 m and up to 3000 m. Affects mossy forest.

Extralimital. Southeastern Tibet, northern Burma and western Yunnan.

GENERAL HABITS. Keeps in pairs or small parties according to the season, sometimes in company with Hoary Barwings or other babblers. Frequents tall trees, feeding mostly on the lichen- and moss-grown trunks and branches, hopping actively along the big boughs, stopping abruptly at intervals, or moving about slowly on the outer branches. On a fleeting glimpse the quick jerky movements may be easily mistaken for a squirrel's. Often sits up in the top branches of a dead tree.

FOOD. Insects and their larvae, seeds and other vegetable matter.

Regularly probes into the blossoms of rhododendron, *Prunus, Magnolia,* etc. for nectar and insects, its forehead and breast sometimes becoming so thickly plastered with pollen as to make it look like a different species.

VOICE and CALLS. 'Often very silent but at times bursts into a bewildering variety of notes, one in particular resembling the jingling of a bunch of keys' (Stanford). Song likened to the call of a Redshank with three other notes added (Cranbrook); also described as a musical phrase of six notes in pairs, with a drop in pitch after each pair and also a drop after the first note (Smythies).

BREEDING. Known only from one nest taken in May, built on a horizontal branch of a small tree; it was cup-shaped, made of moss, lined with rootlets and placed nearly at the end of the branch (Field *apud* Baker, *Nidification* 1: 281). The single egg was a pale blue, unspotted, and measured 23·8 × 17·9 mm. This egg may not have been typical.

MUSEUM DIAGNOSIS. See Field Characters. Ear-coverts may be slaty or brown to black.

Young, like adult but slightly duller.

MEASUREMENTS

	Wing	Bill (from skull)	Tarsus	Tail	
♂♂	101–115	25	*c.* 30	113–118	mm
♀♀	92–100	—	—	100–108	mm
				(SA, SDR, Mayr)	

Weight ♂ ♀ 35–47 g (SDR).

COLOURS OF BARE PARTS. Iris brown. Bill black. Legs and feet brown; soles yellow.

1401. **Longtailed Sibia.** *Heterophasia picaoides picaoides* (Hodgson)

Sibia picaoides Hodgson, 1839, Jour. Asiat. Soc. Bengal 8: 38 (Nepal)
Baker, FBI No. 310, Vol. 1: 295
Plate 80, fig. 16

LOCAL NAMES. *Matcheo-pho* (Lepcha).

SIZE. Bulbul, with a long tail; length *c.* 30 cm (12 in.).

FIELD CHARACTERS. *Above,* dark slate-grey; wings blackish brown with a white patch. Tail very long and graduated, tipped with whitish. *Below,* uniformly grey, paler than back and still paler on belly. Under surface of closed tail appears barred black and grey. Sexes alike.

The slaty colour, extremely long graduated white-tipped tail and white patch on blackish wing make the bird easy to identify.

STATUS, DISTRIBUTION and HABITAT. Resident, locally common; some altitudinal movements with the seasons. The Himalayan foothills from central Nepal (not recorded since 1879 when Scully found it 'tolerably common about Nimboatar [central dun] in winter') east through Darjeeling, Sikkim, Bhutan and Arunachal Pradesh to the Mishmi Hills, thence south to Nagaland; from the base of the hills to *c.* 900 m in winter and probably somewhat higher in summer (up to 2000 m in northern Burma). Affects forest, open scrub with large trees or clearings in evergreen forest.

Extralimital. Extends to northeastern Burma. The species ranges east to Vietnam and south to the Malay Peninsula and Sumatra.

GENERAL HABITS. Keeps in pairs or in parties according to the season, sometimes numbering up to thirty or forty individuals. Frequents mostly the tree-tops. 'A curious blend of tree-pie, drongo and fantail flycatcher, swinging their tails, cocking them up over their backs, then crouching on a branch with outspread tail, or drifting from tree to tree following one behind another' (Stanford).

FOOD. Insects, flower-buds and seeds; feeds regularly on the nectar of *Salmalia, Erythrina, Prunus* and other flowers in company with drongos and Black Bulbuls, the forehead often becoming so thickly coated with whitish pollen as to give the bird an intriguingly novel look.

VOICE and CALLS. Song, a rich, whistling six-note phrase of thrush-like quality ending in *wheet-wheew*, reminiscent in cadence of *Alcippe poioicephala* but much louder (SA). Call-note, a high-pitched but not loud *tsip-tsip-tsip-tsip* uttered rapidly (Smythies).

BREEDING. Known only from a few nests found by Stevens. *Season*, April to June. *Nest*, a deep cup made mostly of moss, lined with rootlets or rhizomorphs and placed high up in pines on a horizontal branch, carefully hidden. *Eggs*, very similar to those of *H. gracilis*, pale grey-green marked with very small blotches of reddish brown, more numerous at the large end. Clutch size not given. Average size of 5 eggs 24·5 × 18·1 mm (Baker).

MUSEUM DIAGNOSIS. See Field Characters.

MEASUREMENTS

	Wing	Bill (from skull)	Tarsus	Tail	
♂♀	109–125	24–26	30–31	205–220	mm
				(Baker, SA)	

Tail 1 ♂ 199; 1 ♀ 188 mm (Mayr).
Weight 1♀ 46 g (SDR).

COLOURS OF BARE PARTS. Iris red or crimson, sometimes brown. Bill black. Legs and feet dusky grey; claws horny brown.

Subfamily MUSCICAPINAE: Flycatchers

For description see Van Tyne, J. & Berger, A. J., 1959: 526 (Family Muscicapidae)
Cf. Mayr, E., in Delacour, J. & Mayr, E., 1945, *Zoologica* 30 (3): 113
  Delacour, J., 1946, *Zoologica* 31 (1): 4
  Deignan, H. G., 1947, *Proc. Biol. Soc. Washington* 60: 165–8
  Vaurie, C., 1953, *Bull. Amer. Mus. Nat. Hist.* 100: 491–521 (Muscicapini)
  Ripley, S.D., 1955, *Auk* 72: 86–8 (Muscicapinae)

Key to the Genera

		Page
1	Rictal bristles very long	
	a Bill well hooked at tip .... *Rhinomyias*	135
	b Bill not hooked .... *Culicicapa*	201
2	Rictal bristles short	
	c Bill wide at base .... *Muscicapa*	136
	d Bill narrow and weak .... *Muscicapella*	200

FLYCATCHERS

Genus RHINOMYIAS Sharpe

Rhinomyias Sharpe, 1879, Cat. Bds. Brit. Mus. 4: 367. Type, by subsequent designation, *Alcippe pectoralis* Salvadori = *Rhinomyias umbratilis* (Strickland)

Olcyornis Baker, 1930, Fauna Brit. Ind., Bds. 7: 137.
Type, by original designation, *Cyornis olivacea* Hume
Cf. Vaurie, C., 1952, *Amer. Mus. Novit.*, No. 1570: 1–36

A rather heavily set flycatcher. Tarsus booted or with a vague trace of one or possibly two anterior scutes. Bill well ridged, notched and hooked at tip. Rictal bristles well developed.

1402. Olive Flycatcher. *Rhinomyias brunneata nicobarica* Richmond

Rhinomyias nicobarica Richmond, 1902, Proc. U.S. Nat. Mus. 25: 295
(Great Nicobar)
Baker, FBI No. 631a, Vol. 8: 627
Plate 81, fig. 13

LOCAL NAMES. None recorded.

SIZE. Sparrow; length *c.* 14 cm (6 in.).

FIELD CHARACTERS. A brown flycatcher of low shrubbery. *Above*, entirely brown; tail more rufous-brown; a pale eye-ring. *Below*, chin, throat and belly whitish; a pale brown band across breast; flanks brown. Sexes alike.

STATUS, DISTRIBUTION and HABITAT. So far known only as a common winter visitor to the Great and Little Nicobar Islands. Not known to occur north of Little Nicobar, but may be expected in the Andamans (see Butler, JBNHS 12: 401). Affects forest, sometimes gardens.

Extralimital. Winter range outside the Nicobars unknown. A migrant has been taken in the Strait of Malacca on 1 November. Breeding range unknown. The species as a whole breeds in eastern China (Guangxi to Zhejiang) and winters, so far as known, in the southern Malay Peninsula. This subspecies has a more rounded wing than the nominate race, suggesting that it may be less of a long-range migrant.

MIGRATION. All specimens from the Nicobars have been taken in March (but see 'Extralimital' above).

GENERAL HABITS. Keeps close to the ground, usually within three metres, in low bushes in heavy forest.

FOOD. Insects.

VOICE and CALLS. Song similar to the opening portion of that of Fantail Flycatcher, *Rhipidura a. albogularis*, but without the ending *to-tea, to-tea, to-tea* (Abdulali).

MUSEUM DIAGNOSIS. See Field Characters.

Specimens taken in March still show some ochraceous spotting of the immature on the sides of nape, tips of tertiaries and secondary coverts.

MEASUREMENTS

	Wing	Bill (from skull)	Tarsus	Tail	
♂♂	74–80	18	17	54–59	mm
♀♀	74–75	18–19	18	52–54	mm

(MD)

COLOURS OF BARE PARTS. Iris brown. Bill: upper mandible black, lower yellowish flesh. Legs and feet yellowish flesh.

Genus MUSCICAPA Brisson

Muscicapa Brisson, 1760, Orn. 1: 32; 2: 357. Type, by tautonymy, '*Muscicapa*', i.e. *Muscicapa striata*, ibid.

Niltava Hodgson, 1837, Ind. Rev. 1 (12): 650. Type, by original designation, *Niltava Sundara* Hodgson

Siphia Hodgson, 1837, Ind. Rev. 1 (12): 651. Type, by monotypy, *Siphia strophiata* Hodgson

Muscicapula Blyth, 1843, Jour. Asiat. Soc. Bengal 12: 939. Type, by subsequent designation, Gray, 1855: 52, *Muscicapa sapphira* Tickell = Blyth

Cyornis Blyth, 1843, Jour. Asiat. Soc. Bengal 12: 940. Type, by subsequent designation, Gray, 1855: 53, *Phoenicura rubeculoides* Vigors

Hemichelidon Hodgson, 1845, Proc. Zool. Soc. London: 32, *ex* Zool. Misc. 1844: 84, nom. nud. Type, by original designation, *Hemichelidon fuliginosa* Hodgson

Anthipes Blyth, 1847, Jour. Asiat. Soc. Bengal 16: 122. Type, by monotypy, *Anthipes gularis* Blyth = *A. moniliger* Hodgson

Ochromela Blyth, 1847, Jour. Asiat. Soc. Bengal 16, pl. 1: 121. Type, by monotypy, *Saxicola nigrorufa* Jerdon

Alseonax Cabanis, 1851, Mus. Hein. 1: 52. Type, by original designation, *Muscicapa undulata* Vieillot

Eumyias Cabanis, 1851, Mus. Hein. 1: 53. Type, by monotypy, *Muscicapa indigo* Horsfield

Bill depressed, wide at base. Rictal bristles usually well developed. Tarsus short. Young of all species with characteristic ochraceous spotting.

Key to the Species

		Page
I Plumage black and white .... *M. westermanni* ♂	164
II Plumage slaty above; throat and breast rufous .... *M. hodgsonii* ♂	163
III Back, rump and tail bright orange-rufous .... *M. nigrorufa* ♂♀	174
IV Throat and breast slaty with a rufous gorget .... *M. strophiata* ♂♀	158
V Throat white bordered with black................. *M. monileger* ♂♀	159
VI Plumage mostly brown................................. VIII	
VII Upper plumage mostly blue (of various shades)	

A Some rufous colour below
  i Throat blue or blackish blue
    1 Rufous restricted to breast .... *M. rubeculoides* ♂ 189
    2 Rufous extending to under tail-coverts
      a Entire throat blackish blue .... *M. sundara* ♂ 178
      b Chin and sides of throat blue; rufous of breast extending as a wedge to centre of throat .... *M. vivida* ♂ 181
  ii Throat rufous
    3 A white supercilium .... *M. hyperythra* ♂ 161
    4 No white supercilium
      c Upper tail-coverts brilliant ultramarine blue ..... *M. sapphira* ♂ 172
      d Upper tail-coverts dull indigo blue
        o¹ Blue of upperparts deeper, indigo .... *M. tickelliae* ♂ 192
        o² Blue of upperparts paler, more glaucous .... *M. banyumas* ♂ 191

FLYCATCHERS

 B No rufous on underparts
 iii A blue patch on side of neck
  5 Size large, wing over 90 mm .... *M. grandis* ♂ 175
  6 Size small, wing under 80 mm .... *M. macgrigoriae* ♂ 177
 iv No blue patch on side of neck
  7 Centre of throat white
   e Upperparts slaty blue, 3rd primary shorter than 4th ....
    .... *M. leucomelanura* ♂ 169
   f Upperparts cerulean blue, usually at least a trace of white supercilium; 3rd primary equal to 4th .... *M. superciliaris* ♂ 166
  8 Throat entirely blue
   g A white patch on tail
    o^3 White almost reaching tip of rectrices .... *M. concreta* ♂ 182
    o^4 White restricted to basal half .... *M. albicaudata* ♂ 198
   h No white on tail
    o^5 Whole plumage including belly verdigris or turquoise........
     .... *M. thalassina* ♂♀197
    o^6 Belly whitish
     a^1 Under tail-coverts barred..,.................. *M. unicolor* ♂ 188
     a^2 Under tail-coverts white
      b^1 Blue parts indigo; 4th primary equal to 5th *M. pallipes* ♂ 183
      b^2 Blue parts with a greyish green tinge; 4th primary shorter than 5th........................ *M. sordida* ♂♀195
VIII Upper plumage mostly brown (of various shades)
 A A blue patch on side of neck
  i Size small, wing under 70 mm .... *M. macgrigoriae* ♀ 177
  ii Larger, wing over 70 mm
   1 A white throat-patch .... *M. sundara* ♀ 178
   2 No white on throat .... *M. grandis* ♀ 175
 B No blue patch on side of neck
 iii A white patch on tail
  3 No throat-patch; whole plumage tinged with blue *M. albicaudata* ♀ 198
  4 A well-defined white throat-patch; white of tail reaching almost to tip .... *M. concreta* ♀ 182
  5 White of throat less pure, less well defined; white of tail more restricted
   a Above paler brown; 2nd primary equal to 6th ....*M. parva* ♀ 153
   b Above darker brown; 2nd primary equal to or shorter than 7th
    .... *M. subrubra* ♀ 156
  6 Throat rufous
   c Throat orange-rufous bordered with ashy ....*M. parva* ♂ 153
   d Throat rufous-chestnut bordered with black ...... *M. subrubra* ♂ 156
 iv No white on tail
  7 Underparts largely rufous or ochraceous
   e Throat whitish or buffish
    a^1 Throat white, tail rusty .... *M. ferruginea* ♂♀151
    a^2 Throat whitish, belly washed with ochraceous
     b^1 Tail rufous-brown .... *M.p. poliogenys* ♂♀185
     b^2 Tail washed with blue .... *M.p. vernayi* ♂ 187
    a^3 Throat ochraceous, belly and under tail-coverts white .....
     .... *M. rubeculoides* ♀ 189

            f Throat orange-rufous like breast
                a^4 Tail blue .... M. tickelliae ♀ 192
            a^5 Size small, wing under 70 mm
                b^2 Upper tail-coverts rusty.................... M. sapphira ♀ 172
                b^4 Upper tail-coverts olive-brown .... M. hyperythra ♀ 161
            a^6 Size medium, wing over 70 mm
                b^5 No grey on breast, tail brown .... M. banyumas ♀ 191
                b^6 Some grey on lower breast, tail chestnut ...... M. pallipes ♀ 183
        8 No rusty or rufous on underparts
            g Underparts olive-brown, throat ochraceous .... M. vivida ♀ 181
            h Tail bright rufous
                a^7 Size small, wing under 70 mm .... M. leucomelanura ♀ 169
                a^8 Size medium, wing over 70 mm .... M. ruficauda ♂♀ 148
            i Tail brown or rufous-brown
                a^9 Forehead and breast streaked .... M. striata ♂♀ 138
                a^{10} Throat olive-buff........................ M. hodgsonii ♀ 163
                a^{11} Throat more or less whitish
                    b^7 Throat pure white .... M. muttui ♂♀ 146
                    b^8 Size small, wing under 65 mm
                        c^1 Upper tail-coverts rusty .... M. westermanni ♀ 164
                        c^2 Upper tail-coverts grey-brown .... M. superciliaris ♀ 166
                    b^9 Size medium, wing over 65 mm
                        c^3 Bill over 16 mm .... M. unicolor ♀ 188
                        c^4 Bill under 16 mm
                            d^1 Underparts mostly brown; inner webs of tertials
                                pinkish buff .... M. sibirica ♂♀ 140
                            d^2 Underparts mostly whitish; inner webs of tertials
                                cream-buff......................... M. latirostris ♂♀ 143

1403, 1404. **Spotted Flycatcher.** *Muscicapa striata sarudnyi* Snigirewski

Muscicapa striata sarudnyi Snigirewski, 1928, Jour. f. Orn. 76: 595, new name for
    Butalis griseola var. *pallida* Zarudny, 1903, *nec Muscicapa pallida* Müller
            (Eastern Iran and Transcaspia)
            Baker, FBI No. 631, Vol. 2: 202 (= *neumanni*)
                    Plate 81, fig. 5

LOCAL NAMES. None recorded.
SIZE. Sparrow; length *c.* 13 cm (5 in.).
FIELD CHARACTERS. A pale greyish brown, characteristically upright-sitting flycatcher. Both in flight and at perch mistakable in the distance for Yellow-throated Sparrow (*Petronia xanthocollis*, 1949). A slight fork in the tail enhances the resemblance.
Above, drab brown, tail and wings darker; crown noticeably streaked with pale and dark brown. A whitish ring round eye. *Below*, whitish with dark streaks on throat and breast. Sexes alike.

× *c.* 1

M. sibirica is much darker, sooty grey on breast; *M. latirostris* has a more distinct whitish eye-ring and lores, and *M. ruficauda* has a rufous tail; none of these have a streaked crown.

STATUS, DISTRIBUTION and HABITAT. Uncommon summer (breeding) visitor in the inner mountains of Pakistan, from Astor, Baltistan, Gilgit and Chitral south to central and northern Baluchistan (Zhob), where it is more common; breeding from *c.* 2100 m to 3300 m. Affects open forest, especially pine. Also passage migrant (chiefly autumn) in some areas (see below).

Extralimital. Extends to Iran, Afghanistan, Transcaspia and Turkestan. The species ranges from western Europe to Mongolia.

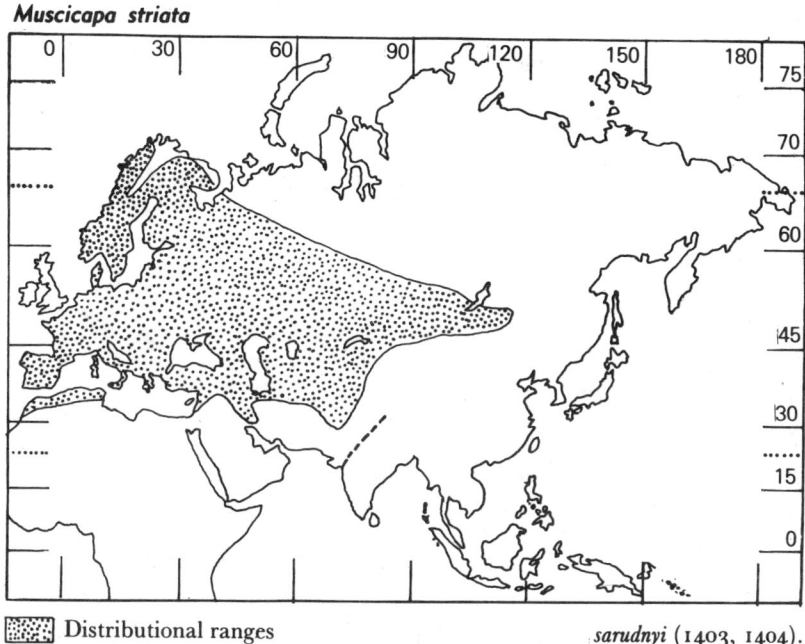

Distributional ranges *sarudnyi* (1403, 1404).
Broken line autumn passage extension

MIGRATION. Winters in east Africa. Recorded on migration as high as 4200 m (Sughet Pass, Kunlun Range) and in southern Sind and southern Baluchistan (Pakistan) and northwestern India as far east as Delhi, 9 15. ix. 1972; A. J. Gaston, JBNHS 75: 123), western Rajasthan, northern Gujarat (Deesa), Kutch and Kathiawar peninsulas. Spring passage migration takes place at the end of April and May (but mostly, if not entirely, west of the Indus, by-passing northwestern India). Occasional birds are sighted in Lower Sind (Pakistan) during February and March (TJR). Arrives on the breeding grounds in the first week of May, leaves in late August and early September. Autumn passage occurs through above areas from the last part of August till October. At this time, affects large trees, and open babool (*Acacia*) and kandi (*Prosopis*) jungle.

GENERAL HABITS. A quiet, solitary bird, usually perching on the lower branches of trees, seldom near top; makes typical sallies with rapid swerves and twists as it pursues a gnat; often returns to the same perch, or to another nearby. Frequently descends to the ground to pick up a crawling insect. Sits upright, head somewhat sunk between shoulders, tail slightly depressed like a bush chat's, and similarly switched loosely up and down from time to time, accompanied by flicks of the wings.

FOOD. Insects, chiefly Diptera.

VOICE and CALLS. Not recorded for this subspecies. Call-note of nominate race is a thin, shrill *tzee*; alarm-note *tzee-tzucc*. Song infrequently heard, half-a-dozen squeaky notes *sip, sip, sree, sreeti, sree, sip* and variants, with pause between each note (Witherby).

BREEDING. Within our limits (Chitral, N. Baluchistan—Ziarat, Quetta), *season* May and June. *Nest*, a cup of grass, moss and roots, lined with hair and rootlets. Placed on the branch of a tree or against the trunk, sometimes in crevices of stumps, a couple of metres from the ground. *Eggs*, 4 or 5, pale sea-green heavily blotched with pale reddish brown; variable (for further details see Baker, *Nidification* 2: 174). Average size of 100 eggs 18.3×14.2 mm (Baker). Building, incubation and care of young by both sexes.

MUSEUM DIAGNOSIS. See Field Characters. For details of plumage see Baker, loc. cit. Differs from *neumanni* in having a yellowish rusty wash on belly.

Young, spotted with buff; much paler than young of *M. sibirica*.

MEASUREMENTS

	Wing	Bill (from skull)	Tarsus	Tail	
♂♂	82–92	14–18	14–16	60–65	mm
♀♀	82–92	14–18	15–16	59–65	mm
				(HW, SA)	

Weight 4 ♂ ♂ 13–16; 1 ♀ 14 g (Paludan—Afghanistan).

COLOURS OF BARE PARTS. Iris brown. Bill black, brownish flesh at base of lower mandible. Legs and feet blackish.

MUSCICAPA SIBIRICA Gmelin: SOOTY FLYCATCHER

Key to the Subspecies

Paler, more grey .... *M.s. gulmergi*
Darker, more brown .... *M.s. cacabata*

1405. *Muscicapa sibirica gulmergi* (Baker)

Hemichelidon sibirica gulmergi Baker, 1923, Bull. Brit. Orn. Cl. 43: 155
(Gulmerg, Kashmir)
Baker, FBI No. 633, Vol. 2: 205

LOCAL NAMES. None recorded.

SIZE. Sparrow −; length *c*. 13 cm (5 in.).

FIELD CHARACTERS. A dark sooty-brown flycatcher, characteristically hawking in the uppermost tree crown by looping aerial sallies, often returning to the same perch and sitting bolt upright.

Adult. *Above*, entirely grey-brown with noticeably large eyes and a pale eye-ring. *Below*, greyish white, the throat patch showing paler and framed at the sides and down the pectoral region by darker ashy grey. Sexes alike.

× *c*.1

Young, sooty-brown spotted with buff and whitish, much darker than *M. striata* of corresponding age. Immatures of other flycatchers cannot be told apart except in certain cases by their behaviour.

M. striata adult is almost entirely white below. The very similar *M. latirostris* shows a bright yellow lower mandible (all dark in *M. sibirica*) and lacks the ashy grey area around the sides of the throat and pectoral region. *M. muttui* is distinguished by the same characters but has a diffuse dusky breast-band and rufous-brown tail. In *M. ruficauda* the tail is a brighter rufous. *M. poliogenys* and *M. ferruginea* have ochraceous underparts.

STATUS, DISTRIBUTION and HABITAT. An altitudinal migrant, generally common. From Tank Zam Valley, S. Waziristan (Hudson, JBNHS 27: 402), the Safed Koh, Kohat, Chitral and Gilgit east through Kashmir (very common), Spiti and Dhaola Dhar to Garhwal; breeds from *c.* 2100 to timber-line (*c.* 3300 m), locally down to 1500 m (Kulu—Whistler, JBNHS 31: 468); optimum zone 2400–3000 m. Affects open forest of pine, deodar, fir and birch, or oak and conifers.

MIGRATION. One of the latest migrants; arrives on its breeding ground in the first half of May and leaves in September. A few birds leave their summer zone immediately after the nesting period and appear at lower levels (*c.* 1200 m) already in the first week of July. Passage at mid-altitudes takes place mostly in April-May and September-October. Recorded as a migrant or straggler as high as 3900 m (Lahul). Winter quarters very imperfectly known (records from November to March are totally lacking); presumably in the foothills below *c.* 1200 m.

GENERAL HABITS. Very similar to those of the Spotted Flycatcher, but it is

Muscicapa sibirica

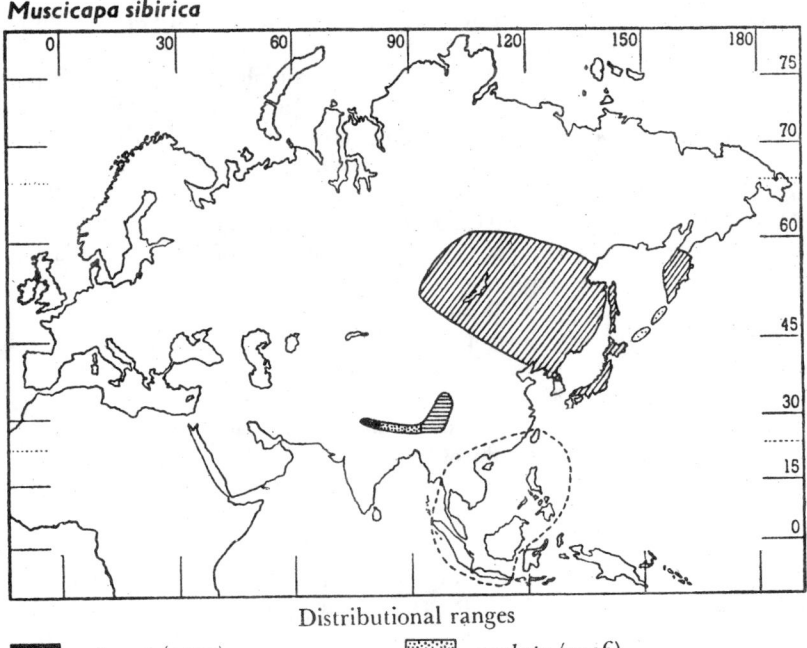

Distributional ranges

■ *gulmergi* (1405).   ▓ *cacabata* (1406).
☰ *rothschildi* (RE)   ▨ *sibirica* (RE).

Broken line: Winter range of *M. s. sibirica*

more of a forest bird. Particularly fond of forest glades and clearings littered with tangled brushwood, with gaunt tree-trunks and stumps standing here and there to serve as hunting bases. Felled or storm-blown trees likewise afford favourite perches. From the tip of a vertical dead branch or upturned root (now telegraph wires where available!) the bird flies out in all directions to take a midge—maybe from as much as ten metres away—by agile twists and turns and circles back, sometimes diverting to seize a second victim before regaining its base. Also launches vertical sorties skyward after the quarry, tumbling back to the perch after each capture. These quick-repeated forays continue throughout the day and often well into the dusk, frequently 3 or 4 individuals hawking within a restricted clearing. May often be seen hunting close to the ground especially after rain, when it has been observed taking food from the ground.

FOOD. Insects, mostly Diptera.

VOICE and CALLS. Usually very silent. Song, very high-pitched, thin, reedy, usually of three notes *tsee-see-see* in descending scale in half- or quarter-tones, with occasional variations; not unlike a thin edition of the song of a Greywinged Blackbird (Lister). Call-note unrecorded.

BREEDING. *Season*, May to July. *Nest*, a compact cup *c.* 8 cm across and nearly as deep, composed almost entirely of moss with lichen on the outside, consolidated with cobweb and lined with fine grass, fur or hair, with an occasional feather. Usually placed on a horizontal branch of a large spruce or Silver Fir where the leafy portion commences. Often in very exposed situations; sometimes against the trunk, tucked behind the out-flaking bark of a birch or similar tree; height above ground varies from two to eighteen metres. Often three or four occupied nests within a 50-metre radius. *Eggs*, 3 or 4, pale green, densely freckled with pale reddish especially around the large end. Average size 16 × 12·1 mm (Baker, Osmaston). Incubating female fed at nest by male whilst hovering momentarily against it. Both sexes share in tending the young; male's share in incubation, and incubation period, unrecorded.

MUSEUM DIAGNOSIS. See Field Characters. For distinction from *cacabata* see 1406.

Young (juvenile). Upperparts, wing-coverts and sides of head as in adult, but crown with central terminal ochraceous streaks; spots elsewhere. Underparts whitish with black margins, obsolete on throat and belly. Under tail-coverts pale ochraceous. Tertiaries with fulvous edges and tip. Tip of tail fulvous. Postjuvenal moult of body-feathers, lesser and median coverts, but here and there juvenal feathers are retained, with ochraceous spots faded to white.

First-year birds recognized by ochraceous tips of greater and primary coverts, fulvous edges of tertiaries and tip of tail.

MEASUREMENTS and COLOURS OF BARE PARTS. As in 1406.

1406. *Muscicapa sibirica cacabata* Penard

Muscicapa sibirica cacabata Penard, 1919, Proc. New Eng. Zoöl. Cl. 7: 22. New name for *Hemichelidon fuliginosa* Hodgson, 1844 (Nepal), preoccupied in *Muscicapa* by *M. fuliginosa* Sparrman, 1787, and *M. fuliginosa* Gmelin, 1789

Baker, FBI No. 632, Vol. 2: 204

Plate 81, fig. 7

LOCAL NAME. *Dang-chin-pa-pho* (Lepcha).

SIZE. Sparrow −; length *c.* 13 cm (5 in.).
FIELD CHARACTERS. As in 1405, q.v.
STATUS, DISTRIBUTION and HABITAT. An altitudinal migrant, locally common. From western Nepal east through Darjeeling, Sikkim to eastern Bhutan (Sakden) and western Arunachal Pradesh (Mago district—Ludlow, *Ibis* 1937: 279). Eastern limit unknown; probably occurs throughout Arunachal Pradesh; thence south of the Brahmaputra and possibly other hill tracts. Breeds from 2400 to 3300 m in Nepal and up to 3600 m in southeast Tibet; recorded at 3900 m at Thangu in Sikkim (Schäfer). Affects open conifer forest or clearings in oak forest.

Extralimital. The species ranges through western China to the Altai, Kamchatka and Japan and winters in the Indochinese countries and southern China eastward to New Guinea (see map, p. 141).

MIGRATION. Arrives on its breeding grounds in the first half of May, leaves in September. Spring passage takes place from the last week of March till May, return in September-October. As for *gulmergi*, winter quarters are poorly known. Winter visitor to the Jalpaiguri district and Meghalaya and, according to Rashid (1967), recorded in northern and northeastern Bangladesh and in the Chittagong region; there are no precise data. Has been obtained in the Khasi Hills, Shillong and Cachar Hills (Godwin-Austen) and near Aimole, Manipur at 1800 m (Hume).

GENERAL HABITS, FOOD and VOICE. As in 1405.
BREEDING. As in 1405, but nest also found in holes in trees. Average size of six eggs 17·1 × 12·2 mm (Baker).

MUSEUM DIAGNOSIS. Like *gulmergi* but rather darker brown; not well differentiated and not every specimen can be separated (HW).

MEASUREMENTS

	Wing	Bill (from skull)	Tarsus	Tail	
♂♂	70–76	11–12	10–12	45–51	mm
♀♀	70–76	11–12	10–12	47–52	mm
				(BB, SA)	

Weight 14 ♂ ♀ 8·5–11·5 (av. 9·6) g—SA. 3 ♂♂ 9–11; 3 ♀ ♀ 9–11 g (GD).
COLOURS OF BARE PARTS. Iris dark brown. Bill, legs, feet and claws black; soles grey.

1407. Brown Flycatcher. *Muscicapa latirostris* Raffles

Muscicapa latirostris Raffles, 1822, Trans. Linn. Soc. London 13: 312
(Sumatra)
Muscicapa grisola var. *daurica* Pallas, 1811, Zoogr. Rosso-Asiat. 1: 461
(Dauria)
Muscicapa Poonensis Sykes, 1832, Proc. Zool. Soc. London: 85 (Poona)
Butalis terricolor Blyth, 1847, Jour. Asiat. Soc. Bengal 16: 120, *ex*
Hodgson MS. (Nepal)
Baker, FBI Nos. 674, 675, Vol. 2: 248, 249
Plate 81, fig. 11

LOCAL NAMES. *Zakki* (Hindi); *Tăvittupăkshi* (Malayalam).
SIZE. Sparrow ±; length *c.* 14 cm (5½ in.).

FIELD CHARACTERS. An ashy brown flycatcher with a conspicuous white ring around the strikingly large eye. *Below,* sullied white, tinged with grey on breast and flanks, and faintly streaked with ashy brown. Throat conspicuously white. In the gloaming, when perched within the dark foliage canopy, glistening white throat often gives the only indication of its position. Sexes alike.

× *c.* 1

M. striata has distinctly streaked throat and crown. *M. sibirica* is much darker below and has the white on throat reduced to a patch. *M. muttui* is very similar but has a diffuse dusky breast-band, rufous-brown tail, and yellow legs (*v.* dark brown); also its tertiaries are edged with rusty (*v.* whitish plus a thin whitish wing-bar).

STATUS, DISTRIBUTION and HABITAT. A partial migrant having a disjunct breeding range. Movements imperfectly understood.

Himalayas: Uncommon summer visitor to Murree hill range (Pakistan) (T. J. Roberts, JBNHS 86: 139). Fairly common summer visitor to the western foothills in Chamba, Dharmsala, Dhaola Dhar and Kulu between 900 and 1800 m. Much scarcer east of these localities where it has been recorded from April to July in Mussooree, Ranikhet, Naini Tal (breeding at Thal, *c.* 1900 m, Almora dist., mid-May 1945—SA), Nepal, Jalpaiguri duars and Bhutan foothills. Said by Ward (JBNHS 17: 480) to be rare in Kishtwar (southwestern Kashmir) and 'probably not to be found west of the Chandra-Bhaga (river)'.

Vindhya Range: A common breeder in the hills around Sehore and Mhow. The breeding population appears to be of summer visitors only.

The Ghats: A scarce breeder in the southern part of the Western Ghats at about 900 m in north Kanara (Davidson), Coorg (Betts), the Palnis (JBNHS

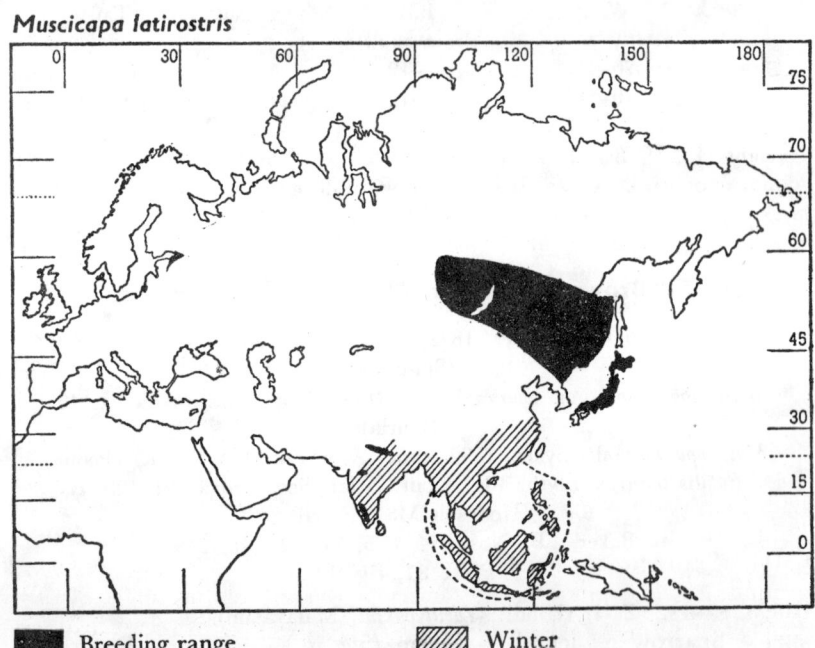

Muscicapa latirostris

Breeding range   Winter

55: 160) and Cardamom Hills (Sparrow, *in epist.* to Whistler; Jackson, 1971, JBNHS 68: 112–13). The resident breeding population is augmented by winter visitors. May breed also in the Chitteri Range (Eastern Ghats) where it has been obtained in June (Whistler, JBNHS 36: 85) and in the Biligirirangan Hills, obtained in July (Sálim Ali, JBNHS 43: 334).

Winters erratically but sparsely over most of the Peninsula but mainly in the Ghats of the southern Peninsula, from the lowlands to *c.* 1500 m (stragglers recorded in Kathiawar and Kutch). Also Sri Lanka (all zones) up to 2000 m where it is most common from October to March. Post-breeding dispersal of southern birds already takes place in August. According to Wright (JBNHS 54: 635) it is a winter visitor to the foothills of Dehra Dun from October to February. Winters also in the Andaman Islands (common), on Car Nicobar and Camorta (Abdulali, JBNHS 64: 183); apparently very scarce in Manipur (Hume), the Cachar Hills (Baker, JBNHS 9: 126) and the Garo Hills (Godwin-Austen). Recorded in winter from all parts of Bangladesh *fide* Rashid (1967, p. 101).

Affects open, mixed deciduous forest, teak plantations, groves about villages and cultivation, margins of jungle, overgrown nullahs, edges of sholas, bamboo jungle, shady gardens and coffee plantations; partial to the vicinity of streams. In the Himalayan foothills typically associated with sub-tropical pine, *Pinus roxburghii* (TJR).

MIGRATION. Arrives on its Himalayan breeding grounds in April, departs in September. May be seen on passage, mainly in September and October, in northern India east of Punjab (Ludhiana) and Gujarat (Sasan and the Gir Forest). Has occurred as a straggler south of Chimre (Ladakh) in September.

Extralimital. Also breeds in southern Siberia from the Yenisey to Sakhalin, Japan and Korea; winters in southeast Asia, the Philippines and Indonesia.

GENERAL HABITS. Closely resembles the Sooty Flycatcher in dress and habits; frequents mostly the lower branches of trees. Makes aerial sallies for insects from a dead branch, often returning to the same perch. Perches upright; flicks wings and depresses tail. More crepuscular than most flycatchers, a habit obviously correlated with its abnormally large eyes (SA). Usually solitary, quiet and very unobtrusive, therefore doubtless often overlooked.

FOOD. Chiefly dipterous insects.

VOICE and CALLS. A feeble *chi-chir-ri-ri-ri* (SA) or *chik, chik-r-r* (Henry). Three or four *chiks* in rapid succession and a small shrill squeak; a weak sub-song full of trills and squeaks heard in winter (Nichols). Primary song: a single, plaintive note slowly repeated; song period in Nepal: mid-April till end of June with a resumption in September (Proud). 'Has a loud note of 3 or 4 syllables and a short song freely uttered' (Whistler).

BREEDING. *Season,* May to July in the north, April to June in the south. Single-brooded. *Nest,* a compact cup made principally of moss, decorated with lichen and lined with rootlets, fine fibres and some feathers. Usually built on a bare, thick branch, some distance from the trunk, at heights varying from two to nine metres, on average about six metres from the ground. *Eggs,* normally 4, very similar to those of *M. rubeculoides* (1440) but more olive-grey, heavily stippled with sienna-brown. Average size of 40 eggs 17·4 × 13·1 mm (Baker). Incubation probably by female only (Dementiev), who is fed on the nest by the male. Incubation period undetermined.

MUSEUM DIAGNOSIS. See Field Characters. For details of plumage and moults see JBNHS 38: 298.

Young (juvenile). Upperparts and sides of head with very bold ochraceous white spots; underparts dull white on belly and under tail-coverts and spangled with dark brown margins elsewhere, heaviest on breast. Lesser and median coverts grey-brown with large ochraceous white tips. Greater coverts and tertials with fulvous white outer edges and tips. Tips of tail narrowly fulvous white. Postjuvenal moult of body, lesser and median coverts. First-year birds distinguished by fulvous white margins and tips of greater coverts. Juvenal plumage not distinguishable from that of *M. striata* (HW).

MEASUREMENTS

	Wing	Bill (from skull)	Tarsus	Tail	
♂♂	69–76	13–15	13–15	46–54	
♀♀	66–73	12–15	13–14	46–51	mm
Wing 1 ♂ 66 mm				(SA, HW, Koelz)	

Weight 50 ♂ ♀ (October) 9–14 (once 16) g—SA. 1 ♂ 11 g, 1♀ 10 g (SDR).

COLOURS OF BARE PARTS. Iris dark brown. Bill horny brown, base of lower mandible fleshy yellow, or cream colour to chrome-yellow; mouth pale yellow or pale yellowish pink. Legs, feet and claws horny brown.

1408. **Brownbreasted Flycatcher.** *Muscicapa muttui muttui* (Layard)

Butalis muttui Layard, 1854, Ann. Mag. Nat. Hist. 13: 127
(Pt. Pedro, Ceylon)
Alseonax muttui khosrovi Koelz, 1954, Contrib. Inst. Regional Exploration,
No. 1: 14 (Aijal, Lushai Hills)
Baker, FBI No. 677, Vol. 2: 251
Plate 81, fig. 6

LOCAL NAME. *Muttupilla* (Malayalam).

SIZE. Sparrow ±; length *c*. 13 cm (5 in.).

FIELD CHARACTERS. Largely as of Brown Flycatcher (1407) but with a diffuse brownish gorget or breast-band rendering white throat even more prominent by contrast. Legs yellow (*v.* horny brown). *Above*, olive-brown becoming rufous-brown on lower rump and tail. Wing-feathers with rusty edges. Lores and ring around the large eye white. *Below*, throat and centre of belly white; breast and sides pale brown. Sexes alike.

M. latirostris is very similar but is greyer both above and on breast, lacks the rusty edges on wing, and has dark legs (not yellow); the lower mandible is brown at tip (largely yellow in *muttui*). *M. sibirica* has a very dark breast. *M. ruficauda* has a rufous tail and lacks the white throat. *M. poliogenys* and *M. ferruginea* have ochraceous underparts.

STATUS, DISTRIBUTION and HABITAT. Breeding range poorly known both in India and outside. Breeds in the Khasi and Cachar hills of Meghalaya and Assam from 1200 m upwards, mostly above 1500 m (Baker). Obtained in Manipur in April (Hume), in the Garo and Mizo hills in March, and in Nagaland (Koelz). May be expected to breed in the eastern Himalayas:

Muscicapa muttui

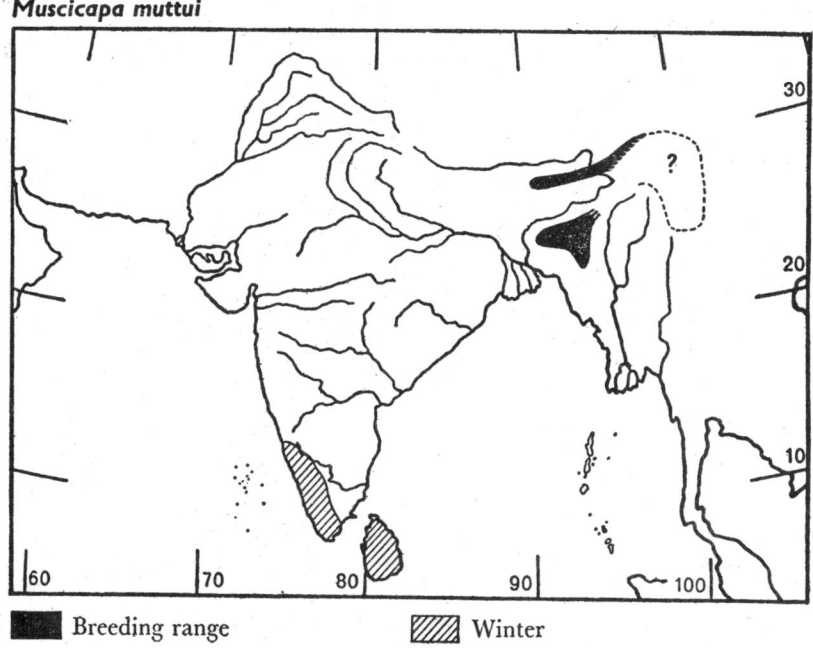

Breeding range   Winter

recorded in Sikkim from August to November (Mandelli, 1875, not noted since). Affects dense evergreen forest.

WINTER RANGE and MIGRATION. A population of this species winters in southwestern India from Londa south through the hills of Karnataka and Kerala, from the low country to *c*. 1500, commonest above 300 m, and in Sri Lanka in all zones up to *c*. 2000 m (scarce). In this season affects evergreen forest, frequenting dense scrub and liana tangles, preferably on the fringe of jungle, cardamon clearings and overgrown rocky streams. Arrives in its winter quarters in October and leaves in March-April, some individuals possibly staying till May (once end May in Kerala at *c*. 1700 m—Primrose, JBNHS 40: 502). Recorded on passage (September and October) at Madhupur (eastern Bihar), Nagpur (eastern Maharashtra) and in Andhra Pradesh (passage only ?). Rashid (1967) lists it as a passage migrant in most parts of Bangladesh.

Extralimital. Northern Burma, northern Thailand, Yunnan and Sichuan; probably breeds above 1500 m in N. Burma.

GENERAL HABITS. A very quiet and secretive flycatcher, thus easily overlooked, frequenting the low vegetation, mostly within a couple of metres above the ground. Affects denser forest than *latirostris*. Often sits motionless in the lowest branches of trees or tangles in shady recesses, every now and then making sallies after winged insects, returning to the same perch or a neighbouring one; occasionally descends to the ground. Usually solitary, but frequently in loose association with other flycatcher species. Appears very territorial and parochial; has been observed hunting in the same spot half an hour at a time. Like the Brown Flycatcher, it is also crepuscular, often hunting till late into the dusk.

FOOD. Chiefly dipterous insects.

VOICE and CALLS. 'In the breeding season it often gives vent to a soft low note, at the same time puffing out its feathers and rapidly vibrating its half-opened wings. Has a pleasant but rather feeble little song, very seldom uttered' (Baker).

BREEDING. The only records of nesting for the species in its entire range are by Baker (N. Cachar, *c.* 1500 m and above). *Season*, from the end of April till June. *Nest*, a very neat, compact cup made entirely of green moss and lined with fine moss-roots. Placed in bushes, creepers or other tangles of vegetation, not far from the ground. *Eggs*, 4 or 5, similar to those of the Brown Flycatcher: pinky brown or grey-blue stippled very closely with reddish brown. Average size of 28 eggs 16·9 × 13·8 mm. Incubation by both sexes; period undetermined.

MUSEUM DIAGNOSIS. See Field Characters. For details of plumage see Baker, loc. cit.

Young (juvenile). Upperparts covered with elongated ochraceous spots, very narrow and pointed on crown. Tips of wing-coverts, upper tail-coverts and edges of tertials and secondaries more rusty ochraceous. Throat and belly whitish; breast spotted like back but less distinctly so. Under tail-coverts buff. First-year birds recognized by rusty spots of greater coverts.

MEASUREMENTS

	Wing	Bill (from skull)	Tarsus	Tail	
♂♂	72–75	16–17	13–14	49–56	mm
♀♀	67–76	16–17	13–14	48–55	mm
				(HW, SA)	

Weight 13 ♂ ♀ (October, in winter quarters) 10–14 (av. 12) g–SA. 2 ♀ ♀ 12, – G (SDR).

COLOURS OF BARE PARTS. Iris dark brown. Bill dark horny brown, lower mandible fleshy with dusky tip; mouth pale lemon-yellow. Legs and feet pale yellowish flesh or yellow, claws dusky.

1409. Rufoustailed Flycatcher. *Muscicapa ruficauda* Swainson

Muscicapa ruficauda Swainson, 1838, Nat. Library, Flycatchers: 251
(India = Kashmir)
Baker, FBI No. 676, Vol. 2: 250
Plate 81, fig. 8

LOCAL NAME. *Chempuvalanpakki* (Malayalam).

SIZE. Sparrow −; length *c.* 14 cm (5½ in.).

FIELD CHARACTERS. A drab brown flycatcher with rufous tail, averaging larger in size than the similar *M. sibirica* (TJR). A noticeable pale eye-ring. *Below*, throat and breast grey-brown; belly whitish. Sexes alike.

Distinguished from *muttui* and *latirostris* by the lack of white on throat; from the other brown flycatchers by its rufous tail, and upper tail coverts (TJR).

STATUS, DISTRIBUTION and HABITAT. A summer (breeding) visitor, common at suitable altitudes in the western Himalayas; from the Safed Koh,

Chitral, Swat and east through Gilgit, Astor and the Indus Valley. It is an occasional winter visitor around Kohat (N.W.F.P.) (TJR). It occurs in Kashmir, east to Lahul and Kulu; much more scarce to the east: Garhwal and Kumaon (breeding), Nepal as far as the Kathmandu Valley [summer records by Lowndes (1955) and Rand & Fleming (1957)]. Breeds between 2100 and 3600 m, optimum zone 2400–3300 m. Affects mixed conifer and deciduous forest, fir or deodar forest, pine and birch.

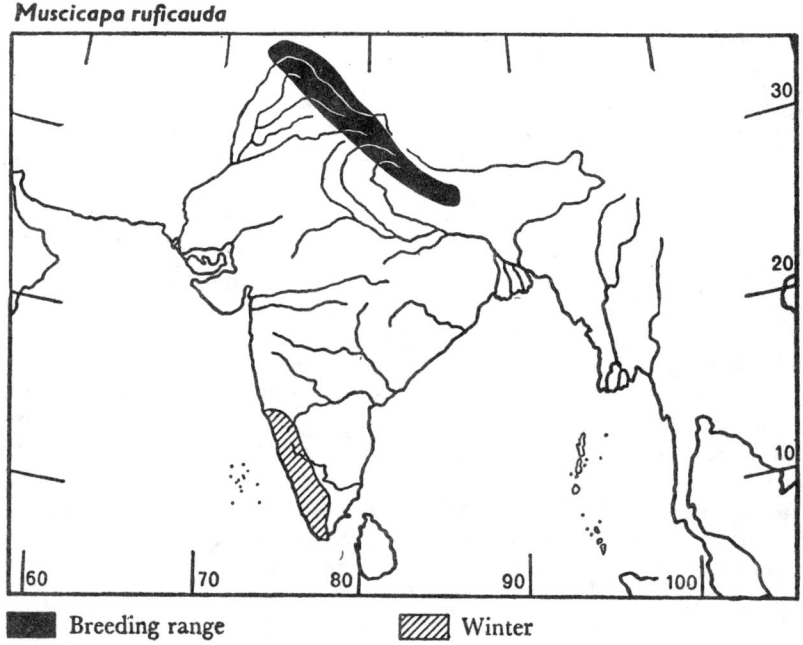

Extralimital. Russian Turkestan and adjacent Afghanistan.

WINTER RANGE and MIGRATION. Arrives on its summer grounds in early May, departs in August or September. A specimen obtained at Sholapur (Maharashtra) in July (Davidson & Wenden, SF 7: 81). Winters (October to April) in southwestern India from north Kanara through the hills of Karnataka and Kerala, from lowlands to *c.* 1000 m. Two records from Cachar, Assam (Baker), and listed by Rashid (authority?) as occuring in the northeast highlands and the Chittagong region of Bangladesh. In this season, frequents evergreen forest, especially along the margins, tracks and in clearings. During autumn migration (September and October) is frequently observed in the Himalayan foothills and has been recorded in southeastern Punjab, Mt Abu, Nagpur and along the Western Ghats. On spring passage in April, recorded at Raipur (eastern M.P.), Madhupur (eastern Bihar) and along the Himalayan foothills. Also obtained at Nellore (no date).

GENERAL HABITS. Solitary and very unobtrusive. Wanders about the crowns of large trees hunting mostly within the canopy rather than by sallies in pursuit outside; insect prey snapped up while flitting restlessly from one perch to another, or while fluttering about amongst the foliage and boughs.

Where it occurs sympatrically with *M. sibirica*, it seems to prefer to forage in the lower forest canopy, frequently descending to the ground (TJR). Flicks wings and bobs its body forward, thus reminding one of a chat. Frequents the edge of forest and does not seem to penetrate far into the woods.

FOOD. Insects, chiefly Diptera.

VOICE and CALLS. Song, three or four loud clear notes repeated at short intervals, varied from time to time in several ways, reminding one of the song of *Monticola cinclorhynchus* (Osmaston); some phrases rendered by Magrath as *tyee-trrirr, trrirr-tee* or *weetititew, ee-willu-willu* and *choi-choi* with an occasional, loud, finch-like *twoink-twoink;* all notes rapidly repeated and with a ventriloquial effect. Usually uttered from high up in leafy trees. While singing, the bird rarely stays in one spot for more than a few seconds. Alarm-note, a plaintive *peup* ceaselessly repeated and followed by a soft *churr* and sometimes a double *te-peup* when excited, as when an intruder approaches the nest (Bates & Lowther).

BREEDING. *Season*, May to July. *Nest*, a solid cup almost entirely made of moss with some bits of bark skin, liberally coated on the outside with lichen and cobweb, and lined with hair and a few feathers. The species is very versatile as to the site; the nest may be placed in bushes between one and two metres from the ground or in trees up to nine metres or more (generally under four metres); it may be built across a horizontal fork in the smaller branches, against the trunk, or in stumps of pollarded trees. It may also be found on the ground, particularly on very steep hillsides where it is sometimes placed at the foot of a small bush or in a depression in a nearly vertical bank (Bates). *Eggs*, 3 or 4, pale sea-green, densely freckled all over with reddish, chiefly at the broad end. Average size of 50 eggs 17·3 × 13·2 mm (Baker). The birds easily desert their nest. Their territory is very large, the male having been heard singing three or four hundred metres from the nest.

MUSEUM DIAGNOSIS. For details of plumage see Baker, loc. cit.

Young (juvenile). Upperparts grey-brown with ochraceous white spots; streaked on crown. Upper tail-coverts paler than in adults. Underparts and sides of head greyish white with blackish fringes, obsolete on throat and belly. Under tail-coverts buff. Median and lesser coverts with indistinct pale tips which are more marked on the greater coverts. Postjuvenal moult of body, lesser and median coverts.

MEASUREMENTS

	Wing	Bill (from skull)	Tarsus	Tail	
♂♂	73–77	14–16	c. 17	57–60	mm
♀♀	75–77	14–16	c. 17	55–58	mm
				(SA)	

Weight 13 ♂ ♀ (Apr.-May, in summer quarters) 12–16 (av. 13·3) g—SA.

COLOURS OF BARE PARTS. Iris dark brown. Bill: upper mandible brown, lower pale flesh-colour; mouth yellow and pink. Legs and feet brownish plumbeous; claws horny brown.

1410. Ferruginous Flycatcher. *Muscicapa ferruginea* (Hodgson)

Hemichelidon ferruginea Hodgson, 1845, Proc. Zool. Soc. London: 32 (Nepal)
Hemichelidon rufilata Swinhoe, 1860, Ibis: 57 (Amoy, Fukien, China)
Hemichelidon ferruginea russata Koelz, 1954, Contrib. Inst. Regional Exploration, No. 1: 13 (Kohima, Naga Hills)
Baker, FBI No. 635, Vol. 2: 206

Plate 81, fig. 9

LOCAL NAME. *Dang-chim-pa-pho* (Lepcha).

SIZE. Sparrow −; length *c.* 13 cm (5 in.).

FIELD CHARACTERS. *Above,* head blackish brown with a conspicuous whitish eye-ring. Back rusty brown. Lower rump ferruginous. Tail chestnut. Wings blackish brown, margins of tertials and tips of greater coverts conspicuous pale rufous. *Below,* chin, throat and centre of belly white. Breast olive-brown with dark centres to the feathers giving a spotted effect. Rest of underparts ochraceous. Sexes alike.

Distinguished from all other brown flycatchers except *M. poliogenys* (1436) by its rusty belly. The latter has an olive-brown back and tail, no rufous edges on wing, a whitish belly, a pale ochraceous breast and a less noticeable pale grey eye-ring.

STATUS, DISTRIBUTION and HABITAT. Probably only a summer visitor or altitudinal migrant in the central and eastern Himalayas from April to October; generally scarce, only noted as common (and breeding) by Osmaston in the Darjeeling district. Occurs from central Nepal (Sheopuri ridge, 2400 m—Proud, JBNHS 61: 534; Bigu, 3300 m east-central Nepal—Diesselhorst, 1968: 253) east through Darjeeling, Sikkim, Bhutan and Arunachal Pradesh (Upper Subansiri—Ludlow, *Ibis* 1944: 193) from 1800 to 3300 m, mostly above 2400 m. Also Meghalaya in the Khasi and Assam in the Cachar hills breeding mostly between 1200 and 1800 m (Baker), the Dibrugarh district (Cripps, SF 11: 106), the Mizo Hills and Nagaland near Kohima (Koelz, loc. cit. above) and Manipur near Aimole, 1800 m (Hume). Winter status unclear. According to Baker, apparently a resident in Assam while 'other birds scatter far and wide during the winter'; these may well be visitors from the Himalayas or extralimital regions. Obtained by Meinertzhagen at *c.* 1350 m in the Sikkim foothills in winter. Affects forest of fir or oak (Himalayas), humid stunted oak forest and dense mixed jungle (Assam).

Extralimital. Southeast Tibet, northern Burma and western China, wintering in the Indochinese countries, Burma, Malaysia and Indonesia.

MIGRATION. No data. Arrives on its breeding grounds in April. Obtained at Nellore in March, probably a straggler.

GENERAL HABITS. Typical of the brown flycatchers; frequents mostly the lower parts of the tall tree canopy, making sallies for insects from a bare branch returning to the same perch; sometimes doing this for hours. Very quiet, retiring and crepuscular.

FOOD. Chiefly dipterous insects.

VOICE and CALLS. Unrecorded. Very silent.

BREEDING. *Season,* May to July. *Nest,* a neat cup of moss and lichen, lined with rootlets; placed on projections, natural or caused by broken branches, near the trunk of a tree, or on a branch, between three and fifteen metres above the ground; very difficult to find as they are generally placed in or on the moss with which they are built. *Eggs,* 2 or 3, very much like those of *M. sibirica* (1405). Average size of nine eggs 17·9 × 13·6 mm (Baker). Incubation by both sexes; period undetermined.

MUSEUM DIAGNOSIS. See Field Characters. For details of plumage see Baker, loc. cit.

Young, upperparts brown with ochraceous spots; head more dusky with paler streaks. Underparts ochraceous with dark margins on breast; throat whitish with dark margins.

MEASUREMENTS

Wing ♂ ♂ 69–72; ♀ ♀ 63–70 mm (Koelz)
Bill from skull 11; tarsus 12; tail 47 mm (MD)
Weight 1♀ 12 g (GD).

COLOURS OF BARE PARTS. Iris brown. Bill blackish brown, flesh to yellow at base of lower mandible. Legs and feet pale flesh.

[The Pied Flycatcher *Muscicapa hypoleuca tomensis* Johansen may be a scarce migrant in Pakistan. According to Briggs (JBNHS 32: 749) 'a few birds indistinguishable from the British species pass through Peshawar almost every spring. Earliest and latest dates seen February 18 and March 30. The only year I have seen the species in autumn was in 1925 when a pair fed in my compound almost every day from October 31 till I left Peshawar on January 11.' No specimen has ever been obtained within our limits.

This subspecies breeds in western Siberia from the Urals to the Yenisey and migrates through southwestern Asia to tropical Africa.

FIELD CHARACTERS. Male in spring: *above* black with a white forehead and a large white wing-patch. *Below* entirely white.

In the field can be told at a glance from male of Little Pied Flycatcher (1419) by its conspicuous white forehead and absence of broad white superciliary stripe.

Female and male in autumn: very similar to *M. parva* (1411) female, but distinguished by the conspicuous wing-patch. Call-note almost constantly uttered on migration, a sharp *whit.*]

MUSCICAPA PARVA Bechstein: REDBREASTED FLYCATCHER

Key to the Subspecies

A Throat orange-rufous
  1 Rufous of throat meeting white of belly .... *M. p. parva*   ♂
  2 Rufous of throat separated from buff belly by
     ashy breast .... *M. p. albicilla*   ♂
B Throat whitish
  3 Breast pale brown .... *M. p. albicilla*   ♀
  4 Breast buffish .... *M. p. parva*   ♀

1411. *Muscicapa parva parva* Bechstein

Muscicapa parva Bechstein, 1794, Allg. über Vög. 2: 356 (Thüringerwald)
Baker, FBI No. 638, Vol. 2: 210
Plate 81, fig. 2

LOCAL NAMES. *Turra* (Hindi); *Yeepidippān* (Tamil).

SIZE. Sparrow —; 13 cm (5 in.).

FIELD CHARACTERS. Male (adult): pale brown above, more blue-grey on the crown, with white patches at either side of basal half of black tail prominently displayed by frequent upward flicks of tail; chin and throat orange-rufous, rest of under-parts white. Tail usually carried erect like a robin's, with the wings partly drooping at the sides.

Female has a whitish throat and buffy breast.

Male *M. p. albicilla* (1412) has a smaller throat-patch separated from the cream-coloured belly by a grey breast. Female and first-winter male not distinguishable in the field.

In adult male Kashmir Redbreasted Flycatcher (1413) the red area on throat and breast is more extensive, and bordered with black, whilst the crown is not grey as in *M. parva* but concolorous with the mantle; adult female and young male darker above, but not separable from those of *M.p. albicilla* in the field.

STATUS, DISTRIBUTION and HABITAT. A widespread winter visitor and passage migrant. Winters from the western Himalayan foothills (Chitral and Kohat to Dehra Dun) south to Baluchistan, Sind, through Gujarat (including Kathiawar and Kutch), Maharashtra and Karnataka; more

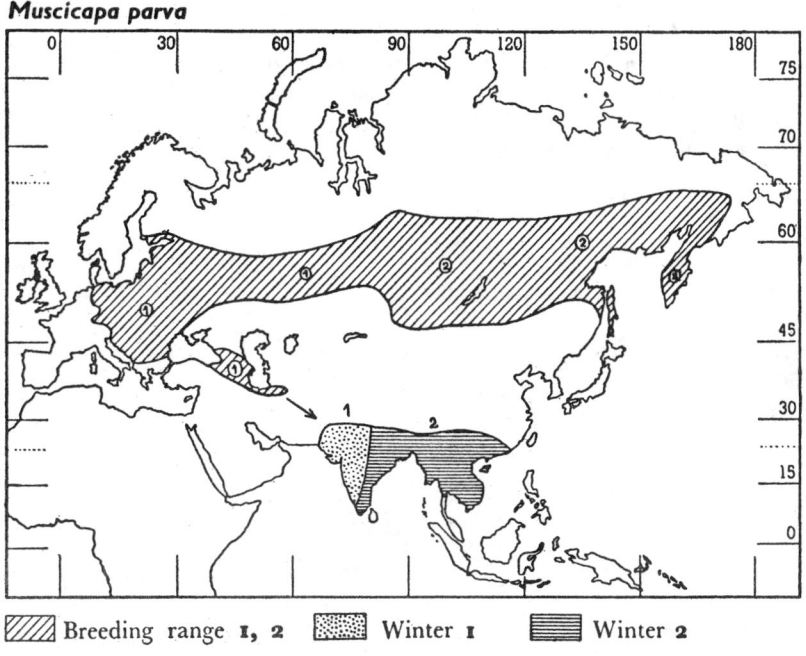

Muscicapa parva

////// Breeding range 1, 2   ▒▒ Winter 1   ≡≡ Winter 2

1 *parva* (1411).   2 *albicilla* (1412).

common in the western and central Peninsula but has been recorded in West Bengal, western Bangladesh (Rashid) and has straggled to the Jalpaiguri duars. In northern India and Pakistan it is more common as a passage migrant and has been recorded only as such in Ladakh, Kashmir and northern Baluchistan. Affects groves, forest plantations, orchards, urban gardens with large trees, etc. May be seen in scrub on migration.

Extralimital. Breeds in eastern Europe and Russia to the Urals and the southern Caspian region. On passage migration through Afghanistan in autumn and spring.

MIGRATION. Arrives in the northwest at the end of August but the bulk of autumn passage takes place in September and October and may last till November. Spring passage starts in March, reaches its peak in April, some individuals still being seen in May. Adult males are the first to leave, females and immatures following two or three weeks later.

GENERAL HABITS. Usually keeps singly, sometimes in company with other species. Makes short darts for insects from the lower branches of trees and flits from bough to bough, sometimes hovering momentarily before a flower or sprig. Frequently descends to the ground from an overhanging branch and with a couple of smart upward flicks of the cocked tail picks up an insect and flits back to its perch. Partially crepuscular, often hawking well into the dusk.

FOOD. Insects, chiefly mosquitoes and midges.

VOICE and CALLS. A double *tick-tick* or *click-click* uttered with a jerking of the tail and twitching of wings; a harsh, jarring alarm-note, and a plaintive *phwee-phwee-phwee* or *weeit, weeit, weeit* as of an unoiled bicycle wheel.

BREEDING. Extralimital; largely as in 1413.

MUSEUM DIAGNOSIS. Adult male differs from *albicilla* in having darker and more extensive rufous on throat. Female distinguished mainly by the dark brown, *v.* black tail.

First-winter male and female have brown ear-coverts; chin and throat whitish, breast and flanks greyish buff. Some first-summer males are similar to females while some have a variable amount of rufous on throat. Young in autumn differs from adult female by buff spots on wing.

MEASUREMENTS

	Wing	Bill (from skull)	Tarsus	Tail	
♂♂	64–72	12–14	17–18	47–54	mm
♀♀	64–71	12–13	*c.* 17	48–54	mm
				(HW, SA)	

Weight 3 oo? (October) 10–10·5 g (SA). 5 ♂ ♂ (spring) 10–13 g (Paludan, Afghanistan).

COLOURS OF BARE PARTS. Iris brown. Bill brown, paler at base of lower mandible. Legs and feet dark brown.

1412. *Muscicapa parva albicilla* Pallas

Muscicapa albicilla Pallas, 1811, Zoogr. Rosso-Asiat. 1: 462, Aves, pl. 1 (Dauria)

Baker, FBI No. 639, Vol. 2: 211

Plate 81, fig. 3

LOCAL NAMES. *Turra* (Hindi); *Chatki* (Bengali); *Yerra eda pitta* (Telugu); *Păttūk kūrūvi, Chūvăppūnencha kūrūvi* (Malayalam, Tamil).

SIZE. Sparrow −; 13 cm (5 in.).

FIELD CHARACTERS. As in 1411 but rufous restricted to throat; breast grey; tail black, also with white patches on either side of basal half. Female and young male not distinguishable.

Male Rustybreasted Flycatcher (1418) is slaty above; its female has the tail pale brown, nearly concolorous with the back, and lacks any white. Female Little Pied (1419) is similar but smaller, and also without any white on base of tail.

STATUS, DISTRIBUTION and HABITAT. A widespread winter visitor and passage migrant. Common in Arunachal Pradesh (Abor, Miri, Dafla and Patkai hills), Bhutan and Sikkim foothills, Nepal up to $c.$ 1500 m; Assam, Nagaland, Manipur, Bangladesh, West Bengal, Bihar, Orissa, eastern Madhya Pradesh, Andhra Pradesh, becoming scarcer farther south and west: recorded in Tamil Nadu, Rameswaram Island, Narcondam I, Andamans (Abdulali, JBNHS 71: 502), Kerala (up to 2100 m), Karnataka, Western Ghats in Maharashtra, Uttar Pradesh and, as a straggler, in the Kangra district. Has bred in the Dachgam Valley (Ladakh, 1800 m) according to Ward (JBNHS 17; 480). Affects groves, orchards, bush-and-scrub jungle.

Extralimital. Breeds from the Urals to Kamchatka. Winters in Burma, Thailand, and the Indochinese countries.

MIGRATION. Migrates mostly through China though some birds may migrate west of central Asia, as it has been recorded in Afghanistan. Arrives in its winter quarters in September, leaves in April; some belated migrants may still be seen in May.

GENERAL HABITS and FOOD. As in 1411.

VOICE and CALLS. As in 1411; a long, somewhat squeaky sub-song may be heard in winter (Nichols).

BREEDING. Said to have nested in Ladakh (Ward, above). The reported nest, taken on 28 May, was a tiny cup of green dry moss, lined with soft hair and placed beside a stone, half hidden in a shallow hollow in a stone wall. *Eggs*, pale sea-green or pale pink more or less profusely covered with pinkish brown. Size, $c.$ 17·5 × 12·8 mm.

MUSEUM DIAGNOSIS. Male differs from *parva* (1411) in having the rufous of throat pale and more restricted; breast ashy, belly buff. Females and first-winter males are distinguished from those of *parva* by the colder, greyer brown of the upperparts, and the black (*v.* dark brown) upper tail-coverts and tail; underparts whiter suffused with grey on breast, instead of creamy white suffused with buff.

Postjuvenal moult does not include tertiaries, greater and primary coverts, and wing-quills which remain from juvenile dress. This feature distinguishes juvenile from adult female (HW).

MEASUREMENTS

	Wing	Bill (from skull)	Tarsus	Tail	
♂♂	66–72	13–14	$c.$ 17	50–53	mm
♀♀	67–71	13–14	$c.$ 17	48–53	mm
				(HW)	

Weight 1 juv. o? (Sept.) 11·5; 1 ♂ 9 g; 1 ♀ $c.$ 11 g (SDR).

COLOURS OF BARE PARTS. Iris hazel-brown. Bill brownish black, lower mandible paler; mouth lemon-yellow and pink. Legs and feet blackish brown.

1413. Kashmir Redbreasted Flycatcher. *Muscicapa subrubra* Hartert & Steinbacher[1]

Muscicapa parva subrubra Hartert & Steinbacher, 1934, Vög. pal. Fauna, Ergänzungsband: 233. New name for *Siphia hyperythra* Cabanis, 1866, Jour. f. Orn. 14: 391 (Ceylon) nec *Muscicapa hyperythra* Blyth
Baker, FBI No. 140, Vol. 2: 212

Plate 81, fig. 4

LOCAL NAME. *Turra* (Hindi).

SIZE. Sparrow −; length *c*. 13 cm (5 in.).

FIELD CHARACTERS. M a l e (adult). *Above*, dark grey-brown; tail blackish brown, white on either side of basal half, this feature rendered conspicuous by constant flicking of the cocked tail. *Below*, throat, breast and upper belly deep rufous-chestnut edged with black; lower belly white.

F e m a l e (adult) darker than that of *M. parva* and with a variable amount of rufous on breast.

The black-edged rufous-chestnut underparts identifies the male. Female distinguished from that of other flycatchers by the blackish brown tail, darker than back, with white on sides of basal half.

STATUS, DISTRIBUTION and HABITAT. Breeds commonly in NW. Himalayas in summer, in Kashmir and the Pir Panjal range between *c*. 1800 to 2400 m; breeding recorded as high as 2700 m (Bates). Affects temperate mixed forest of walnut, cherry, willow, etc. especially where there is dense growth of *Perrottia* or *Corylus*. Winters in Sri Lanka from October to March throughout the hills above *c*. 750 m. October specimens assigned to this species were taken at Bumthang, Bhutan and Point Calimere, Tamil Nadu (Abdulali, 1985, JBNHS 82: 90). In passage, frequents gardens, tea estates, borders of forest, etc.

MIGRATION. Arrives on its breeding grounds in the second half of April, leaves in September. Migrates through the Peninsula in September and October, but very few records: Kangra district (Punjab), Dhulia (northwestern Maharashtra) and Secunderabad (Andhra). On spring passage, recorded at Darbhanga (Bihar), Kathmandu (Nepal—Proud) and near Mussooree. Obtained at Chakrata (Dehra Dun district) on 17 June (Meinertzhagen).

GENERAL HABITS. Very similar to those of *M. parva*. Takes a great part of its food when flitting from one perch to another under the shade of leafy foliage, mostly within about six metres from the ground. Appears to visit the ground more often in its winter quarters, hopping about in search of insects. Jerks its tail well above the back, at the same time flicking the wings and uttering a curious little creaking rattle.

---

[1] We agree with Whistler (1932, JBNHS 36: 81) that the characters of *subrubra* are sufficiently distinct from *parva* to separate the two as species. Desfayes points out (pers. comm.) that the situation is comparable to the species *Prunella immaculata* v. *P. modularis*, or *Erithacus pectoralis* v. *E. calliope* and deserves placing in a superspecies rather than being relegated to subspecific rank.

Muscicapa subrubra, M. westermanni, M. nigrorufa

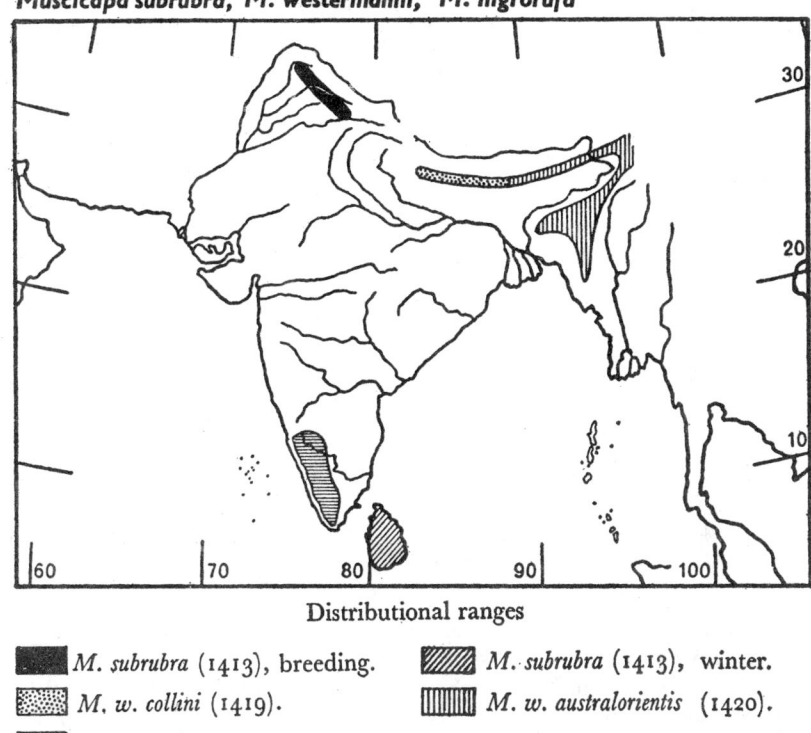

Distributional ranges

▪ *M. subrubra* (1413), breeding.   ▨ *M. subrubra* (1413), winter.
▨ *M. w. collini* (1419).   ▥ *M. w. australorientis* (1420).
▤ *M. nigrorufa* (1427).

FOOD. Insects.

VOICE and CALLS. A sharp *chack* while flitting about; a subdued but harsh *purr* accompanying the flicking of wings and tail. Full song in breeding season, sweet, loud and robin-like but short (Bates); in its winter quarters, in Sri Lanka described as consisting of a single whistled note *chip-chip-chip* followed by a rattling note, uttered quite freely (Henry).

BREEDING. *Season*, second half of May and June. *Nest*, made of dead leaves, moss and strips of bark, and lined with hair and a few feathers; placed in holes. Little holes in small trees, often only about one metre above the ground, are most favoured; other nests may be found up to twelve metres. *Eggs*, 4 or 5, sometimes 3, very pale green, speckled with pale reddish brown, mostly around the large end. Average size of 50 eggs 16·1 × 12·3 mm (Baker) and of 36 eggs 16·6 × 12·5 mm (Osmaston).

MUSEUM DIAGNOSIS. Adult male differs from that of *parva* not merely in the darker colour of upperparts and deeper chestnut of the breast with its black border, but also in the sequence of plumages and wing formula: 2nd primary (as.) equal to or usually shorter than 7th instead of always longer. First-winter male resembles adult male rather than female as in *parva* and *albicilla*. The young female likewise agrees in resembling the adult by having much rufous wash on the breast.

Young, much darker brown than in *parva*, but similarly spotted with buff; tail blackish instead of brown.

MEASUREMENTS

	Wing	Bill (from skull)	Tarsus	Tail	
7 ♂♂	65–70	12–13	18–19	49–54	mm
1 ♀	67	13	18	50	mm
				(HW)	

Weight 15 ♂ ♀ (Apr.-May) 9–12 (av. 10·4) g–SA.

COLOURS OF BARE PARTS. Male. Iris brown. Bill brown, lower mandible yellow; mouth yellow. Legs and feet sooty brown; soles dull olive. Female. Bill dark brown, basal half of lower mandible yellowish. Legs and feet black (HW).

1414. Orangegorgeted Flycatcher. *Muscicapa strophiata strophiata* (Hodgson)

Siphia Strophiata Hodgson, 1837, Ind. Rev. 1 (12): 651–2 (Nepal)
Muscicapa strophiata euphonia Koelz, 1939, Proc. Biol. Soc. Washington 52: 67
(Kulu, Kangra Dt., Punjab)
Baker, FBI No. 636, Vol. 1: 208
Plate 81, fig. 12

LOCAL NAMES. *Siphya* (Nepal); *Phatt-tagrak-pho* (Lepcha).
SIZE. Sparrow; length *c.* 13 cm (5 in.).
FIELD CHARACTERS. Male (adult). *Above*, forehead white; upperparts olive-brown. Tail black, white on either side of basal half. *Below*, an orange-rufous patch on lower throat surrounded by dark slaty on breast, sides of neck and head, and by black on upper throat and chin. Belly ashy, gradually becoming white on vent; flanks olive-brown.

× *c.* 1   Female (adult), frequently wears a plumage identical with that of male; in true female dress, has a paler and smaller gorget; chin and throat ashy.

Similar tail pattern to *parva* (1411) misleading at a flashing glimpse, but combined with white forehead always diagnostic.

STATUS, DISTRIBUTION and HABITAT. Resident, altitudinal and partial migrant, uncommon west of Nepal. The Himalayas from Kangra and the upper Indus Valley (Rupshu) east through Nepal, Sikkim, Bhutan and Arunachal Pradesh, then south through Nagaland and Manipur. Observed in May and June in the Barail Range around 1800 m. Breeds between 2400 and 3600 m in the western Himalayas, possibly higher in Nepal (recorded at 3950 m by Diesselhorst); from 2700 and 3700 m in Bhutan, and between 2500 and 3000 m in the eastern ranges south to Mt Victoria. Winters from the foothills to 2400 m and spreads out sparingly in the hills south of the Brahmaputra to the Chittagong Hill Tracts. Affects forest of oak, rhododendron, conifers, birch and mixed species; may be seen in high forest with little undergrowth as well as thick brushwood in open forest. In winter frequents shady forest, especially along the edges, by tracks and in clearings.

Extralimital. Western and northern Burma, N. Thailand and south-western China; another subspecies in central Vietnam. Winters in the Indochinese subregion.

GENERAL HABITS. Found singly or in pairs, usually flitting from bush to bush or launching aerial sallies from a bare branch or a fallen log. At times may be seen high up in trees or descending to the ground after an insect. Flight and habits likened to a robin's; also very similar to those of Redbreasted Flycatcher (1411, 1412). Constantly jerks up its cocked tail and flicks it open, flashing the conspicuous black-and-white pattern.

FOOD. Insects.

VOICE and CALLS. Alarm-note, a croaking *churr*. Call-note a low *tik-tik* or a single *pink* constantly uttered. Song, short, spirited, typically flycatcher —'a triple noted *tin-ti-ti*, the first syllable metallic and far carrying, the two others soft and audible only a few yards' (Proud).

BREEDING. *Season*, from the beginning of April (Mt Victoria) through May and June. *Nest*, a loose cup of moss and rootlets, lined with finer rootlets, papery bark, and a few feathers. May be placed on the ground on a sloping bank, or in a hole or crevice of a tree up to about three metres, sometimes up to six metres from the ground. *Eggs*, 3 or 4, white. Average size of 23 eggs 18·1 × 13·5 mm (Baker). Other details of breeding biology not recorded.

MUSEUM DIAGNOSIS. See Field Characters. Tail-feathers pointed as e.g. in Orange-flanked Bush Robin (1655). West Himalayan birds are slightly paler clinally.

Young, upperparts with ochraceous spots, narrower about the head; tail as in adult. Breast pale ochraceous with blackish edges, obsolete on throat. Lower belly whitish.

MEASUREMENTS

	Wing	Bill (from skull)	Tarsus	Tail	
♂♂	69–80 (Mostly 72–75)	12–14	19–22	53–63	mm
♀♀	67–77	12–14	19–21	55–58 (BB, SA)	mm

Weight ♂ ♂ 10–14; ♀ ♀ 11–13 g (SDR, SA). 7 ♂ ♂ 11–14; 4 ♀ ♀ 12–15 g (GD).

COLOURS OF BARE PARTS. Iris brown. Bill black. Legs, feet and claws horny brown.

MUSCICAPA MONILEGER (Hodgson): WHITEGORGETED FLYCATCHER

Key to the Subspecies

Supercilium fulvous .... *M. m. monileger*
Supercilium white .... *M. m. leucops*

1415. ***Muscicapa monileger monileger*** (Hodgson)

Dimorpha monileger Hodgson, 1845, Proc. Zool. Soc. London: 26 (Nepal)
Baker, FBI No. 669, Vol. 2: 244
Plate 81, fig. 21

LOCAL NAME. *Phatt-tagrak-pho* (Lepcha).

SIZE. Sparrow −; length *c*. 11 cm (4½ in.).

FIELD CHARACTERS. A short-tailed brown flycatcher with a white gorget bordered with black. *Above*, olive-brown; tail ferruginous; short, broad, fulvous supercilia almost meeting on forehead. *Below*, olive-brown with an almost triangular white throat-patch bordered with black; centre of belly whitish. Sexes alike.

STATUS, DISTRIBUTION and HABITAT. Scarce resident, from central Nepal (Rand & Fleming) east through Darjeeling, Sikkim, Bhutan, and into Arunachal Pradesh probably to the Dihang river; from the foothills to *c.* 2000 m. Affects dense bush jungle, scrubby ravines in broken country or thick undergrowth in tropical forest.

× *c.* 1

GENERAL HABITS, FOOD and VOICE. As in 1416, q.v.
BREEDING. Unknown; probably as in 1416.
MUSEUM DIAGNOSIS. Differs from *leucops* by the fulvous, not white, supercilium.
MEASUREMENTS

	Wing	Bill (from feathers)	Tarsus	Tail	
♂♀	60–65	*c.* 10	*c.* 23	45–50 (Baker)	mm

COLOURS OF BARE PARTS. Iris brown. Bill dark horny brown, paler at base (all black in breeding season). Legs and feet fleshy white; claws paler.

1416. *Muscicapa monileger leucops* (Sharpe)

Digenea leucops Sharpe, 1888, Proc. Zool. Soc. London: 246 (Shillong)
Baker, FBI No. 671, Vol. 2: 245
Plate 81, fig. 20

LOCAL NAME. *Inrephatki* (Kacha Naga).
SIZE. Sparrow −; length *c.* 11 cm (4½ in.).
FIELD CHARACTERS. As in 1415 but supercilia white.
STATUS, DISTRIBUTION and HABITAT. Uncommon resident, Arunachal Pradesh in the Mishmi Hills, south through Nagaland, Meghalaya, Manipur, the hills of Assam and the Chittagong Hill Tracts of Bangladesh; from 750 to 1800 m. Affects undergrowth in deep forest or dense bush-and-bamboo jungle.

Extralimital. Adjacent hills of western Burma. The species extends to northwestern Thailand; *M. solitaris*, often considered a subspecies, complements the range in the Malay Peninsula, Sumatra and the Indochinese countries.

GENERAL HABITS. A retiring, solitary flycatcher feeding in dense undergrowth and thickets, usually within a metre or so from the ground. Captures insects by making little sallies in the air, occasionally taking them on the ground. Often hops on the ground like a shortwing. Spreads and jerks its tail up when perched.

Display. Flies into the air and sails down to its perch, feathers all fluffed out and head thrown back, displaying thus its white gorget (Baker).
FOOD. Insects.
VOICE and CALLS. 'Has a weak but pleasant little song' (Baker) and some chattering notes (Ludlow).

BREEDING. *Season*, end of April and May. *Nest*, globular, loosely built of grass, leaves and sometimes a little moss, and lined with very fine grass stems. Usually placed on a sloping bank overgrown with vegetation, sometimes in dense bushes a few centimetres from the ground. *Eggs*, normally 4, white with small freckles and blotches varying in colour from pinkish red to reddish brown, forming a ring around the large end. Average size of 24 eggs 18·2 × 13·8 mm (Baker). Incubation by both sexes; period undetermined.

MUSEUM DIAGNOSIS. Differs from nominate *monileger* by the white (*v.* fulvous) supercilium.

Young. *Above*, dark brown, streaked with fulvous; *below*, dull fulvous, mottled on the breast with dark brown.

MEASUREMENTS

	Wing	Bill (from skull)	Tarsus	Tail	
3 ♂♂	63–66	—	—	—	mm (Stresemann)
1 ♀	62	14	20	45	mm (MD)

Weight 2 ♂ ♂ 12·3, 13·4 g; 1 ♀ 11 g; 10? 8 g (SDR).
COLOURS OF BARE PARTS. As in 1415.

1417. Rufousbreasted Blue Flycatcher. *Muscicapa hyperythra hyperythra* Blyth

M. (uscicapa) hyperythra Blyth, 1842, Jour. Asiat. Soc. Bengal 11: 885, new name for
D.(imorpha) superciliaris Blyth, 1841, Jour. Asiat. Soc. Bengal 11: 190
(Darjeeling), *nec M. superciliaris* Jerdon, 1840
Baker, FBI No. 643, Vol. 2: 217
Plate 82, fig. 5

LOCAL NAMES. None recorded.

SIZE. Sparrow −; length *c.* 11 cm (4½ in.).

FIELD CHARACTERS. Male (adult). *Above*, slaty blue with a conspicuous white supercilium. Forehead and cheeks black. Tail blue-black, white at base but this character not conspicuous. *Below*, chin black; rest orange-rufous.

May be easily confused with both the Indian Blue Chat (1650), slightly larger, and Rustybellied Shortwing (1636) about same size, which are almost identical in colour-pattern and behaviour but keep more to the ground. Blue Chat differs in having the supercilium longer, extending well behind eye; Shortwing has the supercilium equally short with the flycatcher but less prominent, and its legs are longer.

Female. *Above*, olive-brown; supercilium and eye-ring fulvous. *Below*, orange-rufous, duller than in the male. Some breeding females have no rufous on breast.

Distinguished from other similar-looking brown flycatchers and robins by the fulvous supercilium in conjunction with the rufous underparts.

STATUS, DISTRIBUTION and HABITAT. Resident, subject to vertical movements; the Himalayas from **Kumaon** (JBNHS 32: 796) through Nepal, Sikkim, Bhutan and Arunachal Pradesh to the Burma border, Nagaland,

Manipur, Assam in the Cachar Hills, and Meghalaya in the Khasi and Garo hills. Uncommon in Assam and west of Sikkim. Altitudinal distribution not satisfactorily known. Breeds between 2100 and 3000 m in the Himalayas, occasionally down to 1900 m (Sikkim), perhaps down to 1500 m according to Stevens. Arrives on its summer grounds in April; winters in the foothills up to *c.* 1700 m, in the terai, duars and over the plains of the Brahmaputra. In Assam, breeds mostly above 1200 m though Baker observed it in May, June and July around Margherita at 300–400 m. Affects dense primary forest with luxuriant undergrowth.

Extralimital. Adjacent hills of western and northern Burma, and northernThailand. The species ranges throughout the Indochinese subregion, Indonesia, Timor, Moluccas, Celebes and the Philippines.

GENERAL HABITS. Usually keeps singly or in pairs low down in bushes, scrub, and dense dank thickets—same facies as of *Tesia* and *Pnoepyga* wren-babblers—flits among the branches or runs about mouse-like mounting fallen tree-stems etc. flicking its tail and making sallies after midges. On the ground readily mistaken for a shortwing or chat.

FOOD. Chiefly dipterous insects; once an earthworm (Stevens).

VOICE and CALLS. A characteristic short sweet percussive song of *Cettia* quality (SA). Almost a wheeze, consisting of four notes, the first, second and fourth on the same pitch, third much lower; there is then a little pause followed by two notes resembling *see-saw*, the first note much higher than the second (Smythies).

BREEDING. *Season*, April to June. *Nest*, a small deep cup (*c.* 65 mm diameter × 65 mm deep) of green moss neatly lined with very fine moss roots or with black hair-like fibres and a few feathers; often untidily draped in skeleton leaves, fragments of bark, spiders' egg-cases, etc. looking like a bunch of flotsam rubbish. Placed generally in very wet and dense forest in holes in mossy banks, hollows in tree-trunks or between boulders, or amongst the exposed roots of a tree, usually low down (once *c.* twelve metres up); sometimes within a lump of moss hanging from a branch or amongst a tangle of leaves in a twiner. *Eggs*, normally 4, pale yellowish grey to deep pinkish red, marked with freckles of deep red-brown so numerous that the egg appears blotched or uniformly reddish brown. Average size of 40 eggs 17·5 × 13·8 mm (Baker); of 3 eggs of a c/4 collected in Bhutan 19·6 × 15·2 mm! (SA). Both sexes build the nest, incubate and feed the young; incubation period undetermined.

MUSEUM DIAGNOSIS. See Field Characters. Casually confusable even in the hand with *Brachypteryx hyperythra* (1636), but longer tarsus of latter, *c.* 28–30 mm (*v.* 18–20) is diagnostic.

Young. *Above*, dark brown with ochraceous spots. *Below*, dull ochraceous with dark brown edges to feathers; lower belly dingy white. Tail of young male blackish, with white much more extensive than in adult, covering the basal two-thirds of either side.

MEASUREMENTS

	Wing	Bill (from skull)	Tarsus	Tail	
♂♂	56–63	11–13	18–20	38–45	mm
♀♀	56–60	11–13	18–19	37–41	mm

(BB, SDR, SA)

Weight 12 ♂ ♂ 8–10; 7 ♀ ♀ 8–10 g (SDR, SA).
COLOURS OF BARE PARTS. Iris brown. Bill black. Legs and feet livid grey to fleshy; claws whitish grey.

1418. Rustybreasted Blue Flycatcher. *Muscicapa hodgsonii* (Verreaux)

Siphia hodgsonii Verreaux, 1871, Bull. Nouv. Arch. Mus. Hist. Nat.
Paris 6: 34 (Chinese Tibet = Paohing, eastern Sikang)
Muscicapa amabilis Deignan, 1947, Proc. Biol. Soc. Washington 60: 166.
New name for *S. hodgsonii* Verreaux, 1871, nec *Nemura hodgsoni* Moore, 1854
Baker, FBI No. 642, Vol. 2: 216
Plate 82, fig. 4

LOCAL NAME. *Paon-pali* (Tibetan).
SIZE. Sparrow −; length *c.* 13 cm (5 in.).
FIELD CHARACTERS. Male (adult). *Above,* slaty blue; lores and cheeks velvety black; tail blackish brown, with bases of all but central pair of rectrices white. *Below,* throat, breast and flanks orange-rufous; lower belly and under tail-coverts buffish.
First-winter male similar to female. Often breeds in this 'female plumage'.
Female. *Above,* olive-brown; a pale eye-ring. *Below,* olive-buff, whitish on belly.
May be confused with *M. parva* female but the latter has a whitish throat-patch and a black tail with white bases (*v.* plain brown). *M. leucomelanura* female is smaller and has a rufous tail. *M. westermanni* has a whitish throat and grey back. *M. poliogenys* has a whitish throat and fulvous breast.
STATUS, DISTRIBUTION and HABITAT. Altitudinal migrant, locally common. From central Nepal (Proud, Biswas) east through Darjeeling, Sikkim, Bhutan, Arunachal Pradesh to the Mishmi Hills, Nagaland, Manipur, Meghalaya in the Khasi Hills; breeds between 2100 and 3900 m, mostly 2400–3600 m; in the Khasi Hills above 1200 m. Winters (October to April) in the foothills up to *c.* 2000 m and apparently in the plains of Bangladesh *fide* Rashid. In the breeding season affects mostly pine or fir forest; in Meghalaya, oak and rhododendron; in the non-breeding season keeps to dense forest, thick scrub, forest of large bamboo and tall trees, etc.
Extralimital. Extends to southwestern China and Thailand.
GENERAL HABITS. A quiet and shy flycatcher. Catches most of its food in the air, circling out for insects from the canopy of trees; occasionally or rarely descends to the ground. Perches fairly upright.
FOOD. Insects.
VOICE and CALLS. Has a clear, pleasant, robin-like song (Schäfer); described by Ludlow as 'a constant ripple of whistling notes'.
BREEDING. *Season,* mid-April till July. *Nest,* a cup of green moss mixed with a few dead leaves, scraps of roots and lichen, and lined with fine rootlets or musk-deer hair. Usually placed in hollows between rocks or stones on steep banks covered with moss and ferns, sometimes low down in a moss-covered dead stump, or on the ground under some exposed root. As in *M. hyperythra,* it has also been found within a lump of moss hanging from a branch. *Eggs,* normally 4, rarely 5, pale green to warm buff, stippled all over

with light reddish. Average size of 40 eggs 17·8 × 13·4 mm (Baker). Incubation by both sexes; period undetermined.

MUSEUM DIAGNOSIS. See Field Characters.

Young. *Above*, dark brown with elongated ochraceous spots. *Below*, dull ochraceous with dark edges to the feathers, especially on breast; throat dingy white.

MEASUREMENTS

	Wing	Bill (from skull)	Tarsus	Tail	
♂♂	68–74	c. 12	c. 16	56–58	mm
♀♀	66–68	c. 12	c. 16	50–54	mm

(BB, Kinnear, Stevens)

Weight (one bird in brown plumage) 9 g (SDR). 1 ♂ 11 g (SA).

COLOURS OF BARE PARTS. Iris deep brown. Bill black. Legs and feet dark brown (Yellow in juvenile).

MUSCICAPA WESTERMANNI (Sharpe): LITTLE PIED FLYCATCHER

Key to the Subspecies
(Males not distinguishable)

Rump and edges of rectrices greyish .... *M. w. collini* ♀
Rump and edges of rectrices more reddish .. *M. w. australorientis* ♀

1419. *Muscicapa westermanni collini* Rothschild

Muscicapa collini Rothschild, 1925, Bull. Brit. Orn. Cl. 45: 90, substitute name for *Muscicapa blythi* Rothschild, 1921, Novit. Zool. 28: 48, nec *Muscicapa blythi* Giebel, 1875. New name for *Muscicapula melanoleuca* Blyth, 1843, Jour. Asiat. Soc. Bengal 12: 940 (Nepal, Darjeeling), nec *Muscicapa melanoleuca* Forster, 1817, or *Muscicapa melanoleuca* Güldenstädt, 1775

Baker, FBI No. 649 (part), Vol. 2: 224

Plate 81, fig. 14

LOCAL NAME. *Tuni-ti-ti* (Lepcha).

SIZE. Sparrow −; length c. 10 cm (4 in.).

FIELD CHARACTERS. A striking black and white, small flycatcher.

Male (adult). *Above*, black with a conspicuous white supercilium from lores to nape, a large white wing-patch, and white basal sides of tail.

Female. *Above*, olive-brown; a pale wing-bar. Upper tail-coverts bright rufous-brown. *Below*, throat whitish, rest smoky white.

A passable miniature of *M. ruficauda* (1409). Very similar to female *M. leucomelanura* (1424) which has a rufous tail, buffish breast and no wing-bar. Easily confused also with female *M. superciliaris* (1421) which is distinguished only by the grey upperparts and upper tail-coverts, and lack of wing-bar. Female *M. parva* (1411) has a blackish tail, white at base, and is a winter guest or passage migrant.

STATUS, DISTRIBUTION and HABITAT. Altitudinal and short-range migrant, not common. The Himalayas from Uttar Pradesh, Nepal to Sikkim and Darjeeling. Breeds between 1200 and 2500 m, locally up to 2700 m

(Langtang Valley, Nepal—Polunin), optimum zone probably around 2000 m. Winters (October to March) in the foothills up to *c.* 1800 m, and over the plains as far as (Pittie, 1986, JBNHS 83: 665), Surguja (northeastern M.P.), Manbhum (southern Bihar), Orissa (Simplipal Hills), Midnapore (West Bengal) and Bangladesh mostly west of the Jamuna and Ganges (see map, p. 157). Affects deciduous or evergreen forest on steep hillsides. In winter frequents the vicinity of well-wooded streams.

GENERAL HABITS, FOOD and VOICE. As in 1420.

BREEDING. Unrecorded. Probably as in 1420.

MUSEUM DIAGNOSIS. Male not distinguishable from that of *australorientis* (1420). Female tends to have a hazel-coloured shading on the forehead, lores and around the eye, merging into the brownish grey of the crown, nape and upper back. Rump and edges of rectrices greyish drab, lacking the rufescent tones of the eastern race.

Young. *Above*, with large, pale ochraceous spots. Wing-coverts brown with ochraceous tips, darker in male than in female; edges and tips of tertials fulvous white. Rest of wing and tail as in adult of corresponding sex. *Below*, white, the feathers fringed with black; belly and under tail-coverts white.

Postjuvenal moult includes body and all coverts except primary. First-year birds usually not distinguishable from adults unless there are a few unmoulted greater coverts.

MEASUREMENTS

	Wing	Bill (from skull)	Tarsus	Tail	
♂♂	53–61	12–14	14–16	38–47	mm
♀♀	53–60	12–13	14–16	40–43	mm
				(BB, SA)	

COLOURS OF BARE PARTS. Iris brown. Bill, legs and feet black.

1420. *Muscicapa westermanni australorientis* Ripley

Muscicapa westermanni australorientis Ripley, 1952, Proc. Biol. Soc. Washington 65: 72 (Phou Kobo, Laos)

Muscicapa westermanni indochinensis Ripley, 1952, Jour. Bombay nat. Hist. Soc. 50: 507. *Nom. nud.* (cf. JBNHS 51: 272)

Cyornis westermanni exquisitus Koelz, 1954, Contrib. Inst. Regional Exploration, No. 1: 14 (Karong, Manipur)

Baker, FBI No. 650, Vol. 2: 224

LOCAL NAME. *Dao-put-ti-ti* (Cachari).

SIZE. Sparrow −; length *c.* 10 cm (4 in.).

FIELD CHARACTERS. As in 1419, q.v.

STATUS, DISTRIBUTION and HABITAT. Altitudinal and short-range migrant, locally common, in Bhutan, Arunachal Pradesh, Meghalaya, Nagaland, Manipur and Assam in the Cachar Hills. Breeds between 1200 and 2400 m. Arrives on its summer grounds in March. Winters in the foothills and in the plains of Assam and Bangladesh south to Dacca and the Chittagong region (see map p. 157). Affects dense evergreen forest; in winter also frequents lightly wooded open country, cultivation and orchards.

Extralimital. Extends east to Yunnan and northern Vietnam. The species ranges throughout the Indochinese subregion, the Indonesian islands and the Philippines.

GENERAL HABITS. In the non-breeding season may be seen singly, in pairs or in small parties, often in company with other small insectivorous species. Keeps mostly among the crowns of trees, ever on the move from tree to tree, making little fluttering flights from one branch to another, catching most of its food on the wing but making few sorties outside the canopy; also takes insects from the crevices of bark and from the leaves. Flight easy but less agile than that of many flycatchers.

FOOD. Insects.

VOICE and CALLS. Song, a thin, high *pi-pi-pi-pi* followed by a low rattle *churr-r-r-r-r* or *pi-churr-r-r-r-pi-pi-pi-pi* (Smythies). The song may be heard even in winter (Baker). A sub-song is described as 'very soft warbling notes interspersed fairly frequently by equally quiet grating notes' (Lister). Call-note, a single, mellow *tweet*.

BREEDING. *Season,* April to June. *Nest,* a small cup of moss, moss-roots and stems of maidenhair ferns, compactly interwoven, the moss only showing outside, lined with very fine hair-like rootlets. Placed on the ground, in hollows between stones, or under a rock on steep hillsides or in holes among exposed roots of trees. *Eggs,* 3 or 4, warm buff, so densely covered with minute specks of dark reddish brown that the ground colour can hardly be seen. Average size of 30 eggs 15·1 × 11·9 mm (Baker). Incubation by both sexes; period unrecorded.

MUSEUM DIAGNOSIS. F e m a l e compared with ♀ *collini*, less tinged with hazel on the lores and forehead, and lacking any of this colour on throat and upper breast. Head and nape darker, dark mouse-grey in tone; back washed with light tawny-olive. Rump and outer edges of rectrices vary from russet to chocolate-brown, distinctly brighter and more reddish thus than *collini*.

MEASUREMENTS and COLOURS OF BARE PARTS. As in 1419.

Weight 5 ♂ ♂ 5–7; 3 ♀ ♀ 5–8 g (SA). 1 ♂ 10 g (SDR).

MUSCICAPA SUPERCILIARIS Jerdon: ULTRAMARINE FLYCATCHER

Key to the Subspecies

Base of tail white . . . . *M. s. superciliaris*
No white in tail . . . . *M.s. aestigma*

1421. *Muscicapa superciliaris superciliaris* Jerdon

M.(*uscicapa*) *superciliaris* Jerdon, 1840, Madras Jour. Lit. Sci. 11: 16
(Ajunta, N. Ghats)
Baker, FBI No. 647, Vol. 2: 221
Plate 81, fig. 15

LOCAL NAMES. None recorded.

SIZE. Sparrow −; length *c.* 10 cm (4 in.).

FIELD CHARACTERS. M a l e (adult). *Above,* deep blue with a white supercilium from eye to nape, and a white patch on either side of basal half of tail. *Below*, sides of head and neck deep blue; centre of throat, breast and whole belly white, the glistening white throat especially arresting.

× c. 1

Distinguished from the Slaty Blue Flycatcher (1423) by deeper blue upperparts, the blue on sides of breast forming a broken pectoral band, and—in the western Himalayas—by the conspicuous white supercilium.

Female. *Above*, mouse-grey; tail blackish, edged with blue. *Below*, sides of neck and breast greyish white; centre of throat, breast and whole belly glistening white.

Distinguished from the very similar females of *westermanni* (1419) and *leucomelanura* (1423) by the grey, not olive-brown back; the latter has a rufous, not blackish tail.

STATUS, DISTRIBUTION and HABITAT. Common summer visitor to the western Himalayas; from Kohat (Kurram Valley, but not recorded north of the Kabul river in N.W.F.P.) east through the Murree hills, Kashmir (south of the main range) to Kumaon. Intergrades with *aestigma* in Nepal. Breeds between 1800 and 3200 m, optimum zone 2100–2500 m. Affects open, mixed forests of oak, rhododendron, pine, fir, etc., occasionally orchards. Winters in central India from Delhi south to northern Maharashtra, Goa, southeastern Karnataka, northwestern Andhra Pradesh, Orissa, West Bengal and Bihar. Winter visitors to the last three States mentioned may well be, all or in part, from Nepal or Sikkim since a good percentage of this population also have a white supercilium and white basal portion of tail. A few birds winter in the Himalayan foothills. Also obtained at Bahawalpur, Pakistan (no date). In winter frequents various types of open deciduous forest, village groves, gardens and orchards.

MIGRATION. Arrives on its summer grounds in March and April, departs in September though some birds may still be seen in October while others leave at the end of August. Leaves its winter quarters in March.

GENERAL HABITS. Usually singly, in winter often in the mixed hunting parties. Keeps largely to the middle story of the forest (low trees and bushes) feeding mostly among the foliage canopy, not venturing much into the open. Occasionally descends to the ground. Constantly jerks up its tail, often accompanied by fluffing of head feathers and *trrr* note, especially in proximity of nest.

FOOD. Insects.

VOICE and CALLS. Territorial song, an oft-repeated *che-chi-purr* (Bates). During nesting season males have a very weak but prolonged song comprising short squealy phrases only audible at close range (T. J. Roberts). Call-note, a soft, repeated *tick*. Alarm-note, a *trrr*.

BREEDING. *Season*, middle of April to early July. *Nest*, a soft structure mainly of fine moss with some strips of bark or fine grass, lined with hair and rootlets; placed in holes or clefts in trees, often ivy-covered, at heights up to seven metres, or in a depression in a steep bank. One pair was found to have appropriated a nest of *Troglodytes*. Readily takes to nest-boxes in hill-station gardens. *Eggs*, 3 to 5, usually 4, olive greenish to dull stone-buff, densely freckled all over with reddish brown or, in another type, mostly around the large end, forming a cap. Average size of 100 eggs 16 × 12·2 mm (Baker). Building of nest and incubation by both sexes; period unrecorded. A pair in the Murree Hills occupied the same nest site for four successive years, repairing the old nest each time (T. J. Roberts).

MUSEUM DIAGNOSIS. Distinguished from the eastern population *aestigma* mainly by presence of white supercilium and white patches in the base of outer tail-feathers. See also 1422, under Museum Diagnosis. Owing to much intergrading single winter specimens of females of the two subspecies cannot be separated with certainty.

Young. *Above*, with large fulvous white spots. Wing-coverts blackish tinged with blue (♂) or brownish (♀), with fulvous white tips and edges of tertials. Wing and tail as in adult of corresponding sex. *Below*, whitish tinged with cream on breast, all feathers edged with black, producing a squamated effect.

MEASUREMENTS and COLOURS OF BARE PARTS. As in 1422.

Weight 1 ♂, 1 ♀ (Apr.-May) 8, 8 g—SA.

1422. *Muscicapa superciliaris aestigma* Gray

Muscicapa aestigma Gray, 1846, Cat. Mamms. Bds. Nepal: 90, 155. *Ex* Hodgson *in* Gray, *Muscicapa Astigma* [*sic*], 1844, Zool. Misc.: 84, *nom. nud.* (Nepal)

Cyornis superciliaris cleta Koelz, 1954, Contrib. Inst. Regional Exploration, No. 1: 14 (Mawphlang, Khasi Hills)

Baker, FBI No. 648, Vol. 2: 223

LOCAL NAME. *Tuni-ti-ti* (Lepcha).

SIZE. Sparrow −; length *c.* 10 cm (4 in.).

FIELD CHARACTERS. As in 1421 but lacks white supercilium and white base of tail; however, Nepal and Sikkim birds and those wintering south of these regions are intermediate and may or may not have any supercilium or white on tail. Males may be found breeding before acquiring full adult plumage.

STATUS, DISTRIBUTION and HABITAT. Uncommon summer visitor. Nepal, Sikkim, Bhutan and Arunachal Pradesh, from *c.* 2000 to 2700 m, locally up to 3200 m (Diesselhorst, 1968; Bailey, JBNHS 24: 75); also Meghalaya in the Khasi and Assam in the Cachar hills, Nagaland and Manipur, breeding mostly above 1500 m. Affects fairly open forest (oak and rhododendron, pine, etc.), and open, bush-covered areas, with or without trees. Himalayan birds winter in the plains as far as southern Bihar, West Bengal and all parts of Bangladesh (Rashid); a specimen has been obtained from Darjeeling in January. Movements of Assam populations unknown; probably also descend to the plains.

Extralimital. Extends to southeastern Tibet, Sichuan and Yunnan.

MIGRATION. No data. Probably as in 1421.

GENERAL HABITS, FOOD and VOICE. As in 1421.

BREEDING. As in 1421.

MUSEUM DIAGNOSIS. Adult male separable from the western race (1421) by the absence of white supercilium and white patches in the base of outer tail-feathers. However, there is intergradation over a wide range (the whole of Nepal and Sikkim), and birds from these areas may or may not have any white in tail. Moreover, most males from the Himalayas have at least traces of a white supercilium. Eastern birds are a little darker and particularly duller blue. Birds from Assam (Khasi Hills) show no trace of supercilium and no white in tail.

MEASUREMENTS

	Wing	Bill (from skull)	Tarsus	Tail	
♂♂	60–66	12–14	15–17	42–48	mm
♀♀	59–64	12–13	15–16	42–45	mm
				(BB, SA)	

Weight 1 ♂ 7·5 g; 1 ♀ 7·5 g (SDR).

COLOURS OF BARE PARTS. Iris dark brown. Bill black; mouth pinkish grey. Legs and feet brownish black; claws black.

MUSCICAPA LEUCOMELANURA (Hodgson): SLATY BLUE FLYCATCHER

Key to the Subspecies

		Page
A	Underparts white and grey .... *M. l. leucomelanura*	169
B	Underparts tinged with rufous	
1	Female darker above, more fulvous below .... *M. l. cerviniventris*	171
2	Female paler above, less fulvous below .... *M. l. minuta*	170

1423. *Muscicapa leucomelanura leucomelanura* (Hodgson)

Digenea leucomelanura Hodgson, 1845, Proc. Zool. Soc. London: 26 (Nepal = central hills of Nepal according to Gray & Gray, 1846, and Biswas, 1962, JBNHS 59: 812)
D.(igenea) tricolor Hodgson, 1845, Proc. Zool. Soc. London: 26 (Nepal). Preoccupied by *Muscicapa tricolor* Hartlaub, 1845, and *Muscicapa tricolor* Vieillot, 1818
  Cyornis tricolor notatus Whistler, 1930, Bull. Brit. Orn. Cl. 50: 70
    (Gund, Kashmir). [See Biswas, loc. cit. above.]
    Baker, FBI No. 645 (part), Vol. 1: 219
Plate 81, fig. 17

LOCAL NAMES. None recorded.
SIZE. Sparrow −; length *c.* 10 cm (4 in.).
FIELD CHARACTERS. Male (adult). *Above*, slaty blue, brighter on forehead. Tail black, conspicuously white on either side of base, displayed by constant flicking. *Below*, sides of head and throat dark slaty, almost black; throat white, rest of underparts greyish white.
Distinguished from *westermanni* by slaty blue (*v.* black) upperparts and absence of broad white supercilium.
Female. *Above*, olive-brown; tail rufous. *Below*, buffish.
Separable from the very similar females of *westermanni* and *superciliaris* by the rufous tail. *M. ruficauda* is larger and does not flick its tail. Male often breeds in 'female' dress.
STATUS, DISTRIBUTION and HABITAT. Common altitudinal migrant. Western Himalayas from Swat Kohistan, Indus Kohistan, just west of the Indus, and east to central Nepal. West of the Indus river there is a March record from Kohat. Breeds between 1800 and 4000 m, winters mainly below 1500 m. Affects mixed forest and well-wooded areas with plenty of undergrowth.
MIGRATION. Passage at mid-elevations is leisurely and takes place from the middle of February till the end of April; in autumn from the end of July

till the end of November. Obtained in Darbhanga, Bihar, 31.xii.1903 by C.M. Inglis.

GENERAL HABITS. More secretive and restless than most flycatchers; usually seen singly or in pairs, frequenting low undergrowth and lower branches of trees; takes much of its food on the ground. Both its demeanour and notes are reminiscent of the English Robin. Perches with wings drooping and often flicks its tail upwards.

FOOD. Insects.

VOICE and CALLS. A characteristic *ee-tick* of alarm is a good indication of its presence in undergrowth; also a rapid *ee-tick-tick-tick-tick*, each note accompanied by a tail-flicking. For description of song see 1425.

BREEDING. *Season*, May to July. *Nest*, a small cup of moss well lined with rootlets, hair and an occasional feather; placed in a depression in a bank, in crevices in boulders and trees, in a rift of bark, in a hole in a trunk or wall, occasionally against trunks; when not on the ground, usually within a couple of metres from it, rarely up to six metres. *Eggs*, 3 or 4, pale pinkish cream, minutely but densely speckled with pinkish red, sometimes forming a ring or cap around the large end. Average size of 52 eggs 15·6 × 12·1 mm (Osmaston) and of 100 eggs 15·8 × 12·1 mm (Baker).

MUSEUM DIAGNOSIS. See Field Characters. For distinction from the eastern race see 1424 under Museum Diagnosis.

Young, (juvenile). *Above*, olive-brown with pale ochraceous streaks. Upper tail-coverts rusty. Wing-coverts, except primary, with ochraceous tips. *Below*, dull creamy white with black margins, fainter on throat and belly. Postjuvenal moult takes place in August. First-year birds recognized by fulvous tips of coverts. First-year male resembles female.

MEASUREMENTS

	Wing	Bill (from skull)	Tarsus	Tail	
♂♂	57–63	*c.* 13	*c.* 18	49–56	mm
♀♀	54–60	*c.* 13	*c.* 18	43–55	mm
				(BB, Vaurie)	

Weight 13 ♂♂ (Apr.-May) 7.5–10 (av. 9.1); 11 ♀♀ (Apr.-May) 7–8 (av. 7·5) g—SA.

COLOURS OF BARE PARTS. Iris dark brown. Bill black. Legs and feet blackish brown.

1424. *Muscicapa leucomelanura minuta* (Hume)

Siphia minuta Hume, 1872, Ibis: 109 (Mount Tongloo, Sikkim)
Baker, FBI No. 645 (part), Vol. 2: 219
Plate 81, fig. 18

LOCAL NAMES. None recorded.

SIZE. Sparrow −; length *c.* 10 cm (4 in.).

FIELD CHARACTERS. As in 1423, q.v.

STATUS, DISTRIBUTION and HABITAT. Altitudinal migrant, locally common. Eastern Nepal through Sikkim, Bhutan and Arunachal Pradesh between 2700 and 4000 m, and Meghalaya in the Khasi Hills above 1500 m.

Winters in the foothills (up to 2100 m in central Nepal) and in the plains of the Brahmaputra from the Jalpaiguri district to the Noa Dihing. Khasi Hills population appears to be mostly resident or subject to short vertical movements. Affects forest with plenty of undergrowth. In Nepal it is the only flycatcher to occur regularly in the subalpine *Abies-Betula* forest and occupies a higher zone than *M. superciliaris*. The two species appear to be mutually exclusive in their vertical distribution (Diesselhorst). In the Khasi Hills it affects dense, wet forest of oak and rhododendron, or pine forest if undergrowth is sufficient. In winter often frequents thickets of reeds.

Extralimital. Southeastern Tibet, northern Burma to southwestern Szechuan. The species ranges east to northern Yunnan, north Kansu and Shensi; also Indonesia. Chinese populations winter south to northern Indochinese countries.

GENERAL HABITS, FOOD and VOICE. As in 1423.

BREEDING. As in 1423. *Eggs*, normally 3. Average size of 20 eggs 15·8 × 12 mm (Baker).

MUSEUM DIAGNOSIS. Differs from the western race (1423) in being more saturated and having the underparts tinged with rufous or olive (*v.* white and pure grey). Eastern Nepal population is intermediate. *Cerviniventris* is still darker above and much darker fulvous below.

MEASUREMENTS

	Wing	Bill (from skull)	Tarsus	Tail	
♂♂	59–64	12–13	19–20	47–54	mm
♀♀	55–60	12–13	19–20	45–49	mm
				(Vaurie, SA)	

Weight 4 ♂ ♂ 6–8; 7 ♀ ♀ 7–8; 1 juv. 9 g (SA. GD).

COLOURS OF BARE PARTS. As in 1423.

1425. *Muscicapa leucomelanura cerviniventris* (Sharpe)

Digenea cerviniventris Sharpe, 1879, Cat. Bds. Brit. Mus. 4: 460 (Manipur)
Baker, FBI No. 646, Vol. 2: 220

LOCAL NAMES. None recorded.

SIZE. Sparrow −; length *c.* 10 cm (4 in.).

FIELD CHARACTERS. As in 1423, q.v.

STATUS, DISTRIBUTION and HABITAT. Altitudinal migrant, locally common. Assam in the Cachar Hills, Nagaland, Manipur and the Chin Hills of Burma. Said to have been found breeding at *c.* 1500 m in the Chin Hills (Mackenzie *apud* Baker); however, like the Himalayan subspecies, it is probably a bird of higher altitudes since on Mt Victoria Heinrich found it only above 2500 m and up to over 3000 m in the breeding season. Winters in the foothills and adjacent plains. Affects low bushes along forest edges, especially dense reed-bamboo.

GENERAL HABITS and FOOD. As in 1423.

VOICE and CALLS. Song, a short, unmelodious strophe of three whistled notes, something like *zieh-ti-zietz* (Heinrich). See also 1423.

BREEDING. As in 1423.

MUSEUM DIAGNOSIS. Male hardly distinguishable from *minuta* though averaging darker fulvous below. Female darker above and on sides of head, and more fulvous below.

MEASUREMENTS

	Wing	Bill (from skull)	Tail
♂♂	55–59	11–12	46–51 mm
♀♀	52–54	11–12	44–49 mm

(Vaurie, Stresemann)

COLOURS OF BARE PARTS. As in 1423.

1426. Sapphireheaded Flycatcher. *Muscicapa sapphira* (Blyth)

Muscicapula sapphira Blyth, 1843, Jour. Asiat. Soc. Bengal 12: 939
(Darjeeling)
Cyornis sapphira coelicolor Koelz, 1952, Jour. Zool. Soc. India 4: 42
(Tura, Garo Hills)
Baker, FBI No. 651, Vol. 2: 225
Plate 82, fig. 2[1]

LOCAL NAMES. None recorded.
SIZE. Sparrow −; length *c.* 12 cm (4½ in.).
FIELD CHARACTERS. Male (adult). *Above*, forehead, crown, nape, lower rump and tail bright ultramarine blue. Lores and a line through eye black. Sides of head and back deep purplish blue. *Below*, chin, throat and upper breast orange-rufous; an interrupted breast-band deep blue. Belly ashy.

Male, first-year. Like adult but crown, back and sides of neck and breast dark olive-brown. Lores and eye-ring ochraceous. No breast band. Also breeds in this plumage.

Female. *Above*, rufescent olive-brown, upper tail-coverts rufous-brown. Lores and eye-ring ochraceous. *Below*, chin and throat orange-rufous. Sides of neck and breast olive-brown. Belly whitish, washed with fulvous.

The orange-rufous throat distinguishes the female from all other brown-backed flycatchers except the female of *M. hyperythra* which has a short, fulvous supercilium and the rufous of throat extending to breast. The male of *M. parva albicilla* also has a rufous throat, but ashy breast and black tail with white on base.

STATUS, DISTRIBUTION and HABITAT. Fairly common altitudinal migrant. East-central Nepal, Darjeeling, Sikkim, Bhutan, Arunachal Pradesh, Nagaland, Manipur, Assam in the Cachar and Meghalaya in the Khasi and Garo hills. Vertical distribution little known. Appears to breed between 2100 m (Nepal) and 2600 m (Nagaland—Ripley) in Meghalaya down to 1800 m, locally 1400 m. Winters in the foothills up to 1700 m chiefly under 800 m. Affects evergreen forest.

Extralimital. Extends to Yunnan and northern Laos.

GENERAL HABITS. Found singly or in pairs in high undergrowth and lower branches of trees. Hunts mostly within the foliage canopy, making sallies for

---

[1] First-year male.

insects. Occasionally descends to the ground. Droops wings at sides and flicks up tail from time to time.

FOOD. Insects.

VOICE and CALLS. Only record is a *tick-tick* accompanied by tail-flicking (SA).

Muscicapa sapphira

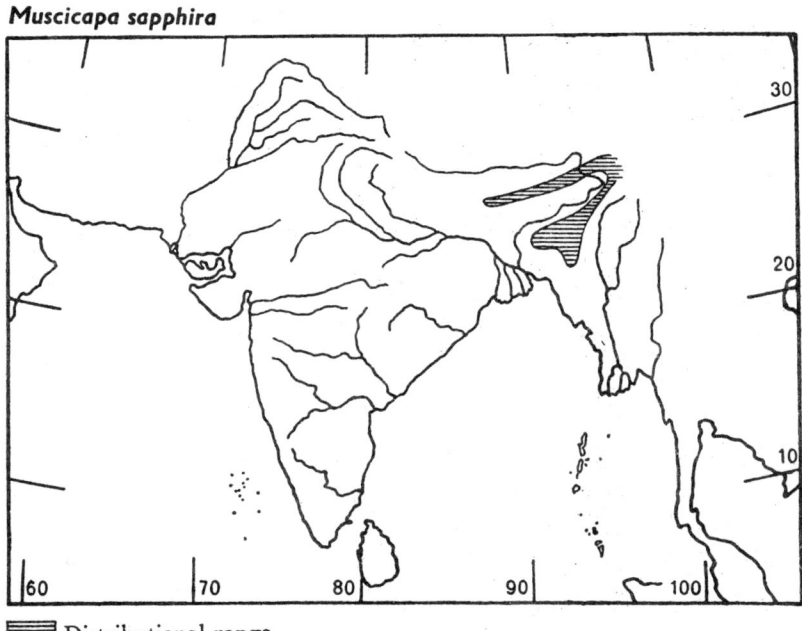

Distributional range

BREEDING. Described only by Baker. *Season*, end of April to June. *Nest*, placed in hollows in tree-stumps, sometimes in banks, rather bulky for the size of the bird, being sometimes as much as 75 or 100 mm in depth. Material principally moss mixed with some roots, lichen, fern-stems, etc., the neat cup within being made of finer moss, moss-roots and a few feathers. *Eggs*, 4, very variable: pale yellowish grey to warm buff; either faintly or densely stippled with reddish brown, sometimes so thickly as to make the egg look all of this colour. Average size of 16 eggs 15·4 × 11·8 mm. Incubation by both sexes; period unrecorded.

MUSEUM DIAGNOSIS. According to Ticehurst (*Ibis* 1939: 753) the males are dimorphic, one phase being that described as 'First-year male' under Field Characters. Males in this plumage appear to be more common than those in full dress.

Young (juvenile) male. *Above*, with ochraceous spots; upper tail-coverts bluish with ochraceous tips; greater coverts edged with blue and tipped with ochraceous. Postjuvenal moult includes body, lesser and median coverts. First-winter birds recognized by ochraceous tips of greater coverts. According to Whistler (MS.) some juveniles moult straight to adult plumage, others to the 'retarded' dress.

MEASUREMENTS

	Wing	Bill (from skull)	Tarsus	Tail	
♂♂	59–65	12–13	16–17	44–51	mm
♀♀	57–61	12–13	16–17	40–44	mm
				(SA, Kinnear)	

Weight 2 ♂ ♂ 7·5, 8 g (SA, SDR).

COLOURS OF BARE PARTS. Iris brown. Bill black. Legs, feet and claws brownish black.

1427. Black-and-Orange Flycatcher. *Muscicapa nigrorufa* (Jerdon)

Saxicola nigrorufa Jerdon, 1839, Madras Jour. Lit. Sci. 10: 266 (Nilgiris)

Baker, FBI No. 678, Vol. 2: 253

Plate 81, fig. 10

LOCAL NAME. *Mēnippǎkshi* (Malayalam).

SIZE. Sparrow −; length *c.* 13 cm (5 in.).

FIELD CHARACTERS. Male (adult). A diminutive orange-rufous flycatcher with slaty black crown, nape and sides of head, and wings.

Female similar but head dark olive-brown or olive-slaty with ochraceous lores and eye-ring.

STATUS, DISTRIBUTION and HABITAT. Resident, fairly common, but patchily distributed and evidently also moving about locally. The southern section of the Western Ghats and associated hills—Nilgiris, Palnis, Anaimalais, and others—from the Wynaad and the Biligirirangans south to the Ashambu hills; from 700 m to the highest summits, more common above 1500 m (see map p. 157). Affects dense, evergreen sholas with a plentiful undergrowth of *Strobilanthes* or *eeta* bamboo (*Ochlandra*), cardamom and edges of coffee plantations and rattan brakes in dank ravines.

× *c.* 1

GENERAL HABITS. Found singly or in pairs, hopping about amongst the seedlings and shady undergrowth, seldom far above the ground; or perches on a low branch or fallen log, making short sallies after insects, occasionally descending to the ground. Actions and behaviour distinctly reminiscent of a small babbler. Usually tame and easy to observe.

FOOD. Insects.

VOICE and CALLS. Alarm-note *zit-zit;* call-note, a melancholy low *pee* (Betts). Song, a somewhat metallic, high-pitched *chiki-riki-chiki* very insect-like, or *chee-ri-ri-ri* uttered every few seconds (SA).

BREEDING. Season, March to July. Nest, a loose, untidy ball of dead *eeta* leaves or coarse grass, sometimes lined with fine grass, sometimes unlined, very similar to that of *Rhopocichla atriceps* (1224). Usually placed in bushes, ferns or other plants within a metre from the ground, sometimes up to a couple of metres. *Eggs*, normally 2, pale greyish white or buffy white, faintly but profusely freckled all over with pale pinky grey or reddish; resembling those of the Verditer Flycatcher more than any others of the family. Average size of 30 eggs 18·4 × 13·1 mm (Baker). In one case observed, the female did all the collection of material but the male was in close attendance (Betts).

MUSEUM DIAGNOSIS. See Field Characters.

Young have never been collected but fledglings described from field observation by Dewar (JBNHS 16: 154) and Betts (JBNHS 50: 42): wings black; rump and tail russet as in adult; head, back and underparts speckled light and dark brown.

MEASUREMENTS

	Wing	Bill (from skull)	Tarsus	Tail	
♂♂	60–63	13–14	19–20	47–51	mm
♀♀	55–59	13–14	19–20	41–46	mm
				(SA)	

Weight 5 ♂♂ 0.35–0.4 oz. (c. 10–11 g); 5 ♀♀ 0.25–0.35 oz. (c. 7–10 g)—Davison.

COLOURS OF BARE PARTS. Iris brown. Bill blackish brown; mouth pale pink or brownish pink. Legs, feet and claws greyish flesh or greyish brown.

1428. Large Niltava. *Muscicapa grandis grandis* (Blyth)

Chaitaris grandis Blyth, 1842, Jour. Asiat. Soc. Bengal 11: 189
(Darjeeling)
Niltava grandis pangpui Koelz, 1954, Contrib. Inst.
Regional Exploration, No. 1: 14 (Sangau, Lushai Hills)
Baker, FBI No. 682, Vol. 2: 257

Plate 81, fig. 19

LOCAL NAME. *Margong* (Lepcha).

SIZE. Bulbul; length c. 20 cm (8 in.).

FIELD CHARACTERS. M a l e (adult). A large, rather sluggish, blacklooking flycatcher. *Above*, forehead, lores and sides of head black; crown, rump and a patch on either side of neck brilliant cobalt blue, but seldom noticeable in the dark shade in which it is usually seen. Rest of upperparts deep purplish blue. *Below*, throat and breast black; belly purplish blue.

Distinguished from all blue flycatchers by its black throat and breast, and lack of any rufous.

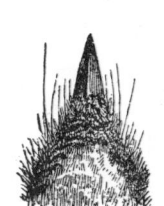

× c. 1

F e m a l e. *Above*, forehead, sides of head and back fulvous-brown. Crown, nape and sides of head olive-brown; a pale blue patch on either side of neck. Wings and tail rufous-brown. *Below*, throat fulvous; rest fulvous-brown.

The similar female of *M. sundara* is smaller and has a white throat-patch.

All species and both sexes of niltava flycatchers are readily diagnosed by the pale blue patch on sides of neck—the hall-mark of the group, as it were.

STATUS, DISTRIBUTION and HABITAT. Fairly common resident, subject to vertical movements. From central Nepal east through Darjeeling, Sikkim, Bhutan and Arunachal Pradesh to the Burma border, thence south through Meghalaya in the Garo and Khasi hills, Nagaland, Manipur, Assam in the Cachar Hills, and Mizoram in the Mizo Hills (Koelz); generally breeds between 1800 and 2700 m (Himalayas and eastern ranges south to Mt Victoria). Winters from the foothills to at least 2000 m. In the hills of Assam, breeds from 900 m to the highest summits. Affects dense, humid forest and secondary jungle on steep hillsides and ravines, especially in the vicinity of streams.

Extralimital. Extends east to the Indochinese countries and south through the Malay Peninsula. Another subspecies in Sumatra.

GENERAL HABITS. Found singly or in pairs in low undergrowth in the thickest and darkest patches of the forest where sunlight rarely filters through. Flits and skulks among low bushes and often feeds on the ground; occasionally takes insects on the wing, but it is less agile than most flycatchers and this is not its common practice.

FOOD. Insects and berries.

VOICE and CALLS. Song: a distinctive, rather mournful, ascending whistle of three or four notes ending interrogatively, thus *whee-whee-wip?* or *tee-ti-tree?* or *tee-ti-tiree?*, repeated unhurriedly and monotonously for several minutes (once timed seven) at a stretch (SA). Fleming considers this bird a beautiful singer and renders its song as *more time to eat* (do, re, re, mi) with other notes like *right here, t-z-z-z-t, ha-ha-ha-ha*.

BREEDING. *Season*, April to July. *Nest*, placed in crevices or holes and conforms to their shape; rather bulky, made chiefly of green moss; the neat cup within is lined with rootlets. Most nests built in among boulders, among the long moss growing on the face of rocks flanking streams and water-falls, against the trunk or among the buttress roots of large trees; occasionally in holes in dead stumps. Situated almost invariably in shady, evergreen forest. *Eggs*, usually 4, sometimes 5 or 3, very similar to those of *Cinclidium leucurum* (1681) but pale buff instead of pale pink, often with a finely speckled pattern of darker buff and a faint indication of a coronal ring on some eggs. Average size of 38 eggs 24·7 × 18 mm (Harrison & Parker, 1966, *Bull. B.O.C.* 86: 71–3). Individual eggs in the same clutch very variable in coloration (D'Abreu). Share of the sexes in the domestic chores, and incubation period, undetermined. Brood-parasitism on it by cuckoo, presumably *Cuculus canorus*, reported (D'Abreu, JBNHS 27: 405).

MUSEUM DIAGNOSIS. See Field Characters.

Young. *Above*, upperparts and sides of head with ochraceous spots and bold black margins to the feathers; lesser and median coverts with ochraceous terminal spot; greater coverts black (edged with blue in male) with tiny ochraceous shaft streaks at tips; primary coverts dull black. *Below*, squamated with large brownish yellow spots and narrow black margins.

MEASUREMENTS

	Wing	Bill (from skull)	Tarsus	Tail	
♂♂	100–112	18–20	23–26	87–100	mm
♀♀	97–105	17–20	23–25	86–91	mm

(BB, SA, Stresemann)

Weight 6 ♂ ♂ 35–40; 6 ♀ ♀ 33.5, 35, 38, 40.8, 42.5, 46 g (SA, SDR).

COLOURS OF BARE PARTS. Iris dark brown. Bill, legs, feet and claws black; soles greyish flesh.

MUSCICAPA MACGRIGORIAE (Burton): SMALL NILTAVA

Key to the Subspecies

Underparts darker, more grey .... *M. m. signata*
Underparts paler, more whitish on abdomen ... *M. m. macgrigoriae*

1429. *Muscicapa macgrigoriae macgrigoriae* (Burton)

Phoenicura macgrigoriae Burton, 1835, Proc. Zool. Soc. London: 152 (Himalayas, restricted to western Himalayas by Rand & Fleming, 1957, Fieldiana 41: 178)
Baker, FBI No. 685 (part), Vol. 2: 260
Plate 81, fig. 16

LOCAL NAMES. None recorded.

SIZE. Sparrow —; length *c.* 11 cm (4½ in.).

FIELD CHARACTERS. Male (adult). *Above*, forecrown, rump and a patch on either side of neck brilliant ultramarine blue. Forehead and lores black. Rest of upperparts deep purplish blue. *Below*, throat deep purplish blue; rest of underparts ashy, whitish on belly.

Female. *Above*, rufescent olive-brown. Wings and tail rusty brown. A pale blue patch on either side of neck. *Below*, throat fulvous; rest of underparts fulvous olive-brown.

Small size and blue patches on sides of neck distinguish both sexes.

STATUS, DISTRIBUTION and HABITAT. Fairly common resident, subject to vertical movements. Himalayas from Mussooree east through Nepal and Darjeeling (where it intergrades with *signata*); from *c.* 1000 to 2100 m in summer and from the foothills to *c.* 1400 m in winter. Affects bushes near streams, in shady glades or alongside roads and tracks through forest, generally not far from water.

GENERAL HABITS and FOOD. As in 1430.

VOICE and CALLS. A sub-song, described as a curious little grating song uttered *sotto voce*, heard from March to May (Proud). See also 1430.

BREEDING. As in 1430.

MUSEUM DIAGNOSIS. Underparts paler and belly more whitish than *signata* (1430).

MEASUREMENTS

	Wing	Bill (from skull)	Tarsus	Tail	
♂♂	62–67	*c.* 12	*c.* 18	49–54	mm
♀♀	61–65	*c.* 12	*c.* 18	46–55	mm
				(BB, SA)	

Weight 3 ♂ ♂ 11–13; 3 ♀ ♀ 12–13 g (GD).

COLOURS OF BARE PARTS. Iris brown. Bill black. Legs and feet brownish plumbeous; soles hoary.

1430. *Muscicapa macgrigoriae signata* (Horsfield)

Leiothrix signata Horsfield, 1840, Proc. Zool, Soc. London: 162
(Assam)
Baker, FBI No. 685 (part), Vol. 2: 260

LOCAL NAME. *Phak-tagrak-pho* (Lepcha).

SIZE. Sparrow —; length *c.* 11 cm (4½ in.).

FIELD CHARACTERS. As in 1429, q.v. See Museum Diagnosis.

STATUS, DISTRIBUTION and HABITAT. Common resident, subject to vertical movements. Sikkim (intergrading with nominate race), Bhutan and Arunachal Pradesh, thence south through the Patkai Range, Nagaland, Manipur, Assam in the Cachar Hills, and Meghalaya in the Khasi and Garo hills. Breeds from *c.* 900 to at least 2000 m; winters from *c.* 1400 m down to the foothills and in the plains of the Brahmaputra. Affects bushes along streams, shady glades and tracks in evergreen forest, and secondary scrub in forest clearings. In the plains, in winter, partial to mixed heavy reed and grass jungle interspersed with trees.

GENERAL HABITS. A retiring flycatcher, usually difficult to observe. Solitary in winter. Very sprightly and active, feeding almost entirely on the wing in typical flycatcher style among the upper bushes or undergrowth. More active in the early mornings and about dusk.

FOOD. Insects (ants, beetles, etc.); also berries in the non-breeding season.

VOICE and CALLS. Song, a four-noted strophe, very high-pitched and remarkably thin *twee-twee-ee-twee* rising in pitch to the second note, then falling (Smythies). Sub-song described as a curious little subdued grating song (Proud). Call-note, a very high-pitched *see-see*, the second note a quarter-tone lower (Lister).

BREEDING. *Season*, end of April to early July. *Nest*, usually placed in clefts in rocks well covered by vegetation, on the banks of streams; made of moss, the inner cup lined with moss and rootlets. *Eggs*, normally 4, sometimes 5 or 3, creamy white to pale greyish yellow, blotched or freckled more or less densely, more so at the larger end where forming an indefinite ring or cap. Average size of 100 eggs 18·1 × 13·6 mm (Baker). Both sexes take part in building though male seems to do little more than bring the materials. Incubation by both sexes; period about 12 days. Brood-parasitized by the cuckoos *Cuculus poliocephalus*, *C. sparverioides* and *C. fugax* (Baker, JBNHS 17: 353, 363, 368).

MUSEUM DIAGNOSIS. See Field Characters. Differs from the nominate race (1429) in having the breast and abdomen greyer, less whitish.

Young. *Above*, rufescent brown with ochraceous spots. Lesser, median and greater coverts with ochraceous tips. *Below*, fulvous brown with ochraceous spots; more whitish on abdomen. Postjuvenal moult of body-feathers and wing-coverts (except primary).

MEASUREMENTS and COLOURS OF BARE PARTS. As in 1429.
Weight 3 ♂ ♂ 11, 11·8, 13·3 g; 1♀ 11.7 g (SDR, SA).

MUSCICAPA SUNDARA (Hodgson): RUFOUSBELLIED NILTAVA

Key to the Subspecies

Male paler below, female paler above .... *M. s. whistleri*
Male darker below, female darker above........... *M. s. sundara*

1431. *Muscicapa sundara whistleri* (Ticehurst)

Niltava sundara whistleri Ticehurst, 1926, Bull. Brit. Orn. Cl. 46: 113
(Murree)
Baker, FBI No. 684 (part), Vol. 1: 259

LOCAL NAMES. None recorded.

SIZE. Sparrow; length *c.* 15 cm (6 in.).

FIELD CHARACTERS. As in 1432, q.v. See Museum Diagnosis.

STATUS, DISTRIBUTION and HABITAT. Fairly common resident, subject to vertical movements. The western Himalayas from the Murree hills to Kumaon. Occasional birds have been recorded in Hazara district in the lower reaches of the Kagan Valley (T. J. R.). Breeds between 1600 and 2700 m, optimum zone 1800–2400 m. Affects undergrowth in dense forest (chestnut, fir, etc.), jungle, and along nullahs. Found in winter from c. 650 m down to the foothills and adjacent plains (Hoshiarpur, Ambala and Rawalpindi). In this season frequents bushes near water and along roads in more open areas.

GENERAL HABITS. Usually keeps singly; in winter frequently with the mixed hunting parties. Bobs its body forward and flicks and spreads its tail every few seconds. Hunts in low undergrowth close to the ground, freely descending to it for food. Both Magrath (JBNHS 19: 148) and Whistler (*Ibis* 1930: 95) note its very Blue Chat-like habits.

FOOD. Insects (ants and beetles recorded). Also berries, especially in the non-breeding season.

VOICE and CALLS. Song described as squeaky and grating (Proud). It is short and monotonous and can be rendered as *sweeee-ch-tri-tr tik*, the last note being emphasized (Roberts), or *s-i-i-i-i-f cha chuck* (Fleming). Alarm-note, a harsh, scolding *tr-r-r-tchik* (SA), and a long-drawn squeak in the neighbourhood of the nest (HW). Other notes, a soft *cha . . . cha* low-pitched and insistent, and a soft, falling *pea . . . pea* (Lister). Also a high-pitched *tzi, tzi, tzi* (Loke).

BREEDING. *Season* April to July, chiefly May and June. Young in nest have been found as late as August. *Nest*, usually placed in a hole in a bank or roadside cutting, in clefts or crevices of rocks, or in cavities in dead stumps,

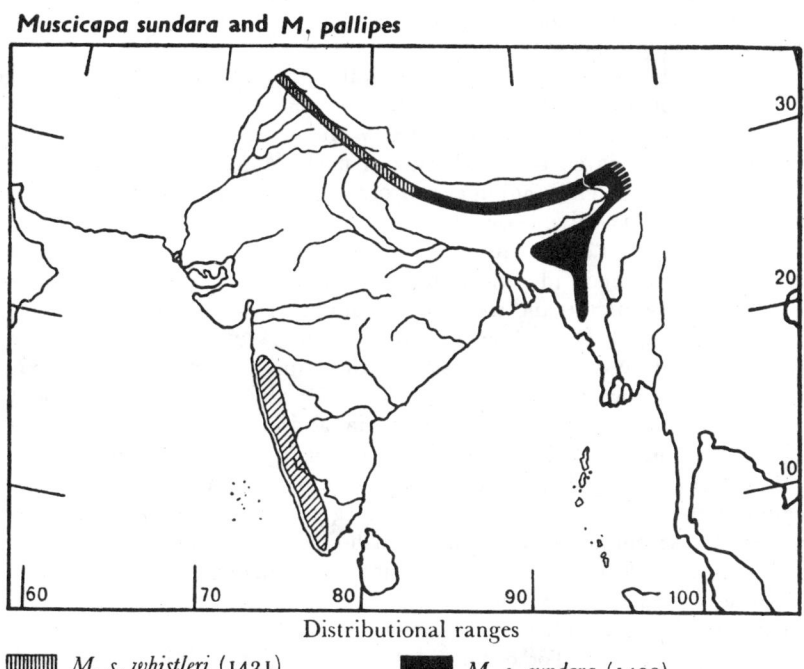

Muscicapa sundara and *M. pallipes*

Distributional ranges

▓▓▓ *M. s. whistleri* (1431)   ■ *M. s. sundara* (1432).

▨▨ *M. pallipes* (1435).

close to the ground. Made mostly of moss, the inner cup lined with rootlets. *Eggs*, normally 4, very similar to those of *M. grandis* (1428). Both nest and eggs are very reminiscent of an English Robin's. Average size of 14 eggs 20·7 × 15·9 mm (Baker). Building and incubation by both sexes; period 12 or 13 days. Brood-parasitized by *Cuculus canorus*, *C. saturatus* and *C. fugax* (D'Abreu, JBNHS 27: 405).

MUSEUM DIAGNOSIS. Male differs from *sundara* (1432) in having the orange-rufous of underparts paler. Female is paler and more olive-grey above; tail paler chestnut; underparts more greyish olive.

MEASUREMENTS and COLOURS OF BARE PARTS. As in 1432.

1432. *Muscicapa sundara sundara* (Hodgson)

Niltava Sundara Hodgson, 1837, Ind. Rev. 1 (12): 650 (Nepal)
Cyanecula fastuosa Lesson, 1840, Rev. Zool.: 266 (Mont. Himal., restricted to Murree by Baker, 1930, Fauna Brit. India, Bds. 8: 632, re-restricted to NE. Himalayas by Ticehurst, 1931, Ibis: 351)
Baker, FBI No. 684 (part), Vol. 2: 259
Plate 82, fig. 14

LOCAL NAMES. *Niltau* (Nepal); *Margong* (Lepcha).
SIZE. Sparrow; length *c*. 15 cm (6 in.).

FIELD CHARACTERS. Male (adult). *Above*, forehead black; crown, rump, shoulders and a patch on either side of neck bright ultramarine blue. Sides of head and back dark purplish blue looking practically black. *Below*, throat as back; rest of underparts orange-rufous.

May be confused with the very similar *M. vivida* which lacks the well-defined blue patch (the niltava hall-mark) on sides of neck; other differences not apparent in the field.

Female. *Above*, olive-brown tinged with ochraceous on rump. A pale eye-ring. A blue patch on either side of neck. Wings fulvous-brown. Tail rusty brown. *Below*, chin and upper throat fulvous-olive. Lower throat white; rest of underparts olive-brown.

The combination of a white throat-patch and the characteristic blue spot on each side of neck identifies the female.

STATUS, DISTRIBUTION and HABITAT. Common resident, subject to vertical movements. Nepal, Darjeeling, Sikkim, Bhutan, Arunachal Pradesh, Nagaland, Manipur, Assam in the Cachar Hills, and Meghalaya in the Khasi Hills. Breeds between *c*. 1800 and 3200 m (Diesselhorst) in the Himalayas and the eastern range south to Mt Victoria, and in Assam from 900 m to the highest summits. Found in winter (November to March) from *c*. 2300 m down to the foothills, in the plains of the Brahmaputra and in the Chittagong region of Bangladesh. Affects dense undergrowth in more open forest than *M. grandis*, secondary growth and brush-covered hillsides.

Extralimital. Extends east to Yunnan and northern Laos.

GENERAL HABITS, FOOD and VOICE. As in 1431.

BREEDING. As in 1431.

MUSEUM DIAGNOSIS. For distinction from *whistleri*, see 1431.

Young, male and female similar to those of *M. grandis* respectively, but with a pale

FLYCATCHERS

ochraceous patch on lower throat. First-winter birds recognizable by ochraceous tips of greater coverts.

MEASUREMENTS

	Wing	Bill (from skull)	Tarsus	Tail	
♂♂	78–87	16–17	21–22	65–73	mm
♀♀	76–83	16–17	21–23	60–68	mm

(BB, SA, Stresemann)

Weight 21 ♂ ♂ 20–24 (av. 21); 7 ♀ ♀ 19–24 (av. 21·5) g—SA.

COLOURS OF BARE PARTS. Iris brown. Bill: ♂ black, ♀ blackish brown; inside of mouth fleshy. Legs and feet plumbeous brown.

1433. Rufousbellied Blue Flycatcher. *Muscicapa vivida oatesi* (Salvadori)

Niltava oatesi Salvadori, 1887, Ann. Mus. Civ. Stor. Nat. Genoa 5 (2): 514 (Muleyit)
Baker, FBI No. 652, Vol. 2: 226
Plate 82, fig. 16

LOCAL NAMES. None recorded.

SIZE. Sparrow +; length *c.* 18 cm (7 in.).

FIELD CHARACTERS. Male (adult). *Above,* forehead and lores black. Crown and rump ultramarine blue. Back and tail dark purplish blue. *Below,* chin, upper throat, sides of head, neck and breast nearly black. Rest of underparts rufous.

Resembles *M. sundara* but lacks the niltava hall-mark—the well-defined blue patch on sides of neck. *M. rubeculoides* also has a dark blue throat but a white belly.

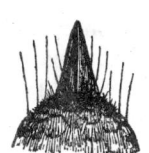

× *c.* 1

Female. *Above,* crown grey-brown; a fulvous eye-ring. Rest of upperparts dark olive-brown. *Below,* throat fulvous, rest olive-brown.

Distinguished from female *M. sundara* and *M. grandis* by lack of the distinctive blue neck-patches; from *unicolor* by fulvous throat and olive-brown breast (*v.* whitish and grey respectively). Females of *rubeculoides* and *banyumas* both have a rufous breast and white belly; female *concreta* has a white throat-patch. Both sexes of *M. poliogenys* have a whitish throat and rufous breast.

STATUS, DISTRIBUTION and HABITAT. Apparently rare. Arunachal Pradesh in the Pachakshiri area (Ludlow, *Ibis* 1944: 194); Assam in north Cachar (one April record—Baker, JBNHS 9: 124); the highest eastern ranges of Manipur (Hume, SF 11: 111), and presumably Nagaland. Obtained in summer from 2100 to 2700 m (Arunachal Pradesh, Sikkim and Mt Victoria); recorded in winter from 1500 to 2000 m in Burma. Affects dense brushwood in evergreen forest.

Extralimital. Extends to Sichuan and northern Vietnam. The species ranges to Taiwan, the Malay Peninsula and Sumatra.

GENERAL HABITS. Found singly or in pairs, skulking in brushwood from which it makes sallies after insects. Behaviour and habitat preference much as in Large Niltava (1428).

FOOD. Insects.
VOICE and CALLS. 'A clear whistle' (Robinson).
BREEDING. Unknown.
MUSEUM DIAGNOSIS. See Field Characters. For details of plumages see Baker, loc. cit. Young undescribed.

MEASUREMENTS

	Wing	Bill (from skull)	Tarsus	Tail	
7 ♂♂	94–102	c. 16	c. 18	74–86	mm
2 ♀♀	93, 94	c. 16	c. 18	—	mm

(Kinnear, Stresemann)

Weight 2 ♂ ♂ 1 ♀ 1·17–1·2 oz. (c. 33 g)—Hume.
COLOURS OF BARE PARTS. Iris brown to reddish chocolate. Bill black. Legs and feet dark blackish brown; soles yellowish.

1434. Whitetailed Blue Flycatcher. *Muscicapa concreta cyanea* (Hume)

Muscitrea cyanea Hume, 1877 (June), Stray Feathers 5: 101 (Muleyit)
Baker, FBI No. 641, Vol. 2: 215
Plate 82, fig. 15

LOCAL NAMES. None recorded.
SIZE. Bulbul; length c. 18 cm (7½ in.).
FIELD CHARACTERS. Male (adult). *Above*, deep blue, crown brighter. Tail dull blue with a larger amount of white on all but the central rectrices. *Below*, throat and breast dark blue fading to ashy on lower breast and flanks, and to white on lower belly and under tail-coverts.

Distinguishable from all dark blue flycatchers by large size (nearly same as of Large Niltava) and by the large amount of white in tail.

× c. 1

Female. *Above*, rufescent brown; a fulvous eye-ring. Tail like male's but brown, also with much white in it. *Below*, rufescent brown with a conspicuous white patch on lower throat. Flanks olive-brown; belly white.

Recognizable by the well-defined white throat-patch and extensive white in tail. Female *sundara* (1432) has blue patches on either side of white throat-patch and no white in tail. *M. monileger* (1415) also has no white in tail, but its white throat-patch is bordered with black.

STATUS, DISTRIBUTION and HABITAT. Little known. Breeds in the Patkai Range at or above 1500 m (exact altitude unknown). Obtained in winter at low altitude in the Margherita-Dibrugarh area in Assam (common). Affects deep forest.

Extralimital. Burma. The species extends east to northern Vietnam and south through the Malay Peninsula to Sumatra and Borneo.

GENERAL HABITS. More sluggish and inactive than most flycatchers. Frequents the lower branches of big trees in dense jungle, searching the leaves and branches for insects; also catches them on the wing but does not descend to the ground (Robinson). Frequently expands its tail laterally, showing the white. Females appear to be much less numerous, or perhaps shyer than males, as they are much scarcer in collections.

FOOD. Insects.

VOICE and CALLS. 'A low whistling song of three notes, in addition to the ordinary twitter' (Robinson).

BREEDING. Little known. A nest brought to Dr H. N. Coltart by Patkai Nagas on 25 June (over 50 years ago) is the only record. It was made of moss and said to have been wedged into a hole in a rocky bank of a ravine running through dense forest. The single egg looked like a very large egg of *M. rubeculoides* (1440): pale buff-stone, stippled all over with dark reddish, forming an ill-defined cap at the large end; it measured 23·9 × 18 mm (Baker).

MUSEUM DIAGNOSIS. See Field Characters. Young, spotted as in other closely related species.

MEASUREMENTS

	Wing	Bill	Tarsus	Tail	
		(from feathers)			
♂♀	91–93	*c.* 18–19	*c.* 23–24	66–72	mm
				(Baker)	
		(from skull)			
1♂	90	21	22	67	mm
				(MD)	

COLOURS OF BARE PARTS. Iris brown. Bill black. Legs and feet fleshy brown.

1435. **Whitebellied Blue Flycatcher.** *Muscicapa pallipes* Jerdon

Muscicapa pallipes Jerdon, 1840, Madras Jour. Lit. Sci. 11: 15 (Coonoor Ghat)
Baker, FBI No. 653, Vol. 2: 228
Plate 82, fig. 9

LOCAL NAME. *Kāttūneeli* (Malayalam).

SIZE. Sparrow ±; length *c.* 15 cm (6 in.).

FIELD CHARACTERS. Male (adult). *Above*, uniformly indigo-blue, forehead and supercilium brighter blue, lores black. *Below*, throat and breast indigo-blue; belly white.

Combination of last two characters identifies it. The Whitebellied Short-wing (1638) is slaty blue rather than indigo-blue, has a paler, bluish white forehead and supercilium, and is more terrestrial with longer legs.

Female. *Above*, rufescent olive-brown; lores white; tail chestnut. *Below*, throat and upper breast orange-rufous; breast greyish fading to white on belly.

Distinguished from female *M. rubeculoides* (1440) by larger size and chestnut tail. Both sexes of *M. parva* have a black-and-white tail. Female *M. tickelliae* (1442) is greyish above and has a blue tail.

STATUS, DISTRIBUTION and HABITAT. Fairly common resident. The Sahyadris (Western Ghats) and associated hills of southwestern India (Nilgiris, Palnis, High Wavys and others) from *c.* 19°N. (Bhimashanker) south through Goa and western Karnataka, Kerela and adjacent hills of Tamil Nadu; from the foothills to *c.* 1500 m; in the Nilgiris up to 1700 m (see map p. 179). Affects undergrowth of lanky seedlings in evergreen forest, sholas, *Strobilanthes* and 'channa' patches on hillsides, and cardamon ravines.

GENERAL HABITS. Found singly or in pairs, often in company with roving bands of insectivorous species. Rather sluggish. Flits unobtrusively among the undergrowth and lower trees, catching insects on the wing in typical flycatcher fashion or dropping to the ground now and again to pick up a morsel. 'When perched on a branch, bolt upright, has a peculiar way of spreading its tail and screwing it from side to side, reminiscent of the Thick-billed Flowerpecker (*Dicaeum agile*)'— SA. Its quiet and retiring habits make it seem rarer than it is. Recent mist-netting, for instance, has shown it to be quite common and breeding at Mahableshwar (c. 18°N.) whence not recorded previously.

FOOD. Chiefly insects; occasionally also berries.

VOICE and CALLS. A low *tsk-tsk* (Betts). A double call of two soft, tremulous notes, the second slightly lower in pitch, a little like the nasal call of the Paradise Flycatcher. Song, high-pitched, sweet and rich though 'a little squeaky, divided into phrases of as many as nine notes' (Nichols).

BREEDING. *Season*, February to September, but chiefly during the monsoon. *Nest*, a rough untidy structure mainly of green moss, the neat inner cup lined with lichen, fine grass or rootlets; generally placed on a ledge of a mossy rock, or in a hole in a dead stump or bank, not far from the ground, in humid forest. *Eggs*, almost invariably 4, resembling those of *Copsychus*: pale sea-green to warm yellowish stone profusely blotched all over with dark brown, more so at the large end. Average size of 45 eggs 20·2 × 15·5 mm (Baker). Other details of breeding biology not recorded.

MUSEUM DIAGNOSIS. See Field Characters. For details of plumage, see Baker loc. cit.

Young. *Above*, olive-brown with ochraceous streaks on head and small spots with dark margins on mantle. Lesser, median, greater coverts and tertials with ochraceous tips. *Below*, ochraceous, breast with dark margins, throat paler, belly whitish. Wings and tail as in adult of corresponding sex. First-winter birds recognized by pale tips of unmoulted greater coverts.

MEASUREMENTS

	Wing	Bill (from skull)	Tarsus	Tail	
♂♂	73–81	16–18	18–19	57–64	mm
♀♀	72–76	16–17	18–19	54–62	mm
				(SA)	

Wing of 15 ♂ ♂ 75–80 (av. 73·5); 16 ♀ ♀ 70–74 mm (Koelz)

Weight 10 ♂ ♂ (Apr.-May) 14–20 (av. 18); 11 ♀ ♀ (Apr.-May) 17–23 (av. 19·9) g—SA.

COLOURS OF BARE PARTS. Iris brown. Bill brownish black; mouth greyish pink or slaty pink. Legs and feet pale horny brown tinged with purplish.

MUSCICAPA POLIOGENYS (Brooks): BROOKS'S FLYCATCHER

Key to the Subspecies

	Page
A Paler. Upperparts more greyish; crown darker than back .... *M. p. poliogenys*	185

B Darker. Upperparts more brownish; crown and back concolorous
.... *M. p. cachariensis* 186
C Upperparts greyer than A, distinctly washed with blue in ♂ ......
.... *M.p. vernayi* 187

1436. *Muscicapa poliogenys poliogenys* (Brooks)

Cyornis poliogenys Brooks, 1879, Stray Feathers 8: 469
(Salbari, Sikkim Terai)
Baker, FBI No. 673 (part), Vol. 2: 247
Plate 82, fig. 8

LOCAL NAMES. None recorded.

SIZE. Sparrow —; length *c.* 14 cm (5½ in.).

FIELD CHARACTERS. *Above*, olive-brown, greyer on crown and sides of head. A pale eye-ring. Tail rufous-brown. *Below*, throat whitish or buff, rest of underparts fulvous, darker on breast. Sexes alike.

This nondescript species may be confused with several other female flycatchers with more or less rufous underparts; these species may be eliminated as follows: *M. hyperythra* has a rufous throat and rufous-brown sides of head. *M. tickelliae* has an orange-rufous throat and blue tail. *M. hodgsonii* has an olive-buff throat and lacks any fulvous on breast and belly. *M. ferruginea* has a more rusty tail, rufous lower belly and a very prominent white eye-ring. *M. rubeculoides* has a less whitish throat, a pure white belly and under tail-coverts. *M. monileger* has a white throat with a well-defined black gorget bordering it. *M. banyumas* has a bright rufous throat and breast.

Female Blue Chat (*Erithacus brunneus*) differs only in having the sides of head olive-brown, not dark grey, and in being more terrestrial with longer legs.

STATUS, DISTRIBUTION and HABITAT. Common resident. From central Nepal down to the terai (Chitawan park-TJR), through Sikkim, Darjeeling, the Jalpaiguri duars and Bhutan foothills (probably intergrading here with *cachariensis*); also Meghalaya in the Garo and Khasi hills, Mizoram and hills of Assam, and adjacent hills of Bangladesh to the Chittagong region; from the edge of the plains to *c.* 1500 m. Affects evergreen and deciduous forest, more open country in winter.

GENERAL HABITS. Little recorded. Affects bushes and undergrowth though generally keeps higher up among trees. Hops over and about fallen stumps and brushwood in forest like a chat (Stevens). See also 1438.

FOOD. Insects.

VOICE and CALLS. Alarm-notes *tik-tik-tik-tik*. Song, a pleasing trill (Stevens); mellow and varied, rendered as '*doe-doe-chi-cha, surani-so-swent, snareeti-do-deee*' (Fleming).

BREEDING. *Season*, middle of April till end of June. *Nest*, a compact cup mostly of green moss mixed with some dead leaves and fine grass, lined with rootlets; placed in a hollow in a bank or among boulders, or in a hole in some dead stump, within a metre from the ground. *Eggs*, 3 to 5, pale olive-green or olive-buff, so densely covered with reddish or red-brown specks as to appear uniformly reddish brown. Average size of 40 eggs 18.5×14.6 mm (Baker). Incubation by both sexes; period unrecorded.

Muscicapa poliogenys and M. albicaudata

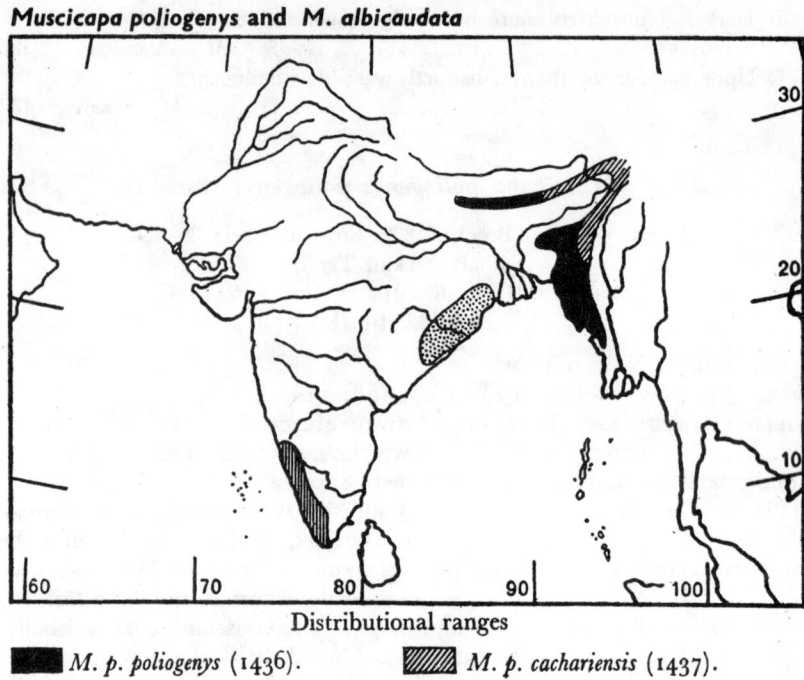

Distributional ranges

■ *M. p. poliogenys* (1436). ▨ *M. p. cachariensis* (1437).
▨ *M. p. vernayi* (1438). ▥ *M. albicaudata* (1446).

MUSEUM DIAGNOSIS. For distinction from *cachariensis* and *vernayi* see 1437 and 1438 respectively (under Museum Diagnosis).

Young similar to that of *M. rubeculoides*.

MEASUREMENTS

	Wing	Bill (from skull)	Tarsus	Tail	
♂♂	72–79	15–17	c. 18	60–65	mm
♀♀	71–78	15–16	c. 18	56–62	mm
				(BB, Rand & Fleming)	

COLOURS OF BARE PARTS. Iris brown. Bill black. Legs and feet pale greyish pink to pale fleshy.

1437. *Muscicapa poliogenys cachariensis* (Madarâsz)

Siphia cachariensis Madarâsz, 1884, Zeitschr. Ges. Orn.: 52, pl. 1, fig. 2
(Dhilkusha, Cachar)
Cyornis poliogenys saturatior Robinson & Kinnear, 1927, Bull. Brit.
Orn. Cl. 48: 43 (Dibrugarh, Assam)
Baker, FBI No. 673 (part), Vol. 2: 247

LOCAL NAME. *Dao-putti* (Cachari).
SIZE. Sparrow −; length *c.* 14 cm (5½ in.).
FIELD CHARACTERS. As in 1436 but throat fulvous as breast.

STATUS, DISTRIBUTION and HABITAT. Common resident. Arunachal Pradesh south through Nagaland, Assam in the Cachar Hills, and Manipur. From the edge of the plains to *c.* 1500 m. Affects various kinds of jungle as well as wet tropical forest.

Extralimital. Adjacent hills of Burma.

GENERAL HABITS, FOOD and VOICE. As in 1436.

BREEDING. As in 1436.

MUSEUM DIAGNOSIS. Differs from nominate race (1436) in being darker and having the ochraceous of breast almost reaching the chin. Above browner, less greyish, the crown not differentiated from the back.

MEASUREMENTS and COLOURS OF BARE PARTS. As in 1436.

1438. *Muscicapa poliogenys vernayi* (Whistler)

Cyornis poliogenys vernayi Whistler, 1931, Bull. Brit. Orn. Cl. 52: 23 (Sankrametta, 3500 ft, Vizagapatam dist., Eastern Ghats)
Not in Baker, FBI

LOCAL NAMES. None recorded.

SIZE. Sparrow −; length *c.* 14 cm (5½ in.).

FIELD CHARACTERS. As in 1436 but male has a bluish wash on upperparts. Looks very like Quaker Babbler, *Alcippe poioicephala* (1389) but keeps singly, not in flocks.

STATUS, DISTRIBUTION and HABITAT. Common resident. The Eastern Ghats from northern Orissa (Mayurbhanj) to northeastern Andhra Pradesh (Visakhapatnam district); from the plains to 1000 m. Affects secondary deciduous and evergreen forest.

Resurveys of the Eastern Ghats failed to turn up a single record of this form. It seems to have been replaced by *Muscicapa tickelliae* (see Price, 1979, JBNHS 76: 416; Ripley *et al.*, 1988, ibid. 85: 96–98).

GENERAL HABITS. Characteristically flycatcher, but frequently descends to the ground and hops about among the debris and undergrowth looking confusingly like a female shortwing or blue chat.

FOOD and VOICE. As in 1436.

BREEDING. Unrecorded. Probably as in 1436.

MUSEUM DIAGNOSIS. Female and first-year male differ from *poliogenys* in the greyer tint of the upperparts. Adult male distinguished by a bluish wash over the whole of the upperparts, more pronounced on the head and nape, becoming a definite bright blue on the longer upper tail-coverts and outer webs of the rectrices. Males breed in both stages of plumage.

MEASUREMENTS

	Wing	Bill (from skull)	Tarsus	Tail	
♂♂	73–78	15–16	18–20	60–68	mm
♀♀	70–74	14–16	18–20	55–60	mm
				(HW, SA)	

COLOURS OF BARE PARTS. As in 1436.

1439. Pale Blue Flycatcher. *Muscicapa unicolor unicolor* (Blyth)

Cyornis unicolor Blyth, 1843, Jour. Asiat. Soc. Bengal 12: 1007 (Darjeeling)
Baker, FBI No. 655, Vol. 2: 230
Plate 82, fig. 11

LOCAL NAMES. None recorded.

SIZE. Sparrow ±; length *c.* 16 cm (6½ in.).

FIELD CHARACTERS. Male (adult). *Above,* blue, brighter on forehead, supercilium and shoulder, deeper on tail. *Below,* throat pale blue, breast blue changing to whitish on belly. Under tail-coverts scalloped grey and white.

May only be confused with the Verditer Flycatcher (1445), which is more blue-green and lacks the whitish on belly.

Female. *Above,* olive-brown, wings and tail browner; pale lores and eye-ring. *Below,* pale grey-brown.

May be confused with female *M. hodgsonii* which is smaller, has an olive-buff throat and whitish belly. Female of *M. vivida* is darker olive below and has a fulvous throat.

STATUS, DISTRIBUTION and HABITAT. Imperfectly known. Uncommon in the western part of its range, common but locally distributed in the east. Garhwal (A. E. Osmaston, JBNHS 28: 148), Nepal (Biswas, JBNHS 59: 815), Sikkim, Bhutan, Arunachal Pradesh, Nagaland, Manipur, Assam in the Cachar Hills, Meghalaya in the Khasi and Garo hills, and Bangladesh in the Chittagong Hill Tracts (Rashid); from the foothills in winter to at least 1800 m in summer; optimum breeding altitude about 1500 m. Affects dense forest, secondary and bamboo jungle, and humid forest on steep hillsides.

Extralimital. Extends east to Laos. Another subspecies in the Malay Peninsula, Sumatra, Java and Borneo.

GENERAL HABITS. Frequents undergrowth as well as high trees. Makes sallies for insects but does not use regular perches. Twitches up its tail half-cocked between the drooping wings.

FOOD. Insects.

VOICE and CALLS. A characteristic *tr-r-r* (of alarm ?) as it twitches its tail (SA). Song described as beautiful, 'arresting and characteristic' (Smythies) and richer than that of most flycatchers (Baker).

BREEDING. Little known. *Season,* April to June. *Nest,* described mostly from material brought in by locals. Said to have been placed in holes between boulders in ravines and stream banks, or in holes in tree-trunks; made mostly of moss and lichen, the inner cup lined with rootlets. *Eggs,* normally 4, deep yellow-buff so densely freckled as to appear uniformly chocolate-brown. Average size of 14 eggs 23·1× 17·5 mm (Baker).

MUSEUM DIAGNOSIS. See Field Characters.

Young. *Above,* spotted with rich ochraceous. Ear-coverts fulvous with darker tips. Greater coverts and tertials with ochraceous tips. Wing and tail as in adult, sex for sex. *Below,* dull ochraceous, paler on belly, with faint dark margins on belly and throat, well marked on breast. First-winter birds recognized by pale tips of tertials.

FLYCATCHERS

MEASUREMENTS

	Wing	Bill (from skull)	Tarsus	Tail	
♂♂	80–85	17–18	17–19	70–76	mm
♀♀	76–84	17–18	17–19	67–72	mm

(BB, Stresemann, SA)

Weight 1 ♂ 21; 1 ♀ 21 g (SA).

COLOURS OF BARE PARTS. Iris dark brown. Bill black, base of lower mandible grey. Legs and feet dark brown.

1440. **Bluethroated Flycatcher.** *Muscicapa rubeculoides rubeculoides* (Vigors)

Phoenicurus rubeculoides Vigors, 1831, Proc. Zool. Soc. London: 35 (Darjeeling)
Baker, FBI No. 657, Vol. 2: 231
Plate 82, fig. 6

LOCAL NAMES. *Ghatki* (Bengali); *Manzil pho* (Lepcha); *Neelachenpan* (Malayalam).
SIZE. Sparrow −; length *c.* 14 cm (5½ in.).
FIELD CHARACTERS. Male (adult). *Above,* dull ultramarine blue, brighter on forehead and supercilium. Lores black. *Below,* throat dark blue; breast rufous; belly and under tail-coverts white washed with buff.

Dark blue throat in conjunction with rufous breast and white belly identifies the male.

Female. *Above,* olive-brown tinged with rufous on rump. Pale lores and eye-ring. *Below,* throat buff; breast ochraceous, rest of underparts white.

× *c.* 1

Several other flycatchers, mostly females, have a rufous breast and are very similar. They may be eliminated as follows: *M. parva* and *subrubra* have white tail-bases; *M. hyperythra, poliogenys* and *sapphira* a more or less ochraceous belly, not white; *M. tickelliae* has a bluish tail; *M. hodgsonii* has a buffish olive throat and breast; *M. banyumas* has both throat and breast rufous, and belly washed with buff, less pure white; *M. pallipes* has a chestnut tail and is larger. *M. poliogenys* is almost identical but has a whiter throat, and an ochraceous wash on belly.

STATUS, DISTRIBUTION and HABITAT. A partial migrant, generally common on its breeding grounds, but very widely scattered in winter. Breeds in the Himalayas from the Murree foothills (Roberts, JBNHS 81: 401), Kashmir (specimens in Brit. Mus.) and Chamba to Bhutan and Arunachal Pradesh; also Meghalaya, Nagaland, Manipur and the hills of Assam and Bangladesh south to the Chittagong Hill Tracts; from the foothills normally to *c.* 1500 m, locally to 1800 m and even 2100 m (Nepal, Diesselhorst). Affects forest with plenty of undergrowth, wooded nullahs, well-wooded gardens and secondary forest.

Extralimital. Adjacent Burma. The species extends to Hupeh and Vietnam.

MIGRATION. A summer (breeding) visitor to the Himalayas arriving in March, leaving in September. Has been recorded in New Delhi and in

Muscicapa rubeculoides

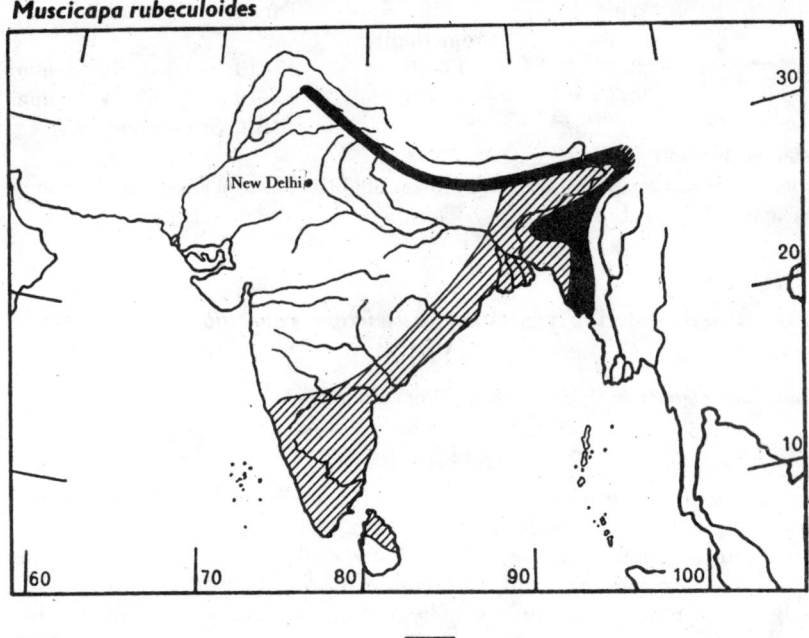

■ Breeding range   ▨ Winter

Sikkim at *c.* 1100 m in winter. A resident or altitudinal migrant in NE. India. Winters scatteredly in the hills of southern India (low country to *c.* 900 m) to Belgaum southwards, Kerala, Tamil Nadu, the Eastern Ghats, Bihar, Orissa, West Bengal (common) and in the plains and foothills of Bangladesh. Also the northern part of Sri Lanka where it arrives in October, up to *c.* 1000 m in the hills. In this season, affects well-wooded country, secondary and bamboo jungle.

GENERAL HABITS.   Keeps to undergrowth of bushes and low trees; makes sallies for insects but rarely returns to the same perch or even the same tree. Frequently drops to the ground momentarily to pick up a morsel, sometimes jerking open and flicking forward the wings to stampede a cricket or grasshopper lurking amongst the mulch. Flicks up partially cocked tail from time to time while uttering its call- or alarm-note.

FOOD.   Insects and grubs.

VOICE and CALLS.   Alarm-note *chr-r, chr-r;* call-note *click, click* or *chik, chik* (SA). Song, a clear but short phrase or metallic trill similar in pattern to that of Tickell's Flycatcher (1442) but somewhat richer. The song is long and elaborate at the beginning of the breeding season, with mimicry of other birds: *Accipiter badius* and *Turdus boulboul,* and consists of repeated plaintive rising and falling phrases, at once recognizable for the species: *Turr-treee-tiih* with a burry timbre (Roberts). A song heard in winter has been rendered as *ciccy-ciccy-ciccy-ciccy-see,* the first two *ciccy* higher in pitch (Smythies). Often given from the low limb of a tree when sitting quietly. Song period in Nepal, mid-April to mid-July (Proud). Also sings sporadically in winter.

BREEDING.   *Season,* April to August. Double-brooded in Assam but apparently single-brooded in the Himalayas. *Nest,* placed in a hollow in a mossy bank or crevices in rocks on stream banks and ravines, holes in dead stumps, holes and clefts in tree-trunks or in hollow bamboo, occasionally among thick clumps of ferns and orchids growing on the trunk of a moss-covered tree; made mostly of green moss, sometimes mixed with some dead leaves, grass bents and lichen, lined with fine rootlets. *Eggs,* normally 4, rarely 5, occasionally 3, pale olive or yellowish stone very densely stippled with olive-brown or reddish olive-brown. Average size of 60 eggs 18·7 × 14·3 mm (Baker). Both sexes share incubation and care of young. Incubation period either eleven or twelve days.

MUSEUM DIAGNOSIS. See Field Characters. For details of plumages see Baker, loc. cit.

MEASUREMENTS

	Wing	Bill (from skull)	Tarsus	Tail	
♂♂	65–77	14–16	15–18	50–60	mm
♀♀	66–74	14–15	15–18	49–55	mm
				(SA, BB)	

Weight 3 ♂ ♂ (March) 12–18; 14 ♂ ♂ (October, in winter quarters) 13–19 (av. 14·9), 14 ♀ ♀ (ditto) 10–16 (av. 13·6) g—SA. 1 ♂ 14·5 g; 1♀ 13·5 g (SDR).

COLOURS OF BARE PARTS. Iris dark brown. Bill brownish black; mouth greyish pink or slaty pink. Legs and feet greyish brown (yellowish fleshy in juvenile).

1441. **Largebilled Blue Flycatcher.** *Muscicapa banyumas magnirostris* (Blyth)

Cyornis magnirostris Blyth, 1849, Jour. Asiat. Soc. Bengal 18 : 814 (Darjeeling)
Baker, FBI No. 663, Vol. 2: 236
Plate 82, fig. 7

LOCAL NAME. *Daogatang* (Cachari).

SIZE. Sparrow −; length *c.* 14 cm (5½ in.).

FIELD CHARACTERS. M a l e (adult). *Above,* indigo-blue, brighter on forehead and shoulders; lores black. *Below,* throat and breast and flanks orange-rufous; belly and under tail-coverts white.

May be confused with *M. sapphira* which is more brilliant ultramarine blue on crown and rump and has a more greyish belly. *M. hodgsonii* is slaty above and has white in tail. The nearly identical *tickelliae* is not likely to occur in the range of *banyumas.*

F e m a l e. *Above,* olive-brown; a fulvous eye-ring. *Below,* throat and breast orange-rufous; belly white washed with buff.

M. sapphira is smaller, more rufescent above and has the breast tinged with olive. *M. poliogenys* has greyer sides of head and whitish throat. *M. rubeculoides* has a buff throat, paler than breast and a purer white belly. See aso 1440 under Field Characters.

STATUS, DISTRIBUTION and HABITAT. A rare and little known species. Central Nepal (one recent record: Lowndes, JBNHS 53: 33, 1955), Sikkim (no recent records), Arunachal Pradesh (obtained in the Miri Hills by Godwin-Austen), the Sadiya frontier tract, Noa Dihing Valley (Saha),

Assam in the Cachar and Meghalaya in the Khasi hills (Baker), Nagaland (Stevens), Manipur (Hume), and possibly the Chittagong region of Bangladesh. Altitudinal distribution little known. The Nepal specimen was collected at *c.* 2600 m in August; in Assam, breeds between 750 and 1800 m, mostly above 1200 m, while in the Margherita area it is found as low as 300 m in summer. Himalayan population appears to winter in Burma (southern Shan States and Tenasserim) while Assam birds spread over the adjacent plains in winter. Affects shady ravines and dense, humid forest with plentiful undergrowth.

Extralimital. The species extends to Yunnan, the Indochinese countries, Java, Borneo and Palawan.

GENERAL HABITS. Unrecorded.

FOOD. Insects.

VOICE and CALLS. 'A cheerful little song.' A song heard by Lister near Darjeeling (JBNHS 52: 43) may belong to this species; it is rendered as *tsea-sea-si-e-e-e-e.*

BREEDING. *Season,* end of April to June. *Nest,* made of moss and lined with moss-roots, usually placed on the ground in some hollow among plants on a bank or between the roots of a tree, sometimes in a hole in a dead stump, or well hidden against a moss-covered trunk, quite low down. *Eggs,* 4 or 5, pale sea-green or buff-stone with small blotches of pale brown or chocolate-brown forming a broad zone at the larger end. Average size of 40 eggs 19·1 × 14·6 mm (Baker). No other details recorded.

MUSEUM DIAGNOSIS. See Field Characters.

Young. *Above,* brown with ochraceous spots. *Below,* breast ochraceous with dark margins, belly paler. First-winter birds recognized by unmoulted greater coverts.

MEASUREMENTS

	Wing	Bill (from feathers)	Tarsus	Tail
♂♀	76–83	*c.* 15	*c.* 19	55–61 mm (Baker)

COLOURS OF BARE PARTS. Iris brown. Bill black. Legs and feet pale fleshy white to light horny brown.

MUSCICAPA TICKELLIAE (Blyth): TICKELL'S REDBREASTED BLUE FLYCATCHER

Key to the Subspecies

Deeper blue above .... *M. t. jerdoni*
Lighter blue above .... *M. t. tickelliae*

1442. *Muscicapa tickelliae tickelliae* (Blyth)

C.(*yornis*) *Tickelliae* Blyth, 1843, Jour. Asiat. Soc. Bengal 12: 941
(Central India = Borabhum)
Baker, FBI No. 660, Vol. 2: 234
Plate 82, fig. 1

LOCAL NAMES. *Ădhărăngă* (Gujarati); *Neelăkkūrūvi* (Malayalam, Tamil).

SIZE. Sparrow −; length *c.* 14 cm (5½ in.).

FIELD CHARACTERS. Male (adult). *Above,* indigo-blue, brighter (azure blue) on forehead, supercilium and shoulders. *Below*, throat and breast orange-rufous; belly white.

Almost identical with *M. banyumas; M. rubeculoides* is very similar but has a blue throat.

Female. Like male but duller and greyer above; lores dull bluish or whitish. Rufous of breast paler. Easily confused with *M. poliogenys vernayi* (1438) but which has a whitish throat (*v.* orange-rufous concolorous with breast).

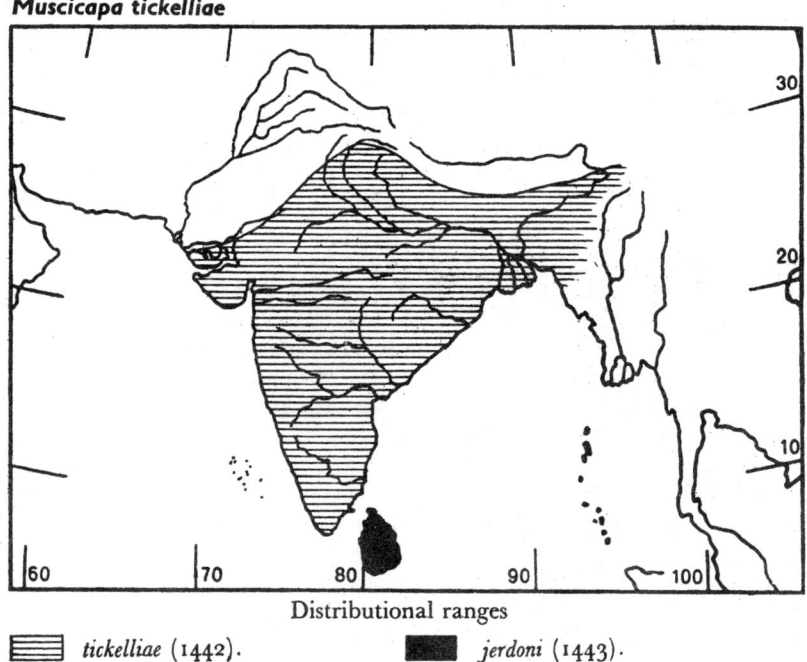

Muscicapa tickelliae

Distributional ranges

▤ *tickelliae* (1442).   ■ *jerdoni* (1443).

STATUS, DISTRIBUTION and HABITAT. Common resident, subject to local movements. The Indian peninsula from Uttar Pradesh (Dehra Dun and the Siwalik foothills [Osmaston], Gonda and Gorakhpur), southern Bihar and southwestern West Bengal, south to the Cardamom Hills (Kerala) and west to a line from Kutch, Mt Abu, Sambhar Lake and Mussooree; thence east through Bangladesh (Rashid) and Assam (N. Cachar, Baker, JBNHS 9: 124). Nesting reported from Murree (Buchanan *in* Whistler's MS.). Breeds mostly in more broken country and hills (in southern India up to *c.* 1500 m) and spreads out over the plains in winter (October to May). Affects thick bushes in or near forest and streams, deciduous scrub and bamboo jungle, overgrown nullahs, village groves, wooded gardens, and orchards. Prefers a drier facies than the Bluethroated Flycatcher.

Extralimital. Extends through Burma, the Indochinese countries, Malaysia, Indonesia and the Philippines.

GENERAL HABITS. Keeps singly, often in mixed hunting parties. Perches bolt upright a couple of metres from the ground, flicking its tail and uttering a sharp *tick, tick* from time to time. Flits about in scrub and undergrowth, catching most of its food in the air, sometimes hovering at a flower or sprig to take an insect.

Display. The male, while leaning on one side, raises the feathers of his head and back and vibrates them as if to attract the female, keeping the body immobile, the tail slightly spread and bent a little sideways (Dharmakumarsinhji).

FOOD. Insects, chiefly diptera.

VOICE and CALLS. Song, a short, metallic trill of six to ten notes, the first half descending, the second ascending the scale (Bates). Unmistakable and often the first indication of the bird's presence; uttered throughout the year, but chiefly in April-May. Call-note *tick-tick*.

BREEDING. *Season* April to August, chiefly May and June. *Nest*, placed in crevices or on ledges of rock, holes in trees or in crevices formed by the twisting aerial roots of large, parasitic fig trees; in the latter case it may be well up above the ground, but usually it is placed within a couple of metres from it. Nests have been found in walls and on window-ledges of disused houses; also under a prickly pear root in a bank and in a rubbish heap of dry leaves (Betham, JBNHS 14: 398). Main material is green moss, loosely packed, often mixed with some dead leaves or grass bents, the inner cup usually lined with fine rootlets. In drier regions, dry bamboo leaves and fine grass are used. *Eggs*, 3 or 4, occasionally 5, similar to those of *M. rubeculoides* (1440). Incubation period 13 days from laying of first egg; fledgling period 21 days (Abdulali, JBNHS 76: 159–61). Average size of 80 eggs 18·4 × 14·2 mm (Baker).

MUSEUM DIAGNOSIS. M a l e differs from the very similar *banyumas* (1441) in being a darker blue above, with the orange-rufous of breast not extending quite as far down on upper abdomen. Some males have extreme point of chin blue-black. Wing of *banyumas* averages somewhat longer.

F e m a l e very similar to male *M. poliogenys vernayi* (1438) but blue of upper tail-coverts a little brighter and blue of forehead and supercilium somewhat more pronounced.

There is no spring moult, and the complete autumn moult takes place from the end of August to October.

Y o u n g (juvenile) shows characteristic spotting but upperparts have a dark bluish tinge; wings and tail blue in both sexes: brighter in male, duller and greyer in female.

MEASUREMENTS

	Wing	Bill (from skull)	Tarsus	Tail	
♂♂	70–77	14–16	17–20·	56–68	mm
♀♀	68–73	14–16	16–19	54–57	mm
				(SA, HW)	

Weight 7 ♂ ♂ 14–16; 3 ♀ ♀ 12–16 g. 19 (unsexed, in ♀ plumage) 12–17 (av. 14·6) g—SA. ♂ ♂ 14.4, 14.5, 14.9 (2), 15 (2) g; ♀ 15 g (SDR).

COLOURS OF BARE PARTS. Iris brown. Bill brownish black; mouth blackish pink, greyish pink or yellowish pink. Legs and feet greyish brown or pinkish slate.

1443. *Muscicapa tickelliae jerdoni* (Holdsworth)

Cyornis jerdoni 'G. R. Gray' = Holdsworth, 1872, Proc. Zool. Soc. London: 442 (few miles from Colombo, Ceylon)
Cyornis tickelliae nesaea Oberholser, 1920, Proc. Biol. Soc. Washington 33: 86 (Walgama, Ceylon)
Baker, FBI No. 662, Vol. 2: 236

LOCAL NAMES. *Kopi-kurullā, Mārāwā* (Sinhala); *Kopi kūrūvi* (Tamil).
SIZE. Sparrow −; length *c.* 14 cm (5½ in.).
FIELD CHARACTERS. As in 1442, q.v. See Museum Diagnosis.
STATUS, DISTRIBUTION and HABITAT. Resident, widely distributed and moderately plentiful. Sri Lanka, from the lowlands to *c.* 1300 m, occasionally higher. Affects forests and well-wooded areas.
GENERAL HABITS, FOOD and VOICE. As in 1442.
BREEDING. *Season*, March to June, occasionally starting in late February in the low country. One nest recorded in October. *Nest* and *eggs* as in 1442; clutch size normally 3. Average size of 11 eggs 19·5 × 14·8 mm (Phillips).
MUSEUM DIAGNOSIS. Male differs from ♂ *tickelliae* (1442) in being a darker, deeper blue above and slightly more rufous on breast, showing less white on abdomen. Female slightly paler. In both sexes the bills average slightly larger and stronger.

MEASUREMENTS

	Wing	Bill (from skull)	Tarsus	Tail	
13 ♂♂	71–76	15–17	16–18	56–63	mm
3 ♀♀	70–75	15–16	16–18	54–61	mm
				(HW)	

Weight 1 ♂ 18 g (SDR).
COLOURS OF BARE PARTS. As in 1442.

1444. Dusky Blue Flycatcher. *Muscicapa sordida* (Walden)

Glaucomyias sordida Walden, 1870, Ann. Mag. Nat. Hist. 5: 218 (Ceylon)
Baker, FBI No. 667, Vol. 2: 241
Plate 82, fig. 12

LOCAL NAME. *Gini-kurullā* (Sinhala).
SIZE. Sparrow −; length *c.* 14 cm (5½ in.).
FIELD CHARACTERS. A dark bluish grey flycatcher with cerulean blue forehead, black lores and whitish belly. Sexes virtually alike; female very slightly duller.
Could be easily confused with ♂ Pale Blue (1439), but for allopatry. Ranges of the two widely disjunct.
STATUS, DISTRIBUTION and HABITAT. Resident, moderately plentiful; Sri Lanka. Confined to the Hill zone from 900 to over 2100 m, occasionally as low as 450 m in the Rattota district. Affects forest and well-wooded ravines.
GENERAL HABITS. Generally perches on low branches, logs or rocks close to the ground.

Muscicapa thalassina and M. sordida

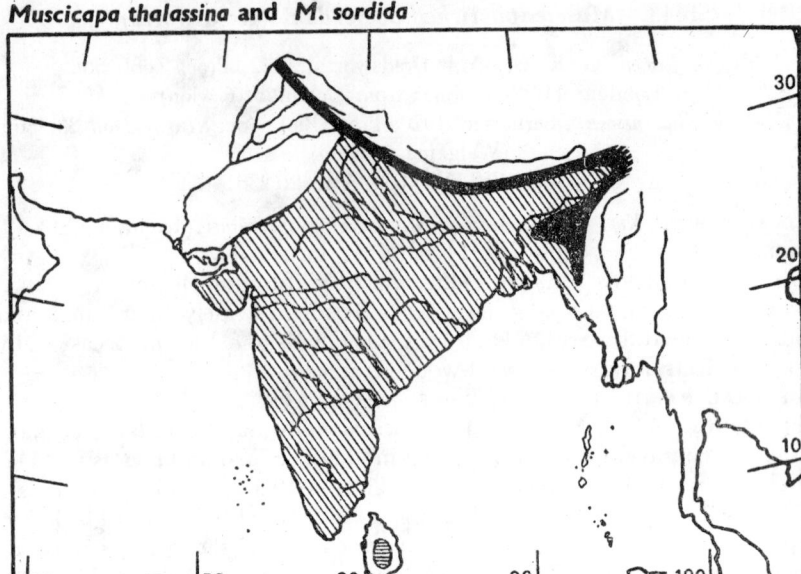

Distributional ranges

■ *M. t. thalassina* (1445), breeding.   ▨ Winter

▤ *M. sordida* (1444).

FOOD. Beetles, caterpillars and flying insects; also berries (*Rubus, Lantana,* etc.).

VOICE and CALLS. A sweet, rather loud song consisting of five or six notes, constantly repeated, uttered from a fairly high perch in a tree during the courting season. Also a subdued sub-song with the same notes and cadence (Henry). Call-note a soft, low *chip-chip*.

BREEDING. *Season*, late February to early June; above *c.* 1500 m nesting takes place again in August and September. *Nest*, placed on ledges and in crevices in moss-covered banks and vertical rock faces in heavy forest, or in cavities in tree-trunks, usually below three metres but sometimes as high as six metres in a hollow branch; made entirely of green moss, the inner cup lined with fine rootlets. Exceptionally a nest may be placed in a low fork; in this case it is a compact, well-built cup of green moss, tendrils and bits of dead leaves, lined with rootlets. *Eggs*, generally 2, exceptionally 3, pale creamy pink to warm salmon-pink, lightly and thinly freckled with pale reddish, the markings forming a ring or a cap around the large end. Average size of 60 eggs 20 × 14·8 mm (Baker).

MUSEUM DIAGNOSIS. See Field Characters. For details of plumages see Baker, loc. cit. Postnuptial moult takes place between August and December.

Young. *Above*, dull olive-brown, each feather with an elongated fulvous apical spot and indistinctly edged with blackish. *Below*, chin and throat fulvous gradually growing paler on breast and white on flanks, abdomen and under tail-coverts, the feathers edged with sooty black. These edges most conspicuous on breast and disappearing on chin and vent and under tail-coverts (Whistler).

MEASUREMENTS

	Wing	Bill (from skull)	Tarsus	Tail	
6 ♂♂	77–82	16–17	18–19	62–65	mm
4 ♀♀	71–79	16–17	18–19	57–63	mm
				(HW)	

COLOURS OF BARE PARTS. Iris red-brown or brown. Bill black. Legs and feet dark lead colour (Wait).

1445. **Verditer Flycatcher.** *Muscicapa thalassina thalassina* Swainson

Muscicapa thalassina Swainson, 1838, Nat. Library, Flycatchers 21: 252 (India)
Muscicapa melanops Vigors, 1832, Proc. Zool. Soc. London: 172,
nec *M. melanops* Vieillot, 1818
Baker, FBI No. 665, Vol. 2: 239
Plate 82, fig. 10

LOCAL NAMES. *Puthir chitta, Nil-kătkătia* (Bengali); *Sibyell-pho* (Lepcha); *Dao-tisha lili gadeba* (Cachari); *Neelamēni* (Malayalam).

SIZE. Sparrow ±; length *c.* 15 cm (6 in.).

FIELD CHARACTERS. Male (adult). Entirely blue-green or verdigris (colour of oxydized copper), brighter on head and throat, darker on wings and tail; a black patch in front of eyes (lores) very prominent. Female duller and greyer but without accompanying male to compare with, difficult to sex.

STATUS, DISTRIBUTION and HABITAT. Common summer (breeding) visitor to the Himalayas, wintering widely spread over the whole Indian peninsula, Assam and Bangladesh.

× *c.* 1

Breeds from the Indus Valley and Kashmir east to Bhutan and Arunachal Pradesh, thence south through Nagaland, Manipur and Assam in Cachar and Meghalaya in the Khasi hills; from 1200 to 2700 m, locally up to 3000 m, optimum zone 1500–2400 m (see map, p. 196). Affects light forest (with a preference for broad-leaved) and bushes along streams; also evergreen and coniferous but avoids dense and tall forest.

Extralimital. Extends east to western China and Vietnam. Another subspecies in Sumatra and Borneo.

MIGRATION. Arrives on its summer (breeding) grounds in early March and April keeping there until October, although a downward movement already starts in July. Abundant in the lower hills in August and September. Winters from the Himalayan foothills (up to *c.* 750 m) south throughout the Indian peninsula east of a line Sambhar Lake—Dwarka [Kutch (JBNHS 61: 449), Kathiawar (JBNHS 60: 456)] south to the hills of southern Kerala, and east to and including the foothills of Assam and Bangladesh to the Chittagong Hill Tracts. Keeps more to broken country and hilly regions up to *c.* 1000 m. Frequents edges of forest, glades, wooded compounds, groves and gardens. In both summer and winter quarters, telephone wires where available provide favourite perches. May be seen in southern India from October to February; in the more northern parts from September to March.

GENERAL HABITS. Keeps in pairs in the breeding season and does not accompany mixed species foraging parties (TJR); at other times, it may be seen singly or in pairs (sometimes several pairs), often in company with *Hypothymis, Terpsiphone* and other flycatchers, ioras (*Aegithina*), leaf warblers and drongos, etc. Perches upright; flicks tail. Makes short aerial sallies for insects from the tips of tall trees as well as bushes, seldom returning to the same perch but flying from branch to branch and tree to tree. Also flutters at flower clusters and leaf sprigs to stampede insects lurking within and occasionally descends to the ground to pick up a morsel.

FOOD. Chiefly tiny winged insects.

VOICE and CALLS. Call-note *tze-ju-jui*. Also a very soft *p'p'pwe . . . p'p'pwe* before giving the song. Song, a pleasant warble rendered as *petititi-wu-pititi-weu* uttered frequently at intervals of several minutes, sometimes in groups of two or three repetitions. Reminiscent of the song of *Prunella modularis* (Lister). A pleasant jingling trill, very like the White-eye's (*Zosterops*) in pattern, slightly louder. It begins almost inaudibly, waxes louder, and soon fades out as it began (SA). Characteristically delivered from tree-tops and quite high above ground. Song period in Nepal mid-March to early July with a resumption in August-October (Proud). Very silent in its winter quarters.

BREEDING. *Season*, April to August. Probably double-brooded. *Nest*, a flat thick-sided cup placed in banks, crevices of rocks, under bridges, sometimes in holes in walls, under the eaves of an occupied hill-station bungalow or rafters of a veranda; sometimes in a hollow among moss or ferns growing on a tree-trunk up to 6 m or so up. Made mostly of green moss, occasionally mixed with other materials, the neat inner cup lined with rootlets, sometimes dry grass. *Eggs*, normally 4, sometimes 3, rarely 5, pale creamy pink or white with a ring of tiny reddish blotches around the large end. Average size of 200 eggs 19·3 × 14·6 mm (Baker). Building, incubation and tending young by both sexes; incubation period undetermined.

MUSEUM DIAGNOSIS. For details of plumage see Baker, loc. cit. Postnuptial moult completed by October.

Young. Grey-brown, more or less tinged with green and spotted with fulvous; the spots smaller above larger below, the edges of the feathers being darker. Spots on head and nape almost white.

MEASUREMENTS

	Wing	Bill (from skull)	Tarsus	Tail	
♂♂	80–90	13–15	16–19	65–75	mm
♀♀	78–84	13–15	16–19	60–71	mm

Bill (skull) ♀ 12 mm (SA, HW, BB)

Weight 20 ♂ ♂ 15–20 (av. 18·25); 7 ♀ ♀ 16–20 (av. 18·0) g (SA). ♂ 20·8 g; ♀ 19·5 g (SDR).

COLOURS OF BARE PARTS. Iris dark brown. Bill, legs, feet and claws black. Mouth yellowish pink.

1446. Nilgiri Verditer Flycatcher. *Muscicapa albicaudata* Jerdon

Muscicapa albicaudata Jerdon, 1840, Madras Jour. Lit. Sci. 11: 16 (Nilgiris)
Baker, FBI No. 668, Vol. 2: 242
Plate 82, fig. 13

LOCAL NAME. *Neelăkkili* (Malayalam).

SIZE. Sparrow ±; length *c.* 15 cm (6 in.).

FIELD CHARACTERS. Male (adult), greenish indigo-blue with bright blue forehead and supercilium, black lores, a white patch at base of outer tail-feathers, and whitish under tail-coverts.

Distinguished from the wintering Verditer Flycatcher by its much darker blue coloration and the white patch at base of tail.

Female, dull grey-brown washed with greenish blue. The white patch at base of tail identifies it.

STATUS, DISTRIBUTION and HABITAT. Common resident. The southern section of the Sahyadris (Western Ghats) from the Bababudan and the Biligirirangan hills of Karnataka south through the Nilgiris, Palnis, and the associated ranges of western Tamil Nadu and Kerala to the Ashambu Hills; from *c.* 600 m to the highest summits, mostly above 1200 m (see map, p. 186). Affects overgrown streams, cardamom plantations, coffee shade, glades and edges of forest, sholas, even gardens and trees near houses. May be a geographical representative of *M. unicolor* along with *M. sordida* of Sri Lanka.

GENERAL HABITS. Usually met with singly, often side by side with *Culicicapa ceylonensis*, *M. muttui* and other flycatchers. Frequents bushes and lower branches of large trees, sometimes also the canopy where it makes short aerial sallies after winged insects. Perches somewhat upright and twitches its tail up and down while calling.

FOOD. Insects, perhaps also some berries.

VOICE and CALLS. General pattern of song that of Verditer Flycatcher; reminiscent also of Pied Bush Chat (SA); 'a sweet phrase of six to eight notes, often with a glide in each note that gives it a somewhat mournful effect' (Nichols); a sweet warble up and down the scale (Betts). Delivered from an exposed perch for long periods at a time. Call-note, a series of four or five sharp *chips* (Nichols). A female also recorded singing (SA).

BREEDING. *Season*, February to June, chiefly March and April. *Nest* a substantial cup of green moss, the inner cup lined with rootlets and rhizomorphs; placed in banks, holes or cracks in trees or among hanging roots; sometimes in walls, under bridges or under the eaves of a house, even an occupied one. Often the same site is used in subsequent years. *Eggs*, usually 3, sometimes 2, pale creamy pink sparsely freckled all over with reddish, more densely at the larger end where the markings tend to form a ring. Average size of 60 eggs 20 × 14·8 mm (Baker).

MUSEUM DIAGNOSIS. See Field Characters. For details of plumages see Baker, loc. cit.

MEASUREMENTS

	Wing	Bill (from skull)	Tarsus	Tail	
♂♂	75–82	14–15	18–19	59–67	mm
♀♀	74–78	13–15	18–19	56–61	mm
				(SA, HW)	

Weight 10 ♂♂ (March–Apr.) 12–18 (av. 15·9); 4 ♀♀ (March–Apr.) 16–19 (av. 17·5) g—SA.

COLOURS OF BARE PARTS. Iris dark brown. Bill horny black; mouth greyish pink or brownish pink. Legs, feet and claws blackish brown.

MUSCICAPINAE

Genus MUSCICAPELLA Bianchi

Muscicapella Bianchi, 1907, Ann. Mus. Zool. Acad. Imp. Sci. St. Petersbourg 12: 14, 43, new name for *Nitidula* Blyth, nec *Nitidula* Fabricius, 1775

Nitidula Blyth, 1861, Proc. Zool. Soc. London: 201.

Type, by monotypy, *N. campbelli* Blyth = *Nemura hodgsoni* Moore

Distinguished by its small size and narrow, slender bill with well-developed hairs over the nostrils.

1447. Pygmy Blue Flycatcher. *Muscicapella hodgsoni hodgsoni* (Moore)

Nemura Hodgsoni Moore, 1854, in Horsfield & Moore, Cat.
Bds. Mus. E. I. Co. 1: 300 (Nepal)
Baker, FBI No. 664, Vol. 2: 237
Plate 82, fig. 3

LOCAL NAMES. None recorded.

SIZE. Sparrow −; length *c.* 8 cm (3 in.).

FIELD CHARACTERS. A diminutive, blue-and-orange flycatcher.

Male (adult). *Above*, dark cyan blue, brighter on crown; forehead, lores and sides of head blue-black. *Below*, orange-yellow.

Female. *Above*, olive-brown, more rufous on rump and upper tail-coverts; *below*, pale yellow.

STATUS, DISTRIBUTION and HABITAT. A scarce resident, subject to vertical movements. From west-central Nepal (Pokhara-TJR) east through Darjeeling, Sikkim, Bhutan and Arunachal Pradesh, thence south through the Patkai Range, Nagaland, Assam in the Cachar Hills and Manipur. Recorded in summer from 2100 to 3500 m (Proud, Rand & Fleming), and in winter from 1800 m (Ripley) down to the foothills. Affects dense, tall forest and secondary scrub at the edge of clearings or along hill streams.

× *c.* 1

Extralimital. Extends to Yunnan and northern Thailand. The species ranges south through the Malay Peninsula, Sumatra and Borneo.

GENERAL HABITS. A very small, lively and restless flycatcher perhaps less scarce than it appears to be but unobtrusive and often overlooked. Keeps singly, sometimes in mixed flocks with *Phylloscopi*, moving slowly from tree to tree. Frequents dense thickets and shrubs as well as tall trees, searching the leaves for insects, sometimes making short sallies within the canopy in their pursuit or fluttering before a sprig to stampede them. Fleming observed them in the tops of tall oaks, fifteen metres or more from the ground. It also descends to the ground to pick up an insect, twitching up and cocking the tail from time to time between half-drooping wings in typical flycatcher style, then looking rather like a shortwing (SA).

FOOD. Small insects.

VOICE and CALLS. A distinctive, high-pitched song *tzit che che che cheeee* (do, do, te, la, te) (Fleming). Call-note, a feeble *tsip* (Stevens).

BREEDING. The only record is a nest taken in Nagaland (altitude ?) by Masson (*apud* Baker) on 20 July. It was a tiny saucer of green moss, lined with rootlets, wedged among the thick stems of a creeper growing over an old stump. The young were old enough to escape.

MUSEUM DIAGNOSIS. For details of plumages see Baker, loc. cit.
Young unknown.

MEASUREMENTS

	Wing	Bill (from skull)	Tarsus	Tail	
♂♂	47–51	c. 11	16–17	34–35	mm
			(SA, SDR, Rand & Fleming)		

Weight 1 ♂ 6·5 g.

COLOURS OF BARE PARTS. Iris brown. Bill brown. Legs and feet bluish grey.

Genus CULICICAPA Swinhoe

Culicicapa Swinhoe, 1871, Proc. Zool. Soc. London: 381.
Type, by monotypy, *Platyrhynchus ceylonensis* Swainson

Bill very wide at base. Rictal bristles numerous and long. First primary (as.) short. Tail square. Young not spotted.

CULICICAPA CEYLONENSIS (Swainson): GREYHEADED FLYCATCHER

Key to the Subspecies

Below, both grey and yellow darker.............. *C.c. ceylonensis*
Below, both grey and yellow paler............... *C.c. calochrysea*

1448. *Culicicapa ceylonensis calochrysea* Oberholder

'*Cryptolopha cinereocapilla* (Vieillot)' Hutton, 1848, Jour. Asiat. Soc. Bengal 17: 689 (Himalayas). *Nom. nud.*
Culicicapa ceylonensis calochrysea Oberholser, 1923, Smith. Misc. Coll. 76: 8 (Quaymoo Choung =Left bank of Thaungyin River, lat. 17°15′N. Amherst Dist., Tenasserim)
Culicicapa ceylonensis orientalis Baker, 1923, Bull. Brit. Orn. Cl. 44: 11 (Szechwan Prov., China)
Culicicapa ceylonensis pallidior Ticehurst, 1927, Bull. Brit. Orn. Cl. 47: 108 (Simla)
Baker, FBI No. 679 (part), Vol. 2: 254

LOCAL NAME. Zărd-phūtki (Bengali).
SIZE. Sparrow −; length c. 9 cm (3½ in.).
FIELD CHARACTERS. Head, neck, throat and breast ashy grey, darker on crown. Back yellowish green, rump yellow. Belly bright yellow. Wings and tail brown edged with yellow. Sexes alike.

STATUS, DISTRIBUTION and HABITAT. Common summer visitor or resident in the hills, spreading widely over the plains in winter. The Himalayas from the Indus Valley and Kashmir (scarce) east to Nepal, Sikkim, Bhutan to Arunachal Pradesh in the Mishmi Hills, thence south through Nagaland, Manipur, Meghalaya and the hills of Assam and Bangladesh south to the Chittagong Hill Tracts. Breeds chiefly between 1500 and 2400 m, locally up to 2700 m or even 3100 m (Nepal - Inskipp & Inskipp); in the eastern Himalayas and in Assam breeds down to the foothills. Also the central

× c. 1

Culicicapa ceylonensis

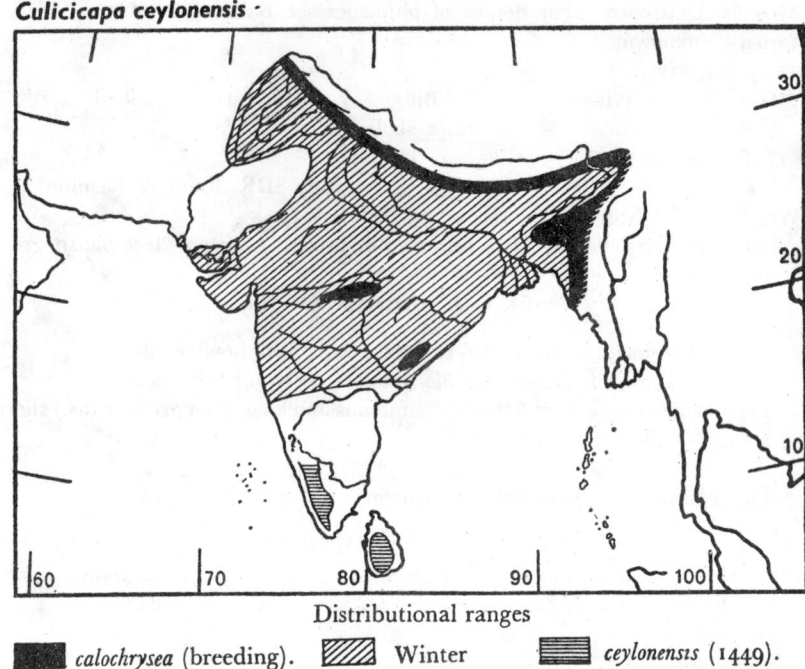

Distributional ranges

■ *calochrysea* (breeding).   ▨ Winter   ≡ *ceylonensis* (1449).

Satpuras above 900 m (Betul and Pachmarhi) but nesting not substantiated, and almost certainly the Eastern Ghats (Visakhapatnam district) Lammasinghi, feeding fledged juveniles, 13th July (Trevor Price, JBNHS 76: 415). Affects open wooded country.

Extralimital. Burma. The species extends to central western China, the Indochinese countries, Palawan, Borneo and the Sunda islands to Timor. A closely related species in the Philippines to Celebes.

MIGRATION. Mostly a summer visitor to the Himalayas above 1500 m from early March till September. Winters in the foothills up to 500 m (around Islamabad, Pakistan) and up to 1800 m in the eastern Himalayas, and in the plains west to Peshawar and Bannu, and throughout the Indus plains to Sind (TJR) and over most of the Indian Peninsula west to Kutch, south to about the Krishna river. In this season affects forests and groves.

GENERAL HABITS. A lively and confiding bird. Keeps in pairs in the breeding season, in family parties thereafter. In winter usually seen singly or in separated twos and threes in the mixed hunting parties of tits, nuthatches, leaf warblers and other small birds, acting as outriders and snapping up any winged insects escaping from the main body. Frequents forest glades, wooded ravines, village groves, upcountry gardens with large trees, etc. Keeps to the lower and middle strata, generally between two and four metres above the ground. Hunts among the inner branches, actively flitting from bough to bough, turning this way and that on its perch, loosely flicking the tail. Makes agile acrobatic dashes after insects which it seizes with a loud snap of the bill, twisting and looping in the air and returning to the same base again and again, all the while uttering its distinctive, sharp twittering note. Telegraph wires through wooded country provide favourite hunting bases. Body pose erect.

Display. A bird of a pair would keep flying without ceasing for nearly ten minutes, going round and round in a rough circle of some thirty metres in diameter, flying deliberately in front of a perched bird, presumably the female, turning and flying out again, then hovering for a moment on rapidly vibrating wings; about this time, the presumed female would fly straight at the male, then return to a perch. All the while both birds called continuously. The performance was repeated at least fifteen to twenty times (Lister).

FOOD. Tiny winged insects.

VOICE and CALLS. Song, a surprisingly loud and lively high-pitched trill heard nearly the whole year—an interrogative *chik* ... *whichee-whichee?* constantly uttered; this usually followed after a second or so by *whi-chichi*, like question and answer. In breeding season these repeated at quick and regular intervals for long periods (once timed 25 minutes at dawn) from the same stance (SA). Some variations of the song rendered as *tee-tata-tei, tyissi-a-tyi,* etc. (for more details see Lister, JBNHS 52: 43). Call-notes (by same observer): A very soft *pit* ... *pit* ... *pit* while foraging; a frequent, clear *kitwik* ... *kitwik* (high-low); and a series of *kui-whi-whi* (rising-low-low); a quiet *chichictrrr* with variations.

BREEDING. *Season,* April to June. *Nest,* a half cup of moss and lichen bound with cobweb, usually unlined, fastened against a moss- or lichen-covered trunk or rock at heights varying from near the ground to about twelve metres; generally composed of the same materials as the substrate, thus remarkably obliterated in its surroundings. *Eggs,* 3 or 4, pale buff, marked with grey blotches and specks, forming a ring around the large end. Average size of 40 eggs 15 × 12 mm (Baker). Share of the sexes in the domestic chores, and incubation period, unrecorded.

MUSEUM DIAGNOSIS. For distinction from *ceylonensis* see 1449.

Young, a slightly duller and paler replica of adult; no spotting.

MEASUREMENTS

	Wing	Bill (from skull)	Tarsus	Tail	
♂♂	63–68	12–14	13–14	51–60	mm
♀♀	57–66	12–14	13–14	50–58	mm
				(BB, SA, HW)	

Weight 20 ♂ ♀ 7–9 g (SA). 1 ♀ 8 g, 2oo? 7 (2) g (SDR).

COLOURS OF BARE PARTS. Iris brown. Bill brown, lower mandible paler or pinkish flesh. Legs and feet yellowish brown.

1449. *Culicicapa ceylonensis ceylonensis* (Swainson)

Platyrhynchus ceylonensis Swainson, 1820, Zool. Ill. 1, No. 3, pl. 13 and text (Ceylon)

Cryptolopha poiocephala Swainson, 1838, *in* Nat. Library, Flycatchers 21: 200. New name for *P. ceylonensis* Swainson

Baker, FBI No. 679 (part), Vol. 2: 254

LOCAL NAME. *Nārāyănpăkshi* (Malayalam).

SIZE. Sparrow −; length *c.* 9 cm (3½ in.).

FIELD CHARACTERS. As in 1448, q.v.

STATUS, DISTRIBUTION and HABITAT. Common resident, wandering to lower elevations in winter. The hills of southwestern India from southern Karnataka to the Ashambu Hills (including the Nilgiris and Palnis), from *c.* 900 m to the highest summits. May also breed farther north in the Western Ghats, but summer data lacking. Also Sri Lanka, generally above 900 m, but down to the foothills in the southwestern Wet zone. Affects evergreen forest, sholas, bamboo facies, wooded ravines, secondary and mixed bamboo forest especially in the vicinity of streams, coffee plantations, and gardens.

GENERAL HABITS, FOOD and VOICE. As in 1448.

BREEDING. *Season*, March to June, mostly April in the Peninsula, and February to May, mostly March in Sri Lanka. *Nest* and *eggs* as in 1448. Clutch size normally 3 in the Peninsula, 3 or often 2 in Sri Lanka. Average size of 100 eggs 15·1 × 12 mm (Baker). Building of nest recorded as by female only.

MUSEUM DIAGNOSIS. Differs from *calochrysea* (1448) in being darker. Birds from the Peninsula are intermediate between those from Sri Lanka and the Himalayas.

MEASUREMENTS

	Wing	Bill (from skull)	Tarsus	Tail	
♂♂	58–66	12–14	12–13	49–59	mm
♀♀	59–63	12–13	12–13	50–54	mm
				(SA, HW)	

Weight 6 ♂ ♀ 6–7 g (SA). 1 ♀ 9 g (SDR).

COLOURS OF BARE PARTS. Iris dark brown. Bill: upper mandible horny brown, lower pale yellowish horn; mouth yellow. Legs and feet brownish orange; claws horny brown; soles bright orange.

Subfamily RHIPIDURINAE: FANTAIL FLYCATCHERS
Genus RHIPIDURA Vigors & Horsfield

Rhipidura Vigors & Horsfield, 1827, Trans. Linn. Soc. London 15: 246.
Type, by subsequent designation, *Muscicapa flabellifera* Gmelin
Leucocirca Swainson, 1838, Nat. Library, Flycatchers 21: 126.
Type, by monotypy, *Turdus leucophrys* Latham
Chelidorhynx 'Hodgson' = Blyth, 1843, Jour. Asiat. Soc. Bengal 12: 930, 936 footnote. Type, by monotypy, *Rhipidura hypoxantha* Blyth

Bill large, about twice as long as wide at base. Rictal bristles numerous and long. Tail longer than wing, graduated. Young not spotted but feathers fringed with rufous.

Key to the Species

		Page
A	Underparts yellow .... *R. hypoxantha*	205
B	Underparts not yellow	
	1 Breast and belly white .... *R. aureola*	206
	2 Breast slaty; belly uniformly slaty, or buffish .... *R. albicollis*	210

1450. Yellowbellied Fantail Flycatcher. *Rhipidura hypoxantha* Blyth

Rhipidura hypoxantha Blyth, 1843, Jour. Asiat. Soc. Bengal 12: 930, 935
(Darjeeling)
Chelidorhynx hypoxantha noa Koelz, 1939, Proc. Biol. Soc. Washington 52: 68
(Naggar, Kulu)
Baker, FBI No. 699, Vol. 2: 275
Plate 83, fig. 1 (Vol. 8)

LOCAL NAMES. *Sitte kloom* (Lepcha); *Pongking-lo, Bang-ho-go* (Naga).
SIZE. Sparrow —; length *c.* 8 cm (3 in.).
FIELD CHARACTERS. A restless, diminutive fantail flycatcher. *Above*, dark greyish olive. Forehead and supercilium yellow. A broad black band from lores through eye and ear-coverts. Tail brown with conspicuously white shafts and white tips. *Below*, bright yellow.
Female differs only in having the eye-band blackish olive-brown (*v.* black).

× *c.* 1

STATUS, DISTRIBUTION and HABITAT. Common altitudinal migrant. The Himalayas from the Chenab river eastward through Nepal, Sikkim and Bhutan, and Arunachal Pradesh to the Burma border; thence south through the Patkai Range, Nagaland, Manipur, Assam in the Cachar and Meghalaya in the Khasi hills (scarce), and Bangladesh in the Chittagong Hill Tracts. Breeds up to tree-line (generally about 3600 m, but in north-central Nepal near 4000 m); lower limit in the western Himalayas not clear, probably about 2400 m; in Bhutan down to 1800 m. In Assam breeds down to *c.* 1000 m (Baker) and on Mt Victoria between 2000 and 2800 m. Winters (October to March) from *c.* 1800 m down to 500 m in the Himalayan foothills and adjacent plains around Islamabad and Sialkot (Pakistan); in Assam and hill ranges to the east, does not appear to descend as low. Affects various kinds of forest and secondary jungle but prefers moist-evergreen biotope. In the breeding season, mostly mixed coniferous and birch or rhododendron forests.
Extralimital. Extends to southeastern Tibet, Yunnan and the Indochinese countries.
GENERAL HABITS. An extremely lively and restless little flycatcher usually met with singly or in pairs, almost invariably in the mixed hunting parties of small insectivorous birds during the non-breeding season. Frequents the lower canopy and higher shrubs, but often hunts from low bushes as well as the canopy of tall trees up to thirty metres or more. Flits and prances among the branches and foliage, pirouettes incessantly with fanned-out tail and partly drooping wings uttering its distinctive *sip, sip* note. Searches for insects in the foliage by fluttering against the sprigs in the manner of a *Phylloscopus*, and launches sprightly twisting and looping sallies after gnats, often springing up and returning to its perch perpendicularly.
FOOD. Tiny winged insects.
VOICE and CALLS. A very thin and high *sip, sip* constantly uttered while foraging. Song, a quick repetition of the call-note (Heinrich), described as 'a feeble little goldfinch-like trill' (Jones). Song period in Nepal, February to at least July (Proud). A male tape-recorded in early April sang for 1.5 minutes without pause with rapid tittering trills interspersed with shorter phrases and tweeping calls (TJR).

BREEDING. *Season*, May (in Assam) to June and July (Himalayas). *Nest*, a deep, thick-walled cup of moss, compactly matted together, plastered with cobweb and lichen, lined with fruit-stems of moss, hair or feathers; placed on a horizontal branch at least as thick as the diameter of the nest making it difficult to spot from below; usually between three and six metres from the ground. *Eggs*, normally 3, cream or pinkish cream, marked with reddish stippling which tends to form a zone around the large end. Average size of 24 eggs 14·4 × 11·3 mm (Baker).

MUSEUM DIAGNOSIS. Female differs from male in having the lores, ocular area and ear-coverts dark olive instead of black.

Young resembles female but has no yellow on forehead. Supercilium and underparts dull pale yellow. Back more grey.

MEASUREMENTS

	Wing	Bill (from skull)	Tarsus	Tail	
♂♂	53–58	9–10	14–15	56–58	mm
♀♀	53–58	9–10	14–15	50–58	mm
				(BB, BA)	

Weight 8 ♂ ♂ 5–6; 2 ♀ ♀ 5, 6 g (GD, SDR, SA).

COLOURS OF BARE PARTS. Iris dark brown. Bill: upper mandible dark brown, lower yellowish brown. Legs and feet horny brown with a yellowish tinge; claws horny brown.

RHIPIDURA AUREOLA Lesson: WHITEBROWED FANTAIL FLYCATCHER

Key to the Subspecies

		Page
A	White on outer pair of rectrices not reaching under tail-coverts .... *R. a. compressirostris*	209
B	White on outer pair of rectrices reaching under tail-coverts	
	1 Darker above, crown blacker than back .... *R. a. aureola*	206
	2 Paler above, crown nearly concolorous with back .... *R. a. burmanica*	209

1451. *Rhipidura aureola aureola* Lesson

Rhipidura aureola Lesson, 1830, Traité d'Orn.: 290 (Bengal)
Baker, FBI No. 700 (part), Vol. 2: 277
Plate 66, fig. 4 (Vol. 5)

LOCAL NAMES. *Shamchiri* (Urdu); *Machharya, Nāchăn, Chakdil* (Hindi); *Nāchăn* (Gujarati).

SIZE. Bulbul −; length *c.* 17 cm (6½ in.).

FIELD CHARACTERS. A cheery restless fan-tailed smoke-brown flycatcher with broad white forehead and white underparts.

Above, crown and ear-coverts black. Forehead and conspicuous broad supercilium white. Back and wings dark grey-brown, the latter with two rows of white spots. Tail blackish, outer rectrices white, others, except central pair, tipped with white. *Below*, throat black, chin and sides of throat whitish. Sides of breast black. Breast and belly white. Sexes alike, but female slightly paler and browner on head.

Distinguished from *R. albicollis* (1455) by its broad supercilium extending to nape and by its pure white belly and breast.

Rhipidura aureola

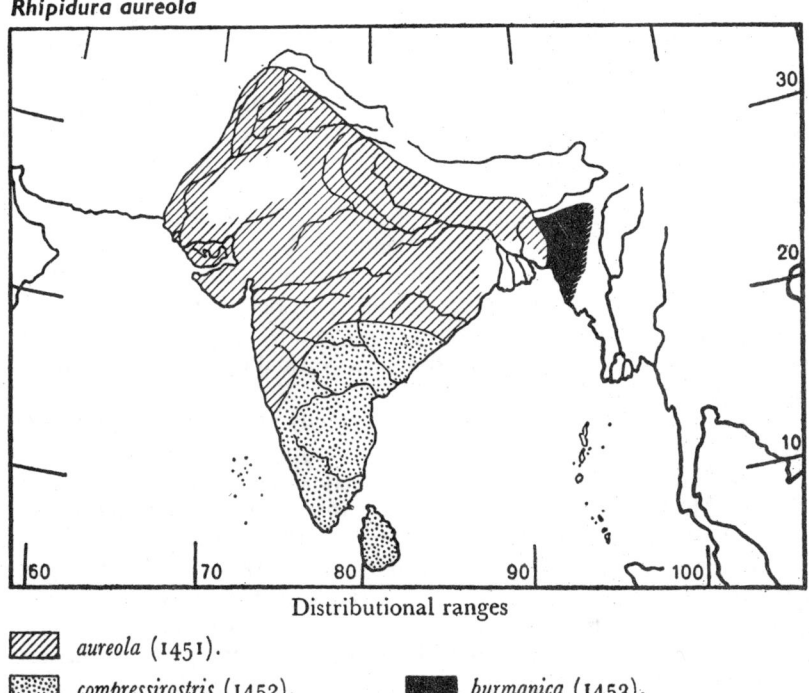

Distributional ranges

▨ *aureola* (1451).

▦ *compressirostris* (1452).   ■ *burmanica* (1453).

STATUS, DISTRIBUTION and HABITAT. Common and widespread resident, subject to local movements in winter. Pakistan and India from the Indus river east along the Himalayan foothills (up to *c*. 600, locally 1000 m) to the Jalpaiguri duars in northern W. Bengal, Bhutan duars and western Bangladesh, including the lowlands east of the Brahmaputra. Widespread and common in the irrigated canal colonies of the Punjab in Pakistan, but much less common and confined to riverain forest in southern Sind (TJR); occurs through Gujarat to southwestern and central Maharashtra (where it intergrades with *compressirostris*); eastward to Madhya Pradesh and Orissa. Absent from the Thar Desert, lower West Bengal and coastal Bangladesh. Affects forest, groves of mango, babool, etc., tamarisks along canals, fallow land with sparse scattered bushes and, especially in winter, wooded compounds and gardens. Prefers more open and drier country than *R. a. albogularis* (1458), but the two are sometimes found side by side in intermediate habitats.

GENERAL HABITS. Generally met with singly or in pairs or family parties, often joining the mixed hunting parties of *Turdoides, Dumetia, Prinia,*

Phylloscopus, etc. Frequents the lower bushes and undergrowth usually near the ground, occasionally descending to the ground. Flits tirelessly from branch to branch, prancing and pirouetting with fanned, erect tail and half-drooping wings, making graceful looping sallies after winged insects which it catches with an audible snap of the mandibles. Sometimes attends grazing cattle, snapping up the tiny insects disturbed by the animals, riding momentarily on the backs of the goats and cows, using them as mobile hunting bases. Often prances lightly in twists and turns up tree-trunks and boughs. From time to time suddenly flicks open wings and tail to stampede winged insects lurking in the crevices of the bark. Makes short aerial darts in their pursuit the moment they attempt to flee, snapping them up in its bill. Usually unafraid of man, and aggressive in mobbing predators such as cats and crows trespassing into proximity of nest, uttering harsh shrike-like 'war-cries'.

FOOD. Insects, chiefly diptera and hemiptera. The jassid mango-hoppers *Idiocerus clypealis* and *I. niveosparsus* identified in stomach contents.

VOICE and CALLS. Song, a delightful clear but rather thin human-like whistle of six to eight tinkling notes, the first part rising, the last falling, often stopping abruptly in the middle of the scale and left unfinished; rendered as *chee-chee-cheweechee-vi* (Dharmakumarsinhji). Normal call-notes a harsh *chuck-chuck* (SA). Alarm-note given as *ch'wch*.

BREEDING. *Season*, February to August, chiefly March to June. *Nest*, a neat compact cup, usually rounded off at the bottom and lacking the untidy 'tail' dangling below that of *albicollis*; inner diameter of cup *c*. 5 cm, depth *c*. 2.5 cm. Made of fine fibres and grasses, profusely plastered exteriorly with cobweb, and lined within with grass fibres. Attached to the upper surface of a small outhanging branch, often in a horizontal fork; usually placed much higher than that of *albicollis*, seldom under a couple of metres and often up to twelve. The same site is commonly used year after year. *Eggs*, normally 3, cream to buff, with a zone of greyish brown specks and spots at the larger end. Average size of 100 eggs 16·8 × 12·8 mm (Baker). Building, incubation and care of young by both sexes. Incubation period undetermined.

MUSEUM DIAGNOSIS. For distinction from *compressirostris* and *burmanica* see 1452 and 1453 respectively.

Young, like adult but upperparts and wing-coverts tipped with rufous. Postjuvenal moult of body-feathers and all coverts except primary.

MEASUREMENTS

	Wing	Bill (from skull)	Tarsus	Tail	
♂♂	79–90	14–16	18–20	84–100	mm
♀♀	72–84	13–16	18–19	83–98	mm
			(HW, SA, BB)		

Weight 1 ♂ 12; 2 ♀♀ 10, 10 g (BB, Roonwal).

COLOURS OF BARE PARTS. Iris dark brown. Bill, legs and feet brownish black; soles dirty white.

FLYCATCHERS

1452. *Rhipidura aureola compressirostris* (Blyth)

Leucocerca compressirostris Blyth, 1849, Jour. Asiat. Soc. Bengal 18: 815
(Ceylon)
Baker, FBI No. 700 (part) and 702, Vol. 2: 277, 279

LOCAL NAMES. *Āttăkkārăn* (Malayalam); *Dărāri-pitta* (Telugu); *Visiri-vāli* (Tamil); *Vali-marittan* (Sri Lanka Tamil); *Endēra-kurullā, Mārāwā, Nătănă-kurullā* (Sinhala).

SIZE. Bulbul —; length *c.* 17 cm (6½ in.).

FIELD CHARACTERS. As in 1451, q.v.

STATUS, DISTRIBUTION and HABITAT. Common resident, subject to local movements in winter. Southern India south of the range of the nominate race (1451); in the plains and hills up to 1000 m. Also Sri Lanka in the lowlands and hills up to *c.* 1500 m. Affects deciduous forest, groves, wooded compounds, orchards, gardens and, to a lesser extent, light and secondary scrub jungle.

GENERAL HABITS, FOOD and VOICE. As in 1451.

BREEDING. *Season*, January to July, chiefly April and May. *Nest* and *eggs* as in 1451.

MUSEUM DIAGNOSIS. Differs from *aureola* (1451) in having *two* central pairs of rectrices black without white tips (*v.* one pair). White tips of outer pair of rectrices less extensive, only about half the length of the feather and not reaching the under tail-coverts (*v.* three-quarters in *aureola*). Upper plumage slightly darker, more sooty when freshly moulted. Tail a little shorter. There are of course complete intergrades and individual variations.

MEASUREMENTS

	Wing	Bill (from skull)	Tarsus	Tail	
♂♂	80–87	15–16	17–18	80–92	mm
♀♀	76–85	14–15	17–18	80–90	mm
				(HW, SA)	

Weight 4 ♂ ♀ 11–13 g (SA).

COLOURS OF BARE PARTS. As in 1451.

1453. *Rhipidura aureola burmanica* (Hume)

Leucocerca burmanica Hume, 1880, Stray Feathers 9: 175, footnote
(Thoungyeen valley)
Baker, FBI No. 701, Vol. 2: 278

LOCAL NAME. *Dao-phari* (Cachari).

SIZE. Bulbul —; length *c.* 17 cm (6½ in.).

FIELD CHARACTERS. As in 1451, q.v.

STATUS, DISTRIBUTION and HABITAT. Common resident. Assam (Cachar) and Bangladesh (east of the range of *aureola*) in the plains and hills up to 1500 m. Not recorded from Arunachal Pradesh, the Brahmaputra valley and Nagaland. Affects dry, open country, parkland, scrub and bamboo jungle.

Extralimital. Burma. The species extends east to the southern Indochinese countries.

GENERAL HABITS, FOOD and VOICE. As in 1451.

BREEDING. *Season*, April to July, chiefly May. *Nest* and *eggs* as in 1451. Average size of 100 eggs 17·2 × 12·8 mm (Baker).

MUSEUM DIAGNOSIS. Differs from *aureola* in being paler above; crown not so black, more concolorous with back. Spots on coverts smaller. White on outer rectrices on average less extensive.

MEASUREMENTS

	Wing	Bill (from skull)	Tarsus	Tail	
♂♂	83–90	14–15	17–18	95–102	mm
♀♀	81–84	14–15	17–18	88–95 (HW)	mm

COLOURS OF BARE PARTS. As in 1451.

RHIPIDURA ALBICOLLIS (Vieillot): WHITETHROATED FANTAIL FLYCATCHER

Key to the Subspecies

		Page
A Breast unspotted		
a Abdomen slaty		
1 Paler, more ashy	*R. a. canescens*	210
2 Darker, more mouse-grey	*R. a. albicollis*	212
3 Still darker, more slaty	*R. a. stanleyi*	213
b Abdomen with a small patch of buff	*R. a. orissae*	213
B Breast spotted, belly buff		
c Buff restricted to belly, flanks grey	*R. a. vernayi*	215
d Abdomen and flanks buff	*R. a. albogularis*	214

1454. *Rhipidura albicollis canescens* (Koelz)

Leucocirca albicollis canescens Koelz, 1939, Proc. Biol. Soc. Washington 52: 68 (Bhadwar, Punjab)

Baker, FBI No. 703 (part), Vol. 2: 279

LOCAL NAMES. *Machharya, Chakdil* (Hindi).

SIZE. Bulbul −; length *c.* 17 cm (6½ in.).

FIELD CHARACTERS. A dark, slaty brown fan-tailed flycatcher with a prominent white band or semi-collar across the throat and short white supercilia; outer rectrices tipped with whitish. Sexes alike.

Distinguished from other members of the genus by its entirely dark belly.

STATUS, DISTRIBUTION and HABITAT. Common resident, subject to vertical movements. The Himalayan foothills from Murree and Kashmir to western Nepal. Breeds commonly from the foothills to 1500 m, more sparsely up to 2000 m; a June record at 3000 m in Lahul (Alexander, JBNHS 49: 610). Winters in the foothills and adjacent plains. Affects shady places in forest (pine, deciduous, evergreen, etc.), wooded nullahs, gardens, groves and secondary scrub.

GENERAL HABITS. Keeps singly or in pairs, often as outriders of the mixed hunting parties of small insectivorous birds, frequenting undergrowth and

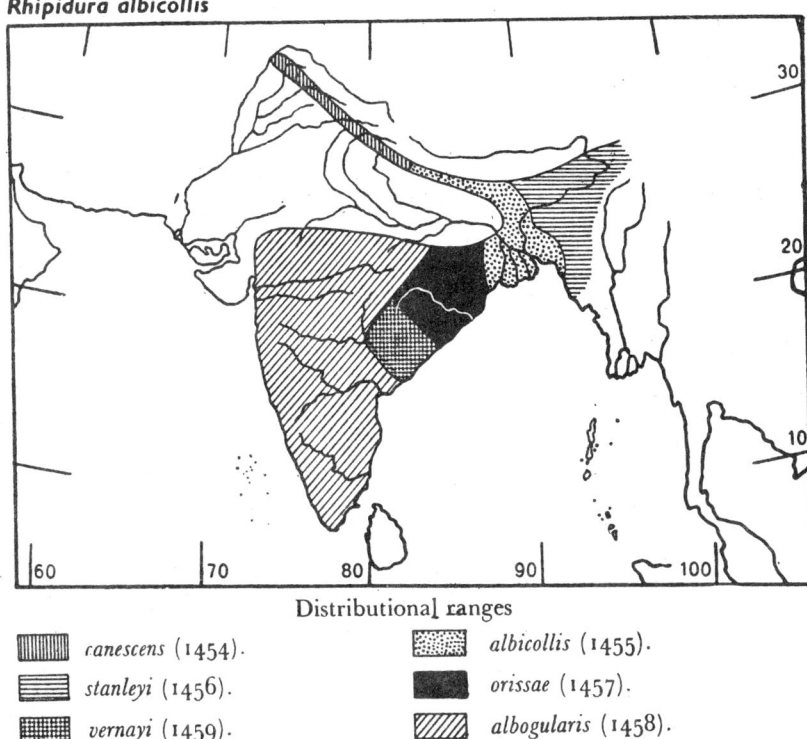

Rhipidura albicollis

Distributional ranges

- canescens (1454).
- albicollis (1455).
- stanleyi (1456).
- orissae (1457).
- vernayi (1459).
- albogularis (1458).

middle story. Flits tirelessly among the branches—tail fanned out and partly erect, wings drooping at the sides—usually close to the main trunk, working up or down the trunk, often moving rapidly out along a bough by a series of rapid, jerky twists and turns, continually flicking open the wings and flirting the tail purposefully. Makes sprightly looping sallies after winged insects thus disturbed, occasionally launching short sorties beyond the canopy. Also hunts near the tops and outside of bushes, tumbling about them as it turns this way and that. Flight from one tree to another fairly direct and slightly dipping, with tail often fanned out.

FOOD. Insects (gnats, flies, etc.).

VOICE and CALLS. Normal note a harsh *chuck* or *chuck-r*. Song, a group of about five thin whistling quarter-notes on a descending scale, rather feeble and jerky, rendered as *tri-riri-riri*, very reminiscent of song of *Phylloscopus magnirostris* (SA).

BREEDING. *Season*, March to August; double-brooded. *Nest*, neat, cup-shaped, rather like an inverted cone, of fine grass stems bound together by a thick external coating of cobwebs; usually without any special lining, and with an untidy 'tail' of strips of nest material dangling below. This feature usually distinguishes it from nest of *R. aureola* (1415). Built in the horizontal fork or elbow of a twig, seldom more than 3 metres from the ground, usually under two. *Eggs*, normally 3, varying from white to a dingy cream colour marked with grey-brown spots and specks almost invariably confined to a zone around the large end; very similar to those of *R. aureola*. Average size of 100 eggs (including those of subspecies *albicollis* and *stanleyi*) 17·3 × 13 mm (Baker). Building of nest and incubation by both sexes. Incubation period 12–13 days; young leave nest on 13th to 15th day.

MUSEUM DIAGNOSIS. Differs from *albicollis* (1455) in being paler, more ashy on back, breast and belly.

Young, like adult but body-feathers, all wing-coverts and tertials tipped with rufous. White of throat less pure and more diffuse. Postjuvenal moult of body-feathers and all coverts except primary.

MEASUREMENTS

Wing: 6 ♂ ♂ 77–81; 4 ♀ ♀ 74–76 mm (Koelz)

COLOURS OF BARE PARTS. As in *albicollis* (1455).

1455. *Rhipidura albicollis albicollis* (Vieillot)

Platyrhynchus albicollis Vieillot, 1818, Nouv. Dict. d'Hist. Nat. 27: 13 (Bengale)

Baker, FBI No. 703 (part), Vol. 2: 279

Plate 66, fig. 3 (Vol. 5)

LOCAL NAMES. *Naklaychara* (Paharia); *Nam-dit-nom* (Lepcha); *Dūmchitri* ('spread-tail'), *Chak-dayal, Chak-dil* (Bengali).

SIZE. Bulbul −; length *c.* 17 cm (6½ in.).

FIELD CHARACTERS. As in 1454, q.v.

STATUS, DISTRIBUTION and HABITAT. Common resident, subject to vertical movements. Western Nepal (where it intergrades with *canescens*) east to Sikkim, Darjeeling and the Jalpaiguri district, and south through the plains of Bengal and Bangladesh and lower West Bengal (but in the west not south of Nepal terai). Breeds from the terai up to 1700 m, locally to 2300 m, and in the plains of Bangladesh and West Bengal to the Sunderbans.

× *c.* 1

GENERAL HABITS, FOOD and VOICE. As in 1454.

BREEDING. As in 1454.

MUSEUM DIAGNOSIS. Differs from *canescens* (1454) in being darker; from *stanleyi* in being paler, less slaty; from *orissae* in being mouse-grey on back (*v.* olive-brown).

MEASUREMENTS

	Wing	Bill (from skull)	Tarsus	Tail	
♂♂	75–84	14–16	19–20	96–109	mm
♀♀	72–80	14–16	19–20	93–102	mm

(SA, BB, Rand & Fleming)

COLOURS OF BARE PARTS. Iris brown. Bill brownish black. Legs, feet and claws horny brown.

FLYCATCHERS

1456. *Rhipidura albicollis stanleyi* Baker

Rhipidura albicollis stanleyi Baker, 1916, Bull. Brit. Orn. Cl. 36: 81. New name for
R. albicollis kempi Baker, 1913 (Sept.), Rec. Ind. Mus. 8: 275 (Abor Hills), preoccupied
by *R. flabellifera kempi* Mathews & Iredale, 1913 (July)
Baker, FBI No. 703 (part), Vol. 2: 279

LOCAL NAMES. None recorded.
SIZE. Bulbul —; length *c.* 17 cm (6½ in.).
FIELD CHARACTERS. As in 1454, q.v.
STATUS, DISTRIBUTION and HABITAT. Common resident, subject to vertical or local movements. Bhutan and Arunachal Pradesh south through the plains and hills of Assam, Meghalaya, Nagaland and Manipur to the Chittagong region of Bangladesh. In the Himalayas breeds up to 2700 m, but most plentiful between 900 and 1500 m; in the Assam hills south of Brahmaputra R. breeds up to 1500 m. Affects various kinds of forest (humid, evergreen, deciduous, dry pine, etc.), as well as secondary jungle and gardens.
Extralimital. Burma. The species extends to Sichuan, Vietnam, Sumatra and Borneo.
GENERAL HABITS, FOOD and VOICE. As in 1454.
BREEDING. As in 1454.
MUSEUM DIAGNOSIS. Differs from *albicollis* (1455) in being darker, more slaty.
MEASUREMENTS and COLOURS OF BARE PARTS. As in 1455.
Weight 6 ♂ ♀ 9–13 g (SDR, SA).

1457. *Rhipidura albicollis orissae* Ripley

Rhipidura albicollis orissae Ripley 1955, Proc. Biol. Soc. Washington 68: 42
(Toda, Bonai, Orissa)
Baker, FBI No. 703 (part), Vol. 2: 279

LOCAL NAMES. None recorded.
SIZE. Bulbul —; length *c.* 17 cm (6½ in.).
FIELD CHARACTERS. As in 1454 but with a patch of buff in centre of belly.
STATUS, DISTRIBUTION and HABITAT. Common resident. Southern Bihar, Orissa and eastern Madhya Pradesh. This population is intermediate between *albicollis* and *vernayi*, and intergrades with the latter. Affects forest, groves, gardens and shrubbery.
GENERAL HABITS, FOOD and VOICE. As in 1454.
BREEDING. As in 1454.
MUSEUM DIAGNOSIS. Differs from *albicollis* in being dark olive-brown above, well demarcated from the black crown, and by the buff patch in centre of belly; from *vernayi* in lacking the pectoral spotting and in having a more reduced buff area on abdomen; however, intermediates commonly occur.
MEASUREMENTS and COLOURS OF BARE PARTS. As in 1455.

1458. *Rhipidura albicollis albogularis* (Lesson)[1]

Muscicapa (Muscylva) albogularis Lesson, 1832, in Bélanger, Voy. Ind.-Orient., Zool.: 264 (le continent de l'Inde et les environs de Pondichéry = Salem district, Madras)
Leucocirca pectoralis Jerdon, 1847, Ill. Ind. Orn., text to plate 2
(Nilgiris)
Baker, FBI No. 705 (part), Vol. 2: 282
Plate 66, fig. 3 (Vol. 5)

LOCAL NAMES. *Măchhăryă* ('mosquito-catcher', Hindi); *Nāchăn* ('dancer', Marathi); *Dāsări pittā* (Telugu).

SIZE. Sparrow with a long tail; length *c.* 17 cm (6½ in.).

FIELD CHARACTERS. As in 1454 but entire belly buff; a slaty breast-band spotted with white. Tail sooty brown with the lateral feathers only diffusing to narrow whitish tips (*v.* broadly tipped with pure white).

R. aureola (1451) has a pure white belly and a large amount of white in its tail.

STATUS, DISTRIBUTION and HABITAT. Common resident, unevenly distributed, subject to restricted local movements. The Indian peninsula (except for the ranges of *vernayi* and *orissae*) from Mt Abu and the Vindhya Range southward to Kerala—coastal plain, central plateau, Western and Eastern Ghats and their associated hills up to *c.* 2000 m (Nilgiris). Affects well-wooded areas—secondary jungle, groves and gardens, even in cities. Prefers more broken country and more humid facies than *R. aureola*, although both species may often be found in the same area.

GENERAL HABITS and FOOD. As in 1454.

VOICE and CALLS. Song of the same general pattern as those of Whitebrowed and Whitethroated Fantails (1451 and 1454): a lively clear whistling of several tinkling notes rising and falling in scale, constantly warbled as the bird waltzes and pirouettes on low branches and brushwood. Also has the normal harsh *chuck* or *chuck-r* calls.

BREEDING. *Season*, March to July occasionally extending into September; often two broods. *Nest* and *eggs* as in 1454. Average size of 50 eggs 16·2 × 12·7 mm (Baker). Both sexes take part in all the domestic chores. Incubation period 12–13 days.

MUSEUM DIAGNOSIS. For distinction from *vernayi* see 1459.

MEASUREMENTS

	Wing	Bill (from skull)	Tarsus	Tail	
♂♂	72–79	14–15	18–19	86–98	mm
♀♀	69–74	13–14	17–19	86–94	mm
				(SA, HW)	

Weight 6 ♂ ♀ 10–13 g (SA).

COLOURS OF BARE PARTS. Iris dark brown. Bill black; mouth pale flesh (mouth and gape yellow in the nestling). Legs and feet slaty brown.

---

[1] Fleming & Traylor (1964) have shown (*Fieldiana*, Zoology 35: 539–40) that *albicollis* and *albogularis* intergrade in the northeastern Peninsula. Their suggestion that they should be treated as subspecies is followed here.

FLYCATCHERS

1459. *Rhipidura albicollis vernayi* (Whistler)

Leucocirca pectoralis vernayi Whistler, 1931, Bull. Brit. Orn. Cl. 52: 40
(Jeypore Agency)
Baker, FBI No. 705 (part), Vol. 2: 282

LOCAL NAMES. None recorded.

SIZE. Bulbul −; length *c.* 17 cm (6½ in.).

FIELD CHARACTERS. As in 1458, q.v.

STATUS, DISTRIBUTION and HABITAT. Common resident. The upper Eastern Ghats (Dandakāranya area) from southern Orissa to the Godavari river. Habitat as in 1458.

GENERAL HABITS, FOOD and VOICE. As in 1454 and 1458.

BREEDING. As in 1458.

MUSEUM DIAGNOSIS. Differs from *albogularis* (1458) by the broader pectoral band continued as a dark wash on the flanks. Pectoral spots reduced in size and number. *Orissae* (1457) lacks any spotting on breast and has a still more reduced pale area on abdomen.

MEASUREMENTS

	Wing	Bill (from skull)	Tarsus	Tail	
5 ♂♂	72–81	14–15	18–19	93–104	mm
3 ♀♀	70–74	13–14	18–19	88–96	mm
Tail 2 ♂ ♂ 88, 90.5 mm				(HW)	

Weight 7 ♂♀ 11–13 g (SA). 2 ♂♂ 7.5, 9.5 g (SDR).

COLOURS OF BARE PARTS. As in 1458.

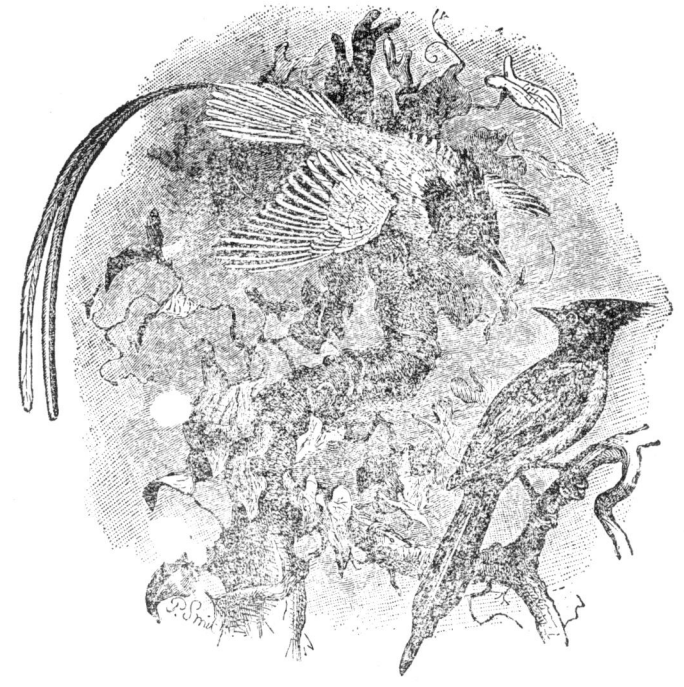

Paradise Flycatcher (1461)

Subfamily MONARCHINAE: Monarch Flycatchers

Key to the Genera

Tail longer than wing; head crested .... *Terpsiphone*
Tail about equal to wing; head not crested .... *Hypothymis*

Genus TERPSIPHONE Gloger

Terpsiphone Gloger, 1827, *in* Froriep's Notizen 16, col. 278. Type, by subsequent designation, *Corvus paradisi* Linnaeus

Tchitrea Lesson, 1830, Traité d'Orn.: 386. Type, by subsequent designation, *Corvus paradisi* Linnaeus

Bill very large, depressed and swollen at base. Rictal bristles long, coarse and numerous. Tarsus short but fairly stout. Central pair of rectrices greatly elongated in male. Young not spotted.

TERPSIPHONE PARADISI (Linné): PARADISE FLYCATCHER

Key to the Subspecies

		Page
A Male not assuming white plumage .... *T. p. ceylonensis*		220
B Male assuming white plumage		
1 Rufous male with greyish throat .... *T. p. saturatior*		221
2 Rufous male with metallic black throat		
a Back with a strong olive wash .... *T. p. nicobarica*		222
b Back cinnamon		
i Paler .... *T. p. leucogaster*		216
ii Darker .... *T. p. paradisi*		218

1460. *Terpsiphone paradisi leucogaster* (Swainson)

Muscipeta leucogaster Swainson, 1838, Nat. Library, Flycatchers 21: 205
(Simla, *fide* Kinnear, 1929, Ibis: 131)
Baker, FBI No. 690, Vol. 2: 268

LOCAL NAMES. *Fhāmbăsīr* = 'cotton flake' (♂), *Răngă būlbūl* (♀), *Lătrāz* (Lolab) [Kashmir]; *Taklal* (Urdu); Peninsular names as in 1461.

SIZE. Bulbul ±; male with very long tail-streamers. Length *c*. 20 cm (8 in.); with streamers up to *c*. 50 cm (20 in.).

FIELD CHARACTERS. As in 1461, q.v.

STATUS, DISTRIBUTION and HABITAT. Common summer (breeding) visitor. Northern Baluchistan, scarce N.W.F.P. north to Chitral, east through the Salt range, Murree foothills (TJR), Kashmir, Punjab, Himachal Pradesh, Uttar Pradesh and Nepal. Breeds in the northern plains (sparingly south to Lahore, Delhi, Lucknow and Ghazipur), and in the hills up to *c*. 1800 m (abundantly in Kashmir), occasionally up to 2100 or even 2400 m. A June record at 3100 m in Lahul (HW). Affects open forest, wooded nullahs, bushes, gardens and groves, especially in the neighbourhood of lakes and streams, e.g. Kashmir Valley.

Terpsiphone paradisi

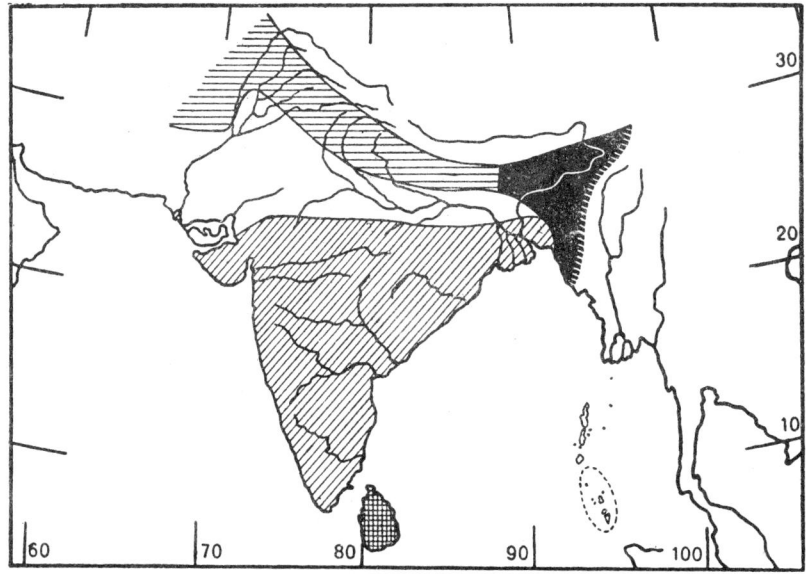

Distributional ranges

▤ *leucogaster* (1460), breeding only. ▨ *paradisi* (1461).

▦ *ceylonensis* (1462). ■ *saturatior* (1463).

☐ *nicobarica* (1464).

Extralimital. Turkestan and Afghanistan.

MIGRATION. Arrives on its summer grounds usually singly in late March (mostly April in the hills); departs in September-October. Appears to prefer night migration. Winters over most of the Peninsula, from the Gangetic plain and from lower Sind where scarce, Kutch and Kathiawar south to Kerala and Tamil Nadu; also the plains of Bangladesh (Rashid). Movements difficult to elucidate because of the presence of the very similar peninsular race. May not be present in the same area every winter especially in the more northern parts. Recorded in Gilgit as a migrant (April, May and August).

GENERAL HABITS, FOOD and VOICE. As in 1461.

BREEDING. *Season*, April to August. *Nest* and *eggs* as in 1461. Average size of 50 eggs 20·1 × 14·9 mm (Baker) and of 10 eggs 20·8 × 15·4 mm (Osmaston). Favourite nest-sites in Kashmir are willow and mulberry groves around lakes and inundations, orchards, and chenar (*Platanus*) trees near habitations. A breeding pair accompanied by a first/second summer ♂ (Kashmir Valley, 12.vii.1983) suggests co-operative breeding (Holmes & Parr, JBNHS 85: 471).

MUSEUM DIAGNOSIS. White males with streamers not distinguishable from nominate *paradisi* but *leucogaster* tends to have a narrower bill.

Rufous males with streamers paler above than nominate *paradisi* (1461); in about 80 per cent of individuals white of lower parts extends right up to black throat contrasting and clear cut, without any intermediate grey area on breast; wings largely but irregularly variegated with white on primaries and secondaries.

Females and immature males also paler on upperparts; dark portions of vanes of primaries and secondaries usually far more extensive; tertials usually have a dark shaft-streak.

MEASUREMENTS

		Wing	Bill (from skull)	Tarsus	Tail	Streamers (from base of tail)
White	♂♂	88–101	23–26	16–18	104–125	316–412 mm
Rufous	♂♂	93–100	24–26	16–18	82–116	267–332 mm
	♀♀	82–92	22–25	15–18	88–106	— mm
						(HW)

Weight 3 ♂♂ (Apr.-May, in summer quarters) 18–22; 2 ♀♀ (ditto) 18, 20 g—SA.

COLOURS OF BARE PARTS. Iris dark brown; eye-rim blue of various shades. Bill dark bluish; mouth bright greenish yellow. Legs and feet bluish grey; claws dusky.

1461. *Terpsiphone paradisi paradisi* (Linné)

Corvus paradisi Linnaeus, 1758, Syst. Nat., ed. 10, 1: 107
(in India = Chandernagor, see Stresemann, 1952, Ibis 94: 517)
Baker, FBI No. 688 (part), Vol. 2: 264
Plate 66, fig. 2 (Vol. 5)

LOCAL NAMES. Rufous plumage: *Shāh būlbūl, Hūsaini būlbūl*. White plumage: *Sūltān būlbūl* (Hindi); *Dūdhrāj, Tărwārio* ('swordsman') .[Gujarati]; *Tōkā pigilipittā* ('long-tailed bird', Telugu); *Wāl kōndā lāthi* (= 'tail crest shaker'), *Vāl kūrūvi* (Tamil, Malayalam).

SIZE. Bulbul ±; male with very long tail-streamers. Length *c.* 20 cm (8 in.); with streamers up to *c.* 50 cm (20 in.).

FIELD CHARACTERS. Old male (4 years or more): entire head and throat black with a conspicuous crest (at close range a metallic blue-black). Bill and a narrow eye-rim blue. Wings black and white. Rest of plumage silvery white, the tailfeathers with black outer webs and black shaft-streaks.

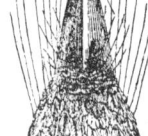

× *c.* 1

The black head with upstanding crest, glistening white plumage and extremely elongated central pair of rectrices ('ribbons') render the bird unmistakable.

Young, male (up to 3 years): as in adult but back, tail and streamers rufous (these however are not fully assumed until second autumn moult). First-year birds have ashy breast. Intermediate individuals with more or less white in plumage, or white with rusty traces, frequently seen.

Female and first-winter male : crown bluish black with a shorter crest than adult male. Rest of upperparts rufous. Throat, sides of head and nuchal collar ashy. Belly white. No streamers. General effect of a bulbul.

STATUS, DISTRIBUTION and HABITAT. Common but unevenly distributed. The Indian peninsula from Kathiawar (Gir forest), western Rajasthan, Madhya Pradesh (Neemuch, Jhansi) south to the hills of Kerala and the Chitteri range, and east through lower West Bengal and the lowlands of

Bangladesh. There are only spring records in Bihar and Orissa but presumably also breeds there. Rare and very sporadic in the Deccan. Southern populations resident, subject to local movements; northern populations mostly migratory but probably replaced in part by wintering Himalayan birds. Breeds mostly in broken foothills country and peninsular hills, up to the highest summits. Affects well-watered and shady forest, bamboo-clad nullahs, plantations and village groves, spreading in winter to gardens and scrub.

MIGRATION and winter movements little understood. It is known to winter in Sri Lanka from October to May. Large numbers pass through Pt Calimere (Tamil Nadu) *en route* in October. Two individuals have been recorded on a ship 25 km off the coast of Malabar and the species has been sighted in the Maldives. Northern birds are probably more migratory, and southern birds prone to erratic winter movements. These populations, however, are greatly increased by wintering Himalayan birds.

GENERAL HABITS. Usually seen in pairs, by themselves or with the roving mixed hunting parties. Frequents shady trees, keeping mostly to the higher branches perched upright or flitting gracefully from bough to bough or tree to tree. Makes short aerial sallies after insects, looping back after the capture to the same tree or flying on to another. Occasionally also descends close to the ground to flutter and dislodge insects from the low herbage. Flight swift and undulating, the long streamers of the male floating gracefully behind. In the nimble aerial contortions after insect prey these looping fluttering 'ribbons' present a spectacle of exquisite charm. When perched the male often flirts his tail, opening and shutting the feathers in a scissors-like movement, making play with the long ribbons. Bathes regularly by splashing at the surface of a pool or stream from a bush on the bank, spinning round abruptly on touching the water, raising a spray, then regaining perch to shuffle the plumage and preen. This repeated quickly 3 or 4 times.

Display. During the breeding season, male sings and courts with wings beating, tail raised and streamers arching gracefully. May also be seen singing in flight, 'fluttering his wing pugnaciously' (Dharmakumarsinhji). Nuptial flight is thus described by Baker: the male launches himself very slowly from some high twig, rises and drops every few yards with slowly beating wings, his long tail undulating with each rise and fall as he flies round in small circles before alighting on the perch he just left.

FOOD. Chiefly winged insects. Where available prefers large insect prey and Sphingid (Hawk) moths (TJR). Recorded items include flies and gnats, dragon-flies, jassid hoppers, small bugs and beetles. Commonly butterflies and moths; occasionally spiders.

VOICE and CALLS. Ordinary call-note, a nasal grating *chē* or *chēchwē* (SA) or *sqeenk* (Nichols), not unlike the *pench* of a snipe, uttered by both sexes. Alarm, a harsh *tst* and some harsh mobbing notes rendered as *weep poor willie weep—poor willie* (B.T. Phillips). Song, surprisingly soft and low-pitched in comparison with its harsh contact calls; comprises 5 to 6 rapidly enunciated yodelling whistles in a rising and falling scale, rendered as *qwee-twe-twuh-twoi* (TJR).

BREEDING. *Season*, March to August, chiefly May and June. *Nest*, a neat, deep, inverted cone of grasses, fine roots, bast fibre and a few leaves compactly bound together with cobweb and so thickly plastered on the outside with the last mixed with spiders' egg-bags, etc. as to appear quite white; lined with rootlets and silky vegetable down. Placed in horizontal, occasionally upright, forks of trees or saplings from one to fifteen metres above the ground, often in a bare fork of a down-sloping branch. Has a predilection for the vicinity of streams, and frequently builds in partly submerged bushes or on branches overhanging water. *Eggs*, generally 4, often 3, especially in the south, pink to nearly white, marked with reddish brown specks and small spots, usually forming a well-defined ring around the large end, usually with secondary marks of lavender. Average size of 100 eggs 20·2 × 15·1 mm (Baker). Building, incubation and care of young by both sexes; even the conspicuous white male may be seen sitting on the nest though most of incubation as well as construction is done by female. Period of incubation 15–16 days; young leave the nest about 12 days after hatching. No second brood if first is successful. Male often breeds in rufous plumage and even before acquiring the streamers. He is very intolerant of intruders and often gives away the position of a nest by his excited calls as he dashes at all comers in his characteristic, wobbly flight.

MUSEUM DIAGNOSIS. White-plumaged and rufous-plumaged males about equal in numbers. For details of plumages see Baker, loc. cit. Young, not spotted but showing obsolete pale centres and dark fringes on breast-feathers. For distinction from *leucogaster* and *saturatior* see Key to the Subspecies; also 1460 and 1463 respectively.

MEASUREMENTS

		Wing	Bill (from skull)	Tarsus	Tail	Streamers (from base of tail)
White	♂♂	90–99	23–27	16–18	94–146	290–412 mm
Rufous	♂♂	92–99	22–26	16–18	100–118	299–405 mm
	♀♀	85–92	22–24	16–17	86–110	— mm

(HW)

Weight 5 ♂♂ 20–22; 5 ♀♀ 16–22; 50 ♂♀ (October, on passage) 16–21 (av. 18·5) g—SA. 2 ♂♂ 16, 16·5 g (SDR).

COLOURS OF BARE PARTS. Iris brown; bare fleshy eye-rim slaty blue. Bill very dark greyish blue, blackish at tip. Gape slaty blue; mouth sulphur-yellow or bright yellow. Legs and feet greyish blue; claws brown.

1462. *Terpsiphone paradisi ceylonensis* (Zarudny & Härms)

Tchitrea paradisi ceylonensis Zarudny & Härms, 1912
Orn. Monatsb. 20: 60 (Ceylon)
Baker, FBI No. 688 (part), Vol. 2: 264

LOCAL NAMES. White-plumaged: *Redi-horā* ('cotton thief'), *Lainsu-horā, Kadde-horā*. Rufous-plumaged: *Gini-horā* ('fire thief') [Sinhala]; *Pirāmana-kūrūvi, Vedivāt-kūrūvi* (Tamil).

SIZE. Bulbul ±; male with very long tail-streamers. Length *c.* 20 cm (8 in.); with streamers up to *c.* 35 cm (16 in.).

FIELD CHARACTERS. As in 1461. Males of the Sri Lanka subspecies never assume white plumage. Breast of Sri Lanka males paler ashy than in immature Indian males. The white males seen in Sri Lanka are wintering birds from India.

STATUS, DISTRIBUTION and HABITAT. Plentiful resident, subject to local movements. Sri Lanka in the low-country Dry zone and lower hills up to c. 900 m. In the northeast monsoon may be seen anywhere in the well-wooded areas of the lowlands including the Wet zone, and in the hills occasionally up to 1500 m. A familiar bird of the country roadsides, gardens, surroundings of irrigation reservoirs, rivers and watercourses.

GENERAL HABITS, FOOD and VOICE. As in 1461.

BREEDING. *Season*, April to July, chiefly May. *Nest*, as in 1461. *Eggs*, normally 3, sometimes 2, exceptionally 5; colour as in 1461. Average size of 10 eggs 20·2 × 15·3 mm (Baker). Nesting behaviour as in 1461.

MUSEUM DIAGNOSIS. Differs from the Indian subspecies only by the fact that males never assume a white plumage, and by slightly paler breast.

MEASUREMENTS

	Wing	Bill (from skull)	Tarsus	Tail	Streamers (from base of tail)	
10 ♂♂	92–96	25–26	16–17	97–115	245–313	mm
1 ♀	92	25	16	116	—	mm
					(HW)	

COLOURS OF BARE PARTS. As in 1461.

1463. *Terpsiphone paradisi saturatior* (Salomonsen)

Tchitrea affinis saturatior Salomonsen, 1933, Ibis: 732
(Buxa Duars, Bhutan)
Baker, FBI No. 689, Vol. 2: 267

LOCAL NAME. *Dao-rajah-gophu* (Cachari).

SIZE. Bulbul ±; male with very long tail-streamers. Length c. 20 cm (8 in.); with streamers up to 50 cm (20 in.).

FIELD CHARACTERS. As in 1461 but crest more rounded. Chestnut males and females have greyish throat and breast. White males have feathers of upperparts with fine black shaft-streaks.

STATUS, DISTRIBUTION and HABITAT. A partial migrant or resident subject to local movements, locally common. Sikkim, Darjeeling and Jalpaiguri districts, east through Bhutan and Arunachal Pradesh, south through Nagaland, Manipur, the hills of Assam, Meghalaya and Bangladesh to the Chittagong region. Most plentiful along the base of the Himalayan foothills, scarce in the plains of the Brahmaputra and in Nagaland and Manipur. Breeds up to c. 800 m, occasionally 1000 m. Affects thin forest, secondary growth, gardens, bamboo and scrub jungle.

Extralimital. Northern Burma. The species in many geographical races extends from Turkestan and Afghanistan, east to the Indochinese countries, north to Mongolia and Manchuria, and south to the Lesser Sunda Islands.

MIGRATION. Unclear. Himalayan population appears to leave the area almost entirely in winter, returning in March-April. Birds of this subspecies are known to winter in Burma from the southern Shan States to Tenasserim, but some may also come from Yunnan. Also winters in the lowlands of Bangladesh.

GENERAL HABITS, FOOD and VOICE. As in 1461.

BREEDING. As in 1461.

MUSEUM DIAGNOSIS. White-plumaged males predominate (80–86 per cent). Rufous males and females have a distinct olive wash on upperparts, giving the back a brownish appearance. Rump, and edges of remiges and rectrices bright cinnamon. Abdomen yellowish; under tail-coverts pale chestnut. Males have a greyish throat and breast, and lack the sharply defined metallic black of other Indian subspecies.

MEASUREMENTS

	Wing	Streamers (White birds)	
♂♂	89–96	224–390	mm
♀♀	87–92	—	mm
		(Salomonsen)	

COLOURS OF BARE PARTS. As in 1461.

1464. *Terpsiphone paradisi nicobarica* Oates

Terpsiphone nicobarica Oates, 1890, Fauna Brit. Ind., Bds. 2: 48 (Nicobars)
Baker, FBI No. 691, Vol. 2: 269

LOCAL NAMES. None recorded.

SIZE. Bulbul ±; male with very long tail-streamers. Length *c.* 20 cm (8 in.); with streamers up to *c.* 50 cm (20 in.).

FIELD CHARACTERS. As in 1461, q.v.

STATUS, DISTRIBUTION and HABITAT. Common resident (and winter visitor ?) in the Nicobar Islands. Rare in the South Andaman Islands and perhaps only a winter visitor (Abdulali, JBNHS 64: 183–4 and footnote).

GENERAL HABITS, FOOD and VOICE. As in 1461.

BREEDING. Unrecorded.

MUSEUM DIAGNOSIS. Adult males have a glossy bluish black throat (*v.* greyish in *saturatior*), abdomen darker, buff-coloured; under tail-coverts a darker chestnut. Upperparts also darker, the back more olive, the rump brown rather than cinnamon. Size smaller. Abdulali (loc. cit. p. 184) suggests that the resident Nicobar population may not assume a white plumage and that white birds may be winter visitors.

MEASUREMENTS

		Wing	Bill (from skull)	Tarsus	Tail	Streamers (from base) of tail	
White	♂♂	88–96	20–24	14–16	92–108	140–358	mm
Rufous	♂♂	88–94	20–25	—	92–102	165–253 (HW)	mm

Black-chinned ♂ ♂ (tail) 93–106 mm
Grey-chinned ♂ ♀ (tail) 82–88 mm

(Abdulali)

COLOURS OF BARE PARTS. Unrecorded.

MONARCH FLYCATCHERS

Genus HYPOTHYMIS Boie

Hypothymis Boie, 1826, Isis, col. 973. Type, by monotypy,
Muscicapa caerulea Gmelin = *M. azurea* Boddaert
Monarcha Vigors & Horsfield, 1827, Trans. Linn. Soc. London 15: 254.
Type, by monotypy, *Muscipeta carinata* Swainson = *Muscicapa melanopsis* Vigors
Bill as in *Terpsiphone* but smaller, densely covered with plumules at base. Rictal bristles numerous and long. Sexes dissimilar. Young not spotted.

HYPOTHYMIS AZUREA (Boddaert): BLACKNAPED MONARCH FLYCATCHER
Key to the Subspecies

			Page
A	No throat-bar	*H. a. ceylonensis*	225
B	A black bar across throat		
	1 Abdomen white	*H. a. styani*	223
	2 Abdomen blue	*H. a. tytleri*	226
	3 Abdomen white, tinged with blue		
	a Larger; wing averaging more than 70 mm	*H. a. idiochroa*	226
	b Smaller; wing averaging less than 70 mm	*H. a. nicobarica*	227

1465. *Hypothymis azurea styani* (Hartlaub)

Siphia styani Hartlaub, 1898, Abh. Nat. Ver. Bremen 16: 248
(Harhow, Hainan)
Hypothymis azurea sykesi Baker, 1920, Bull. Brit. Orn. Cl. 41: 8. New name for
Muscicapa caeruleocephala Sykes, 1832, nec. *M. caeruleocephala* Scopoli, 1786
Hypothymis azurea similis Koelz, 1939, Proc. Biol. Soc. Washington 52: 68
(Londa, Bombay Presidency)
Baker, FBI Nos. 692 and 693, Vol. 2: 270, 271
Plate 66, fig. 1 (Vol. 5)

LOCAL NAMES. *Kālā kătkătiā, Kālā māthā kătkătiā* (Bengali); *Yeepidippăn* (Tamil); *Vēl-neeli* (Malayalam).

SIZE. Sparrow; length *c.* 16 cm (6 in.).

FIELD CHARACTERS. Male. A partially fan-tailed azure-blue flycatcher, duller on the wings, with a velvety black patch on nape and thin black crescent across throat. Belly whitish.

Female. *Above*, crown blue, rest of upperparts brown; *below*, ashy blue fading to whitish on belly. No black nape-patch or necklet.

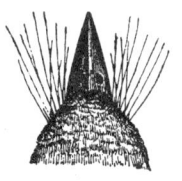

× *c.* 1

STATUS, DISTRIBUTION and HABITAT. Widespread resident, subject to erratic local and winter movements. India south and east of a line running from the Gulf of Kutch to Gorakhpur (eastern U.P.); occasional birds turn up in winter in Karachi (Pakistan) environs, a nest reported at Malir in April 1971 by Jack Coles (T. J. Roberts, JBNHS 78: 74); south to Kerala and Tamil Nadu (including the Palnis—JBNHS 55: 159); east through Bangladesh, Assam, Nagaland?, Meghalaya and Manipur and north along the Hima-

Hypothymis azurea

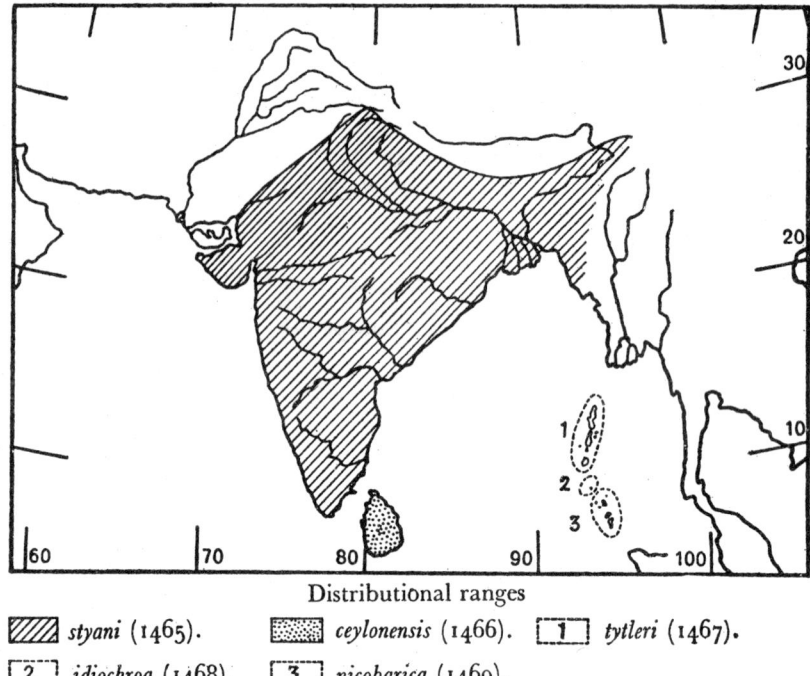

Distributional ranges

▨ *styani* (1465).   ▦ *ceylonensis* (1466).   [1] *tytleri* (1467).
[2] *idiochroa* (1468).   [3] *nicobarica* (1469).

layan foothills west to Dehra Dun (Osmaston, 1935) and east through Nepal, Sikkim and Bhutan to Arunachal Pradesh including the Sadiya frontier tract. In the hills generally up to *c.* 900 m, less often 1200 m (Assam), 1300 m (Sikkim), or 1500 m (Kerala, Maharashtra). More widespread in winter and has straggled to Lucknow, Kutch and Karachi. Affects well-wooded country, evergreen or mixed deciduous forest, secondary jungle, cultivation, coffee, cardamom and teak plantations, etc.; partial to bamboo facies. In drier areas prefers heavy foliage, usually along streams.

Extralimital. Ranges east through southern China and the Indochinese countries to Hainan. The species extends to the Philippines and Indonesia.

GENERAL HABITS. Keeps singly or in pairs, often in company with *Muscicapa tickelliae, Rhipidura albicollis, Hemipus picatus*, etc. As a rule, keeps to taller trees than other flycatchers but also frequents undergrowth. A very active species, always on the move, pivoting on its perch from side to side, prancing and flitting about, wings drooping and tail partly fanned out and cocked, after the manner of *Rhipidura* though to a less degree. Makes agile looping sallies after winged insects, sometimes flutters in front of a sprig to disturb those hiding within and occasionally descends to the ground to pick up one. Under excitement, the black nape-tuft is erected to a point.

FOOD. Winged insects, including butterflies, moths, cicadas, small beetles and bugs. If prey too big to swallow, held under foot and torn piecemeal.

VOICE and CALLS. On the whole a silent bird but commonly utters a distinctive high-pitched rasping, interrogative *sweech-which?* (*sweech-which-*

which? when alarmed) or *chē-chwē?* much like that of the Paradise Flycatcher; reminiscent also of *Aethopyga siparaja* or *Parus major* (SA). Song, a ringing *chew chew chew*, resembling Tailor Bird, being identically the same in the Eastern Ghats and Bombay region of Maharashtra (Trevor Price, JBNHS 76: 45).

BREEDING. *Season*, March to August. *Nest*, a neat, conical cup of *Rhipidura* pattern though a little more massive; sometimes also with an untidy 'tail' of loose material dangling below. Made of grass stems and shreds of bark, lined with finer grass, coated with green moss when available, and so thickly plastered on the outside with cobweb and spiders' egg-cases as to appear quite white; firmly bound into position with cobwebs in a slender fork or elbow of outer branches of trees; also built in saplings, coffee bushes, bamboo clumps, etc., sometimes in fully exposed situations. Usually placed between three and five metres above ground; recorded as low as 60 cm and as high as nine metres. *Eggs*, normally 3 in southern India and Assam, often 4 in the north; very similar to those of *Terpsiphone* (1461) though smaller, and with the same range of colour variations. Average size of 110 eggs 17·4 × 13·3 mm (Baker). Female does most of the nest-building though male accompanies her closely. Incubation and care of young by both sexes; incubation period *c.* 12 days.

MUSEUM DIAGNOSIS. See Key to the Subspecies. For chick (in down) see Biswas, JBNHS 59: 807.

MEASUREMENTS

	Wing	Bill (from skull)	Tarsus	Tail	
♂♂	64–75	14–17	16–18	65–76	mm
♀♀	66–76	14–17	16–18	65–74	mm

(Wing mostly over 70 mm)

(HW, SA, BB)

Weight 16 ♂ ♂ 10–14; 16 ♀ ♀ 9–13 g (SA). 1 ♂ 8.5 g, oo? 9·5–12·5 g, 5 ♀ ♀ 9, 9·7, 10, 10·2, 10·6 g (SDR).

COLOURS OF BARE PARTS. Iris brown; bare eye-rim blue. Bill: adult ♂ dark greyish blue, black at tip; ♀ horny brown; mouth (both sexes) greenish or sulphur yellow. Legs and feet: ♂ slaty blue; ♀ greyish brown; claws dark brown.

1466. *Hypothymis azurea ceylonensis* Sharpe

Hypothymis ceylonensis Sharpe, 1879, Cat. Bds. Brit. Mus. 4: 277
(Cotta, Ceylon)
Baker, FBI No. 694, Vol. 2: 272

LOCAL NAMES. *Nil-kurullā, Mārāwā* (Sinhala).

SIZE. Sparrow; length *c.* 16 cm (6 in.).

FIELD CHARACTERS. As in 1465 but black collar absent and nuchal spot indistinct.

STATUS, DISTRIBUTION and HABITAT. Resident, moderately plentiful. Sri Lanka in the lowlands and lower hills to *c.* 900 m, occasionally to 1600 m in the drier Uva hills. Affects evergreen forests, well-wooded ravines, etc.

GENERAL HABITS, FOOD and VOICE. As in 1465.

BREEDING. *Season*, March to May in the hills; also recorded in September in the low-country Wet zone. *Nest*, as in 1465. *Eggs*, 2 or 3, similar to those of *Terpsiphone* (1461) but smaller. Average size of 13 eggs 17·2 × 13·2 mm (Baker). When sitting on nest, if uneasy or slightly alarmed, has a curious habit of bending back the head and pointing the bill straight up; this habit also noted in the Paradise Flycatcher (Phillips).

MUSEUM DIAGNOSIS. Differs from the Indian race (1465) in lacking the black crescent on throat (sometimes present but obsolete). Nuchal patch reduced; general coloration a little darker, more purplish.

Female differs in having the head, neck and breast a brighter and purer blue; brown of upperparts more leaden brown and more washed with blue. Postnuptial moult about August-September.

MEASUREMENTS

	Wing	Bill (from skull)	Tarsus	Tail	
9 ♂♂	70–74	14–16	15–16	67–73	mm
7 ♀♀	67–71	14–15	15–16	65–70	mm
				(HW)	

Weight 1♀ 12 g (SDR).

COLOURS OF BARE PARTS. Iris black. Bill: ♂ light plumbeous blue, inner surface chartreuse green; ♀ dull plumbeous. Legs and feet light plumbeous; soles yellow.

1467. *Hypothymis azurea tytleri* (Beavan)

Myiagra tytleri Beavan, 1867, Ibis: 324 (Port Blair, Andamans)
Baker, FBI No. 695, Vol. 2: 273

LOCAL NAMES. None recorded.

SIZE. Sparrow; length *c*. 16 cm (6 in.).

FIELD CHARACTERS. As in 1465 but blue on underparts extending to vent.

STATUS, DISTRIBUTION and HABITAT. Common resident. Andaman Islands including Great and Little Coco Islands. Affects wooded areas.

GENERAL HABITS, FOOD and VOICE. As in 1465.

BREEDING. *Season*, April to June. *Nest* and *eggs*, as in 1465. Clutch size usually 3, sometimes 2. Average size of 54 eggs 18·3 × 16·9 mm (Baker).

MUSEUM DIAGNOSIS. Like the Indian *styani* (1465) but underparts bluer and extending to vent. Axillaries blue.

MEASUREMENTS

	Wing	Tail	
3 ♂♂	71–75	70–73	mm
2 ♀♀	71–73	69–72	mm
		(Abdulali)	

COLOURS OF BARE PARTS. Unrecorded.

1468. *Hypothymis azurea idiochroa* Oberholser

Hypothymis azurea idiochroa Oberholser, 1911, Proc. U.S. Nat. Mus. 39: 604 (Car Nicobar)
Baker, FBI No. 697, Vol. 2: 274

LOCAL NAME. *Kalong tesa* (Car Nicobar).

SIZE. Sparrow; length *c*. 16 cm (6 in.).

FIELD CHARACTERS. As in 1465, q.v.
STATUS, DISTRIBUTION and HABITAT. Restricted to Car Nicobar.
GENERAL HABITS, FOOD and VOICE. As in 1465.
BREEDING. As in 1467.
MUSEUM DIAGNOSIS. As in the Andaman *tytleri* but belly white. Differs from *styani* (1465) in having the axillaries light blue (*v.* white).
MEASUREMENTS

	Wing	Tail	
1 ♂	76	72	mm
3 ♀♀	72–77	68–70	mm
	(Abdulali)		

COLOURS OF BARE PARTS. Iris blackish brown. Bill blue, tip blackish. Legs and feet dull blue.

1469. *Hypothymis azurea nicobarica* Bianchi

Hypothymis azurea nicobarica Bianchi, 1907, Ann. Mus. Zool. St. Petersburg
  12: 76 (Nicobars = Nancowry)
Hypothymis azurea calocara Oberholser, 1911, Proc. U.S. Nat. Mus. 39: 610
  (Nankauri I., Nicobar Is.)
Baker, FBI No. 696, Vol. 2: 273

LOCAL NAMES. None recorded.
SIZE. Sparrow; length *c.* 16 cm (6 in.).
FIELD CHARACTERS. As in 1465, q.v.
STATUS, DISTRIBUTION and HABITAT. Common resident. Nicobar Islands except Car Nicobar.
GENERAL HABITS, FOOD and VOICE. As in 1465 and other races.
BREEDING. As in 1467.
MUSEUM DIAGNOSIS. Male, like the Andaman *tytleri* (1467) but slightly smaller, axillaries white, and more white on belly.
Female differs from *idiochroa* of Car Nicobar in being slightly smaller, browner and less grey above; blue of head darker, clear-cut from back and not grading into it.
MEASUREMENTS

	Wing	Tail	
15 ♂♂	63–73	56–78	mm
9 ♀♀	65–70	57–67	mm
	(Abdulali)		

COLOURS OF BARE PARTS. As in 1468.

Subfamily PACHYCEPHALINAE : Thickheads or Shrikebilled Flycatchers

Genus PACHYCEPHALA Vigors

Pachycephala Vigors, 1825, Trans. Linn. Soc. London 14: 444.
  Type, by original designation, *Muscicapa pectoralis* Latham
Muscitrea Blyth, 1847, Jour. Asiat. Soc. Bengal 16: 121.
  Type, by monotypy, *Muscitrea cinerea* Blyth
Bill strong, laterally compressed, notched at tip. Three strong rictal bristles, numerous smaller ones overhanging nostril. Wing long. Young not spotted.

1470. **Grey Thickhead** or **Mangrove Whistler. *Pachycephala grisola***
(Blyth)

Tephrodornis grisola Blyth, 1843, Jour. Asiat. Soc. Bengal 12: 180. Female described under *Tephrodornis superciliosus* Swainson, v. *Lanius Keroula* Hardwicke & Gray (*nom. nud.*), 1842, Jour. Asiat. Soc. Bengal 11: 799 (Neighbourhood of Calcutta).
Cf. Mukherjee A. K., 1970, *Jour. Bombay nat. Hist. Soc.* 67: 112
M.(uscitrea) cinerea Blyth, 1847, Jour. Asiat. Soc. Bengal 16: 122
(Island of Ramree, Arracan)
Baker, FBI No. 887, Vol. 2: 484
Plate 81, fig. 1

LOCAL NAMES. None recorded.
SIZE. Sparrow +; length *c.* 17 cm (6½ in.).
FIELD CHARACTERS. General aspect very like woodshrike (*Tephrodornis pondicerianus*) but lacking whitish supercilium and white in the tail-feathers.

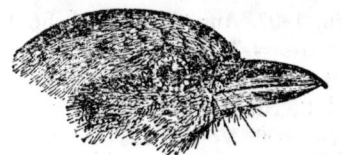

× *c.* 1

Above, olive-brown, more ashy on crown. Underparts white with a darker, very pale grey-brown pectoral band. Sexes alike.
STATUS, DISTRIBUTION and HABITAT. Resident. The Sundarbans of West Bengal and Bangladesh eastward from the Calcutta environs and down the Chittagong coast. Andaman Islands. Confined to a narrow zone fringing the shores. Affects mangroves and other small trees.
Extralimital. Extends along the eastern coast and islands of the Bay of Bengal, east to Vietnam and south through the Malay Peninsula. Other subspecies in Indonesia and the Philippines.
GENERAL HABITS. A quiet and unobtrusive bird usually seen alone or in pairs. Catches insects on the wing as well as on the branches and trunks of trees. May be seen among the roots of mangroves and higher up in trees.
FOOD. Insects.
VOICE and CALLS. Song, a loud and clear whistle, repeated three or four times in a rising scale, or prolonged and drawn out, followed suddenly by a higher or lower note, somewhat reminiscent of the call of *Aegithina tiphia* (Osmaston).
BREEDING. *Season*, April to July. *Nest*, a thin, flimsy, cup-shaped structure attached by means of cobwebs to the twigs supporting it. Placed in small trees between one and four metres above the ground. *Eggs*, 2, creamy buff or pale *café-au-lait* marked with small spots ranging from sepia to blackish brown, and secondary spots of lavender, forming a zone around the large end, sparse elsewhere. Average size of 26 eggs 21·7 × 15·7 mm (Baker).
MUSEUM DIAGNOSIS. For details of plumages see Baker, loc. cit. Nestling. Back and wings warm reddish brown; breast pure white, spotless (Osmaston, JBNHS 17: 159).

MEASUREMENTS

	Wing	Tarsus	Tail	
♂♀	81–89	c. 23	55–61	mm
			(Baker)	

Bill from skull c. 18 mm

Weight 1 ♂ 0.75 oz. (21 g); 3 ♀ ♀ 0.75 + oz. (21 + g)—Hume.

COLOURS OF BARE PARTS. Iris dark reddish brown. Bill blackish brown. Legs and feet slate or greyish brown.

REFERENCES CITED

Ali, Sálim (1933-4): 'The Hyderabad State ornithological survey' (with notes by Hugh Whistler). 5 parts. *J. Bombay nat. Hist. Soc.*, Vols. 36-37.
Ali, Sálim (1935-7): 'The ornithology of Travancore and Cochin' (with notes by Hugh Whistler). 8 parts. *J. Bombay nat. Hist. Soc.*, Vols. 37-39.
Ali, Sálim (1945): *The Birds of Kutch*. Oxford University Press, Bombay.
Ali, Sálim (1949): *Indian Hill Birds*. Oxford University Press, Bombay.
Ali, Sálim (1954-5): 'The birds of Gujarat'. 2 parts. *J. Bombay nat. Hist. Soc.*, Vol. 52.
Ali, Sálim (1962): *The Birds of Sikkim*. Oxford University Press, Madras.
Ali, Sálim (1968): *The Birds of Kerala*. Oxford University Press, Madras.
Baker, E. C. Stuart (1922-31): *Fauna of British India*. Birds. 8 vols. Taylor & Francis, London.
Baker, E. C. Stuart (1932-5): *The Nidification of Birds of the Indian Empire*. 4 vols. Taylor & Francis, London.
Bates, R. S. P. & E. H. N. Lowther (1952): *Breeding Birds of Kashmir*. Oxford University Press, Bombay.
Deignan, H. G. (1964): Subfamily Timaliinae in Peters's *Check-list of Birds of the World*, Vol. 10.
Dementiev, G. P., N. A. Gladkov, E. S. Ptushenko, E. P. Spangenberg & E. M. Sudilovskaya (1951-4): *Ptitsy Sovietskogo Soyuza*. 6 vols. Moscow. [English translation as *Birds of the Soviet Union*. Smithsonian Institution and National Science Foundation, Washington, D. C. (Jerusalem, Israel) 1966-9]
Dharmakumarsinhji, R. S. (1954): *Birds of Saurashtra*. Times of India Press, Bombay.
Diesselhorst, Gerd (1968): 'Beitrage zur Ökologie der Vögel Zentral und Ost-Nepals'. *Khumbu Himal* 2, edited by Prof. Dr Walter Hellmich. Universitätsverlag Wagner Ges. M.B.H., Innsbruck-München.
Hartert, Ernst (1910-22). *Die Vögel paläarktischen Fauna*. R. Friedländer u. Sohn, Berlin.
Heinrich, G. *vide* Stresemann, E. & G. Heinrich.
Henry, G. M. (1955): *A Guide to the Birds of Ceylon*. Oxford University Press, Bombay.
La Touche, J. D. D. (1931-4): *A Handbook of the Birds of Eastern China*. 2 vols. Taylor & Francis, London.
Ludlow, Frank (1944): 'The birds of southeastern Tibet'. 3 parts. *The Ibis*, Vol. 86.
Mason, C. W. & H. Maxwell-Lefroy (1912): 'The food of birds in India'. *Mem. Agr. Dept. India*, Entomological Series, Vol. 3.
Peters, J. L. *see* Deignan, H. G.
Phillips, W. W. A. (1939-45): 'Nests and eggs of Ceylon Birds'. *The Ceylon Journal of Science*, Section B, Zoology, Vol. 21, pts. 2 and 3, Vol. 22, pt. 2, and Vol. 23, pt. 1.
Phillips, W. W. A. (1953): *A (1952) Revised Checklist of the Birds of Ceylon*. The National Museums of Ceylon, Natural History Series (Zoology).
Phillips, W.W.A. (1958): '1956 Supplement to the 1952 revised checklist of the birds of Ceylon'. *Spolia Zeylanica* 28 (2): 183-92.
Rashid, Haroun er (1967): *Systematic List of the Birds of East Pakistan*. The Asiatic Society of East Pakistan, Dacca.
Ripley, S. D. (1961): *A Synopsis of the Birds of India and Pakistan*. Bombay Natural History Society, Bombay.

REFERENCES

Schäfer, E. (1938): 'Ornithologische Ergebnisse zweier Forschungsreisen nach Tibet'. *J. f. Orn.* 86, Sonderheft.

Smythies, B. E. (1953): *The Birds of Burma*. Oliver & Boyd, Edinburgh.

Stresemann, E. & G. Heinrich (1940): 'Die Vögel des Mount Victoria'. *Mitt. Zool. Mus. Berlin*, Vol. 24, Heft 2: 151–264.

Ticehurst, C. B. (1922–4): 'The birds of Sind'. 8 parts. *The Ibis*.

Ticehurst, C. B. (1926–7): 'The birds of British Baluchistan'. 3 parts. *J. Bombay nat. Hist. Soc.*, Vols. 31, 32.

Vaurie, Charles (1965): *The Birds of the Palaearctic Fauna* (Passeriformes). H. F. & G. Witherby, London.

Wait, W. E. (1931): *Manual of the Birds of Ceylon*. 2nd ed. Ceylon Journal of Science.

Whistler, Hugh (1930–7): 'The Vernay scientific survey of the Eastern Ghats. Ornithological Section'. 16 parts. *J. Bombay nat. Hist. Soc.*, Vols. 34–39.

Whistler, Hugh (1944): 'The avifaunal survey of Ceylon'. *Spolia Zeylanica* 23 (3 & 4): 119–321.

Whistler, Hugh. Extracts from literature and MSS. notes in British Museum (Nat. Hist.), London.

Witherby, H. F., F. C. R. Jourdain, N. F. Ticehurst & B. W. Tucker (1938–41): *The Handbook of British Birds*. 5 vols. H. F. & G. Witherby, London.

(Other references in the text)

INDEX

Actinodura, 81
Aenobarbulus (Pteruthius), 79
aenobarbus *see* aenobarbulus (Pteruthius)
aestigma (Muscicapa), 168
affinis (Garrulax), 55
aglae (Siva), 94
ahomensis (Gampsorhynchus), 80
albicaudata (Muscicapa), 198
albicilla (Muscicapa), 154
albicollis (Ixulus, Yuhina), 100
albicollis (Platyrhynchus, Rhipidura), 212
albilineatus (Alcippe, Lioparus), 110
albogularis (Garrulax, Ianthocincla), 5
albogularis [Muscicapa (Muscylva), Rhipidura], 214
albosuperciliaris (Garrulax), 39
Alcippe, 108
Alcippornise, 108
Alseonax, 136
amabilis (Muscicapa), 163
annectans [*sic*] (Lioptila), 126
annectens (Heterophasia), 126
Anthipes, 136
ardosiaca (Lioptila), 131
argentauris (Leiothrix, Mesia), 64
assamensis (Garrulax, Ianthocincla), 29
Astigma [*sic*] (Muscicapa), 168
atrovinacea (Yuhina), 98
aureigularis (Leiothrix, Mesia), 66
aureola (Rhipidura), 206
austeni (Alcippe, Proparus), 115
austeni (Garrulax, Trochalopteron), 50
austeni (Grammatoptila), 13
australorientis (Muscicapa), 165
azurea (Hypothymis, Monarcha), 223

badius (Garrulax), 7
baileyi (Yuhina), 100
bakeri (Garrulax, Trochalopteron), 63
bakeri (Yuhina), 98
banyumas *see* magnirostris (Cyornis, Muscicapa)
bayleyi (Heterophasia, Lioptila), 130
belangeri *see* leucolophus (Garrulax)
Belangeri (Garrulax), 1
bethelae (Garrulax), 55
bilkevitchi [Garrulax, Trochalopteron (Ianthocincla)], 44
biswasi (Dryonastes), 35
blythi (Muscicapa), 164
brahmaputra (Garrulax), 13
brucei (Alcippe), 121
brunnea *see* mandelli (Alcippe, Minla)

brunneata *see* nicobarica (Rhinomyias)
brunneicauda *see* wagstaffei (Alcippe)
brunneicauda (Minla), 111
burmanica (Leucocerca, Rhipidura), 209

cacabata (Muscicapa), 142
cachariensis (Muscicapa, Siphia), 186
cachinnans (Crateropus, Garrulax), 40
caerulatus (Cinclosoma, Garrulax), 33
caerulea (Muscicapa), 223
caeruleocephala (Muscicapa), 223
caeruleotincta (Lioptila), 132
calipyga (Bahila, Leiothrix), 67
calocara (Hypothymis), 227
calochrysea (Culicicapa), 201
campbelli *see* Muscicapella
canescens (Leucocirca, Rhipidura), 210
capistrata (Heterophasia), 128
capistratum (Cinclosoma), 128
carinata (Muscipeta), 223
castaneceps (Alcippe, Minla), 111
castaniceps (Ixulus, Yuhina), 98
certus (Schoeniparus), 120
cerviniventris (Digenea, Muscicapa), 171
ceylonensis (Culicicapa, Platyrhynchus), 203
ceylonensis (Hypothymis), 225
ceylonensis (Tchitrea, Terpsiphone), 220
Chelidorhynx, 204
chinensis *see* nuchalis (Garrulax)
chrysaeeus (Proparus ?), 108
chrysoptera (Ianthocincla), 60
chrysopterus (Garrulax), 60
chrysotis [Alcippe, Pr. (oparus)], 109
chumbiensis (Alcippe, Fulvetta), 114
cineraceum (Trochalopteron), 24
cineraceus (Garrulax), 24
cinerea (Alcippe, Minla), 111
cinerea [M. (uscitrea)], 228
cinereiceps (Alcippe), 116
cinereiceps (Siva), 108
cinereifrons (Garrulax), 3
cinereigenae (Minla, Siva), 93
cinereocapilla (Cryptolopha), 201
cinnamomeum (Garrulax), 40
cleta (Cyornis), 168
coelicolor (Cyornis), 172
collaris (Alcippe), 119
collini (Muscicapa), 164
commoda (Alcippe), 124
compressirostris (Leucocerca, Rhipidura), 209
concreta *see* cyanea (Muscicapa, Muscitrea)
conjuncta (Staphida), 97

233

cranbrooki (Garrulax, Grammatoptila), 13
cuculopsis *see* Heterophasia
Culicicapa, 201
Cutia, 70
cyanea (Muscicapa, Muscitrea), 182
cyanouroptera (Minla, Siva), 94
Cyornis, 136

daflaensis (Actinodura), 86
daurica (Muscicapa), 143
delacouri (Alcippe), 111
delesserti (Crateropus, Garrulax), 19
dorsalis (Leioptila), 131
Dryonastes, 1

egertoni (Actinodura), 81
Erpornis, 95
erythrocephalum (Cinclosoma), 57
erythrocephalus (Garrulax), 57
erythrolaema (Garrulax, Trochalopteron), 61
erythropterus (Lanius), 73
Eumyias, 136
euphonia (Muscicapa), 158
exquisitus (Cyornis), 165

fairbanki (Garrulax, Trochalopteron), 42
fastuosa (Cyanecula), 180
ferruginea (Hemichelidon, Muscicapa), 151
flabellifera (Muscicapa), 205
flavicollis (Yuhina), 100
flaviscapis *see* validirostris (Pteruthius)
flemingi (Garrulax), 55
fuliginosa (Hemichelidon), 142
Fulvetta, 108
furcatus (Parus), 64
fusca (Alcippe), 123

galbanatus (Garrulax), 18
galbanus (Garrulax), 18
Gampsorhynchus, 80
garoensis (Pseudominla), 111
Garrulax, 1
gertrudis (Leiothrix), 64
gilgit (Garrulax, Ianthocincla), 46
glauconotus (Pteruthius), 73
godwini (Garrulax, Trochalopteron), 61
gracilis (Heterophasia, Hypsipetes), 131
Grammatoptila, 1
grandis (Chaitaris, Muscicapa), 175
gratior (Garrulax), 21
griseata (Ianthocincla), 52
griseicauda (Garrulax), 31
griseotincta (Yuhina), 103
grisescentior (Ianthocincla), 46
grisola (Pachycephala, Tephrodornis), 228
grosvenori (Garrulax), 26

gularis (Anthipes), 136
gularis (Garrulax, Ianthocincla), 21
gularis (Yuhina), 103
gulmergi (Hemichelidon, Muscicapa), 140

hardwickii (Garrulax), 16
harterti (Ixulus), 102
Hemichelidon, 136
henrici (Garrulax, Trochalopteron), 54
Heterophasia, 126
Hilarocichla, 71
hodgsoni (Muscicapa, Siphia), 163
hodgsoni (Muscicapella, Nemura), 200
hybrida (Pteruthius, Pterythius [*sic*]), 77
hyperythra (Muscicapa), 161
hyperythra (Siphia), 156
hypoleuca *see* tomensis (Muscicapa)
Hypothymis, 223
hypoxantha (Rhipidura), 205

Ianthocincla, 1
idiochroa (Hypothymis, Monarcha), 226
ignotincta (Minla), 88
imbricatus (Garrulax), 48
imprudens (Garrulax), 59
indigo (Muscicapa), 136
indochinensis (Muscicapa), 165
intermedius *see* aenobarbulus (Pteruthius)
Ixops, 81
Ixulus, 95

jerdoni (Cyornis, Muscicapa), 195
jerdoni (Garrulax), 42

kali (Garrulax), 59
kangrae (Alcippe, Fulvetta), 113
kempi *see* stanleyi (Rhipidura)
Keroula (Lanius), 228
khasiana (Actinodura), 83
khasiensis (Alcippe), 124
khasium (Trochalopteron), 63
khosrovi (Alseonax), 146
kumaiensis (Leiothrix), 66

latirostris (Muscicapa), 143
Leioptila, 126
Leiothrix, 64
lepida (Leiothrix), 94
leucolophus (Corvus, Garrulax), 14
Leucocirca, 204
leucogaster (Muscipeta, Terpsiphone), 216
leucomelanura (Digenea, Muscicapa), 169
leucophrys (Turdus), 204
leucops (Digenea, Muscicapa), 160
lewisi (Actinodura), 83
lineatum (Cinclosoma), 46
lineatus (Garrulax), 46

INDEX

Lioparus, 108
livingstoni (Garrulax), 35
ludlowi (Alcippe, Fulvetta), 116
lutea (Leiothrix), 66
lutea (Sylvia), 64
luteola (Leiothrix), 67

macgrigoriae (Muscicapa, Phoenicura), 177
magnirostris (Cyornis, Muscicapa), 191
mandelli (Alcippe, Minla), 120
manipurensis (Alcippe, Proparus), 116
maximus (Garrulax, Pterorhinus), 29
McClellandi (Garrulax), 8
melanoleuca (Muscicapa), 164
melanoleuca (Muscicapula), 164
melanops (Muscicapa), 197
melanops (Pteruthius), 77
melanopsis (Muscicapa), 223
melanotis (Garrulax), 8
melanotis (Pteruthius), 77
meridionale (Garrulax, Trochalopteron), 43
meridionalis see pectoralis (Garrulax, Ianthocincla)
merulinus (Garrulax), 37
Mesia, 64
Minla, 88
minuta (Muscicapa, Siphia), 170
Monarcha, 223
monileger (Dimorpha, Muscicapa), 159
monilegera (Cinclosoma), 6
monilegerus (Garrulax), 6
montivaga (Actinodura), 83
Muscicapa, 136
Muscicapella, 200
Muscicapidae, 1
Muscicapula, 136
Muscitrea, 227
muttui (Butalis, Muscicapa), 146
Myzornis, 68

nagaensis (Cutia), 70
nesaea (Cyornis), 195
neumanni (Muscicapa), 138
nicobarica (Hypothymis, Monarcha), 227
nicobarica (Rhinomyias), 135
nicobarica (Terpsiphone), 222
nigriceps (Heterophasia, Sibia), 130
nigrimenta (Yuhina), 105
nigrimentum (Garrulax, Trochalopteron), 59
nigroaurita (Lioptila), 132
nigrorufa (Muscicapa, Saxicola), 174
Niltava, 136
Nipalense (Cinclosoma), 81
nipalensis (Actinodura, Cinclosoma), 85
nipalensis (Alcippe, Siva), 124
Nipalensis (Cinclosoma ?), 81
nipalensis (Cutia), 70

Nipalensis (Sibia), 81
Nitidula, 200
noa (Chelidorhynx), 205
nocrecus (Pteruthius), 73
notatus (Cyornis), 169
nuchalis (Garrulax), 17

oatesi (Muscicapa, Niltava), 181
occidentalis (Garrulax, Ianthocincla), 26
occidentalis (Pteruthius, Pterythius [sic]), 75
occipitalis see bakeri (Yuhina)
occipitalis (Siva), 98
occipitalis (Yuhina), 104
ocellatum (Cinclosoma), 32
ocellatus (Garrulax), 32
Ochromela, 136
Olcyornis, 135
olivacea (Cyornis), 135
orientalis (Culicicapa), 201
orissae (Rhipidura), 213

Pachycephala, 227
pallida (Butalis), 138
pallida (Malacias), 128
pallida (Muscicapa), 138
pallidior (Culicicapa), 201
pallipes (Muscicapa), 183
pangpui (Niltava), 175
paradisi (Corvus, Terpsiphone), 218
parva (Muscicapa), 153
patkaicus (Garrulax), 16
pectoralis (Alcippe), 135
pectoralis (Garrulax, Ianthocincla), 8
pectoralis (Leucocirca), 214
pectoralis (Muscicapa), 227
perstriata (Alcippe), 115
phoenicea (Ianthocincla), 62
phoeniceus (Garrulax), 62
picaoides (Heterophasia, Sibia), 133
picaoides (Sibia), 126
plumbeiceps (Staphida, Yuhina), 97
poiocephala (Cryptolopha), 203
poioicephala (Alcippe, Thimalia), 122
poliogenys (Cyornis, Muscicapa), 185
poliotis (Actinodura, Ixops), 88
Poonensis (Muscicapa), 143
Proparus, 108
Pseudominla, 108
Pteruthius, 71
pulchella (Heterophasia, Sibia), 132
pyrrhoura (Myzornis), 68

querulum (Trochalopteron), 49

rama (Siva), 94
Rhinomyias, 135
Rhipidura, 204
ripleyi (Pteruthius), 73
ripponi (Actinodura), 84
rouxi (Ixulus, Yuhina), 102

rubeculoides (Muscicapa, Phoenicurus), 189
rubeculoides (Phoenicura), 136
rufiberbis (Garrulax, Ianthocincla), 28
ruficauda (Muscicapa), 148
ruficollis (Garrulax, Ianthocincla), 36
rufifrons (Garrulax), 1
rufigenis (Ixulus, Yuhina), 96
rufilata (Hemichelidon), 151
rufimenta (Cinclosoma), 27
rufitincta (Ianthocincla), 28
rufitinctus (Garrulax), 28
rufiventer (Pteruthius), 72
rufogularis (Alcippe, Minla), 118
rufogularis (Garrulax, Ianthocincla), 27
rufulus (Gampsorhynchus), 80
russata (Hemichelidon), 151

sannio *see* albosuperciliaris (Garrulax)
sapphira (Muscicapa, Muscicapula), 172
sarudnyi (Muscicapa), 138
saturatior (Cyornis), 186
saturatior (Tchitrea, Terpsiphone), 221
Schoeniparus, 108
setafer [Cinc.(losoma), Garrulax], 48
Sibia, 81
sibirica (Muscicapa), 140
signata (Leiothrix, Muscicapa), 177
sikkimensis (Garrulax, Grammatoptila), 12
simile (Trochalopteron), 22
similis (Garrulax), 22
similis (Hypothymis), 223
simlaensis (Minla, Siva), 90
Siphia, 136
Sittiparus, 108
Siva, 88
sordida (Glaucomyias, Muscicapa), 195
squamata (Ianthocincla), 51
squamatus (Garrulax), 51
Stactocichla, 1
stanfordi (Alcippe), 126
stanleyi (Rhipidura), 213
Staphida, 88
striata *see* sarudnyi (Muscicapa)
striata (Muscicapa), 136
striaticollis (Alcippe, Siva), 117
striatus (Garrulax, Garrulus), 10
strigula (Minla, Siva), 91
strophiata (Muscicapa, Siphia), 158
styani (Hypothymis, Monarcha, Siphia), 223
subcaerulatus (Garrulax), 34
subrubra (Muscicapa), 156
subsquamatum (Trochalopteron), 51
subunicolor (Garrulax, Trochalopteron), 52

sundara (Muscicapa, Niltava), 180
superciliaris [D. (imorpha)], 161
superciliaries (Muscicapa), 166
superciliosus (Tephrodornis), 228
sykesi (Hypothymis), 223

Tchitrea, 216
Terpsiphone, 216
terricolor (Butalis), 143
thalassina (Muscicapa), 197
thalia (Siva), 94
tickelliae [C.(yornis), Muscicapa], 192
titania (Yuhina), 105
tomensis (Muscicapa), 152
torqueola (Siva), 88
toxostomina (Stactocichla), 38
toxostominus (Garrulax), 38
tricolor [D.(igenea)], 169
tricolor (Muscicapa), 169
Trochalopteron, 1
turensis (Alcippe), 124
tytleri (Hypothymis, Myiagra), 226

umbratilis (Rhinomyias), 135
undulata (Muscicapa), 136
unicolor (Cyornis, Muscicapa), 188
uropygialis (Garrulax), 8

validirostris (Pteruthius), 73
variegatum (Cinclosoma), 23
variegatus (Garrulax), 23
vernayi (Cyornis, Muscicapa), 187
vernayi (Leiothrix, Mesia), 64
vernayi (Leucocirca, Rhipidura), 215
vibex (Garrulax), 12
victoriae (Siva), 93
vinctura (Actinodura), 85
vinipectus (Alcippe, Siva), 113
virgatum (Trochalopteron), 49
virgatus (Garrulax), 49
vivax (Yuhina), 102
vivida *see* oatesi (Muscicapa, Niltava)

waddelli (Garrulax), 8
wagstaffei (Alcippe), 111
waldeni (Actinodura), 87
westermanni (Muscicapa), 164
whistleri (Garrulax), 4
whistleri (Muscicapa, Niltava), 178

xanthochlorus (Pteruthius), 76

yangpiensis (Yuhina), 103
Yuhina, 95
yunnanensis (Minla, Siva), 93

zantholeuca (Erp. [ornis], Yuhina), 106
ziaratensis (Ianthocincla), 44

PLATES

Synopsis *numbers in brackets* (SE) = *Species Extralimital*

PLATE 77

1 *Garrulax subunicolor*, Plaincoloured Laughing Thrush (1320)
2 *Garrulax lineatus*, Streaked Laughing Thrush (1314)
3 *Garrulax e. nigrimentum* (1326), ssp of 1324
4 *Garrulax erythrocephalus*, Redheaded Laughing Thrush (1324)
5 *Garrulax squamatus*, Bluewinged Laughing Thrush (1319)
6 *Garrulax austeni*, Browncapped Laughing Thrush (1318)
7 *Garrulax virgatus*, Manipur Streaked Laughing Thrush (1317)
8 *Garrulax merulinus*, Spottedbreasted Laughing Thrush (1304)
9 *Garrulax rufogularis*, Rufouschinned Laughing Thrush (1294)
10 *Garrulax striatus*, Striated Laughing Thrush (1279)
11 *Garrulax moniligerus*, Necklaced Laughing Thrush (1275)
12 *Babax lanceolatus*, Chinese Babax (1270)
13 *Garrulax pectoralis*, Blackgorgeted Laughing Thrush (1277)
14 *Babax waddelli*, Giant Tibetan Babax (1271)
15 *Garrulax o. maximus* (1297), ssp of 1299
16 *Garrulax ocellatus*, Whitespotted Laughing Thrush (1299)

INDEX TO THE VOLUMES BY FAMILY
English names (for Latin names see front endpaper)

Accentors, 9
Avadavat, 10
Avocet, 2

Babblers, 6–7
Barbets, 4
Barn Owls, 3
Bee-eaters, 4
Bitterns, 1
Boobies, 1
Broadbills, 4
Bulbuls, 6
Buntings, 10
Bustards, 2
Button and Bustard-Quails, 2

Chaffinch, 10
Chats, 8–9
Coots, 2
Cormorants, 1
Coursers, 3
Crab Plover, 2
Cranes, 2
Creepers, 9
Crested Swift, 4
Crows, 5
Cuckoo-Shrikes, 6
Cuckoos, 3
Curlews, 2

Darter, 1
Dippers, 9
Divers, 1
Doves, 3
Drongos, 5
Ducks, 1

Egrets, 1

Fairy Bluebird, 6
Falcons, 1
Finches, 10
Finfoot, 2
Flamingos, 1
Flowerpeckers, 10
Flycatchers, 7
Frigate Birds, 1
Frogmouths, 4

Geese, 1
Goatsuckers, 4
Goldfinches, 10

Grebes, 1
Grey Creeper, 9
Gulls, 3

Hawks, 1
Hedge Sparrows, 9
Herons, 1
Honeyguides, 4
Hoopoes, 4
Hornbills, 4
House Sparrows, 10
Hypocolius, 5

Ibisbill, 2
Ibises, 1
Ioras, 6

Jaçanas, 2
Jaegers, 3
Jays, 5

Kingfishers, 4

Larks, 5
Laughing Thrushes, 7
Leaf Birds, 6
Leaf Warblers, 8
Longtailed Titmice, 9
Loons, 1

Magpies, 5
Megapodes, 2
Minivets, 6
Monarch Flycatchers, 7
Munias, 10
Mynas, 5

Nightjars, 4
Nuthatches, 9

Orioles, 5
Owls, 3
Oystercatchers, 2

Painted Snipe, 2
Parrots, 3
Partridges, 2
Pelicans, 1
Penduline Titmice, 9
Petrels, 1
Phalaropes, 2
Pheasants, 2